18th International Conference on Stability, Handling and Use of Liquid Fuels (IASH 2024)

Stability, Handling and Use of Liquid Fuels

Louisville, Kentucky, USA
8-12 September 2024

Editor:

Robert E. Morris

ISBN: 979-8-3313-0830-8

Printed from e-media with permission by:

Curran Associates, Inc.
57 Morehouse Lane
Red Hook, NY 12571

International Association for Stability, Handling and Use of Liquid Fuels (IASH)
3525 Piedmont Road
5 Piedmont Center
Suite 300
Atlanta, GA 30305
USA

Phone: +1 404.760.2829

www.iash.net

Additional copies of this publication are available from:

Curran Associates, Inc.
57 Morehouse Lane
Red Hook, NY 12571 USA
Phone: 845-758-0400
Fax: 845-758-2633
Email: curran@proceedings.com
Web: www.proceedings.com

IASH ®, The International Association for

Stability, Handling and Use of Liquid Fuels

www.iash.net

Proceedings of the 18th International Conference on Stability Handling and Use of Liquid Fuels

Edited by

Dr. Robert E. Morris
IASH Secretary
Nova Research, Inc.
1900 Elkin Street, Suite 230
Alexandria, VA 22308

Louisville, Kentucky USA
September 8-12, 2024

PREFACE

Theme: **Towards 100% Sustainable Fuels**

Location: **Louisville, Kentucky, US**
Dates: **September 8-14, 2024**

The Hyatt Regency in Louisville, USA was the venue for the 18th International Conference organized by the International Association for the Stability, Handling and Use of Liquid Fuels. The conference attracted a large number of participants, with a technical level comparable to that of previous conferences. It should be noted that there was a larger turnout than in Dresden, which was still marked by the end of the pandemic and some travel restrictions. There was also an unusually high number of first-time attendees, which is good for the continuity of the conference.

There were 152 delegates including 62 first time attendees and 14 spouses/guests in attendance from 18 countries.

The Keynote address was presented by John Pendray, Senior Technical Advisor-Technology, Development, Cummins, Inc. The title of his presentation was "Rapid Transportation Decarbonization with Biomass Derived Liquid Fuels".

There were seven (7) paper submissions for the Chevron Research Award of Excellence in Honor of John Bacha from the following individuals:

- Aimen Laalam, Department of Petroleum Engineering, Colorado School of Mines, USA
- Amanda M. Arts, US Air Force / University of Dayton Research Institute, USA
- Danika Nicoletti, LuminUltra Technologies Ltd., Canada
- Mark Romanczyk, US Naval Research Laboratory, Washington, DC, USA
- Robert Parker, Trinity College Dublin; Ireland
- Ryma Benrabah, Université de Lorraine, France
- Victoria Kompanijec, Air Company, USA

All of the submissions were of high technical quality and merit. The award was made to Mark Romanczyk from the US Naval Research Laboratory. His presentation and paper were entitled:

"Characterization of Oxygen-Containing Compounds in Stable and Thermally Unstable Jet Fuels by Using High-Resolution Mass Spectrometry: Insights into Composition-Thermal Instability Relationships".

All of the other paper submissions were accepted for presentation in the General Sessions.

A total off forty-five (45) technical presentations were given in eleven sessions that included the following topics:

- **Session 1: Fuel Production, Optimization and Certification**
  Session chairs : Zainab Al-Saihati (Saudi Aramco) and Krege Christison (Chevron)
- **Session 2: Analytical Technique Development**
  Session Chairs: Melanie Thom (Baere Aerospace) and Georg Eckel (DLR)

- **Session 3: Fuel Supply Chain, Storage, Handling and Logistics**
  Session Chairs: Jordan Metz (ExxonMobil Technology and Engineering) and Joseph L. Stark (Innospec Inc.)
- **Session 4: Fuel Quality Control**
  Session Chair: Victoria Kompanijec (Air Company) and Paul P. Wells (ExxonMobil Technology and Engineering)
- **Session 5: Fuel Properties and Effects**
  Session Chairs: Robert Parker (Trinity College Dublin) and Jack Burgazli (Innospec Inc.)
- **Session 6: Panel Discussion**
- **Session 7: Thermal Stability**
  Session Chairs: Mark Romanczyk (US Naval Research Laboratory) and Zach West (UDRI)
- **Session 8: Fuel Chemistry, Research and Development**
  Session Chairs: Amanda Mae Arts (UDRI) and David Evans (Defence Science & Technology Group)
- **Session 9: Microbiological Contamination / Monitoring**
  Session Chair: Danika Nicoletti (LuminUltra Technologies Ltd.) and Graham Hill (ECHA Microbiology Ltd.)
- **Session 10: Ground Fuels (Gasoline, Diesel and Biodiesel) and Marine Fuels**
  Session Chairs: Nathalie Brassart (IFPEN) and Joanna Bauldreay (Bauldreay Jet Fuel Consulting Limited)
- **Session 11: Fuel Testing and Test Methods**
  Session Chairs: Ryma Benrabah (CNRS-LRGP) and Dietmar Posselt (Aviation Fuel Projects Consulting)

The poster session was comprised of 16 technical presentations

Conference Proceedings are available to members on-line through the IASH Library

Following the success of the first panel discussion organized in Dresden, it was decided to organize one again in Louisville. This year, the topic of the discussion was "Towards 100% Sustainable Fuels". The panel was moderated by Joanna Bauldreay (Bauldreay Jet Fuel Consulting Limited) and comprised of Gurhan Andac (GE Aerospace), Patrick Bosmans (CEPS - Central Europe Pipeline System), Amy Carico (Airlines for America) and Rick Kamin (Naval Air Systems Command). The moderator and panelists discussed various issues relating to the possibility of using alternative fuels at high levels of up to 100%. The certification process and current discussions within ASTM were presented. The cost of these fuels and the problems associated with logistics were also addressed. The civilian sector is making rapid progress in this area. However, we must not forget the military sector. Indeed, there is a very strong link between the two through fuel distribution and standards. The military sector is also particularly confronted with problems of interoperability.

The IASH Awards Dinner was held on Wednesday evening at Churchill Downs. Guests had the opportunity to visit this equestrian complex, famous for hosting the annual Kentucky Derby. They were able to appreciate the architecture of the building, recognizable by its twin towers with pointed roofs, and visit the museum. Three prizes were awarded at the conference, in addition to the chevron award. The Nahum Por Award was presented to Steven Zabarnick, for his leadership, support, and promotion of the Association over the past 30+ years. IASH Lifetime Achievement Awards were given to the following individuals for their multifaceted and dedicated efforts in Research, Development and Use of hydrocarbon-based fuels and related technologies throughout their careers:

- Dietmar Posselt, Aviation Fuel Projects Consulting
- Edwin Corporan, U.S. Air Force Research Laboratory

The Conference Chairman and the IASH Board of Directors are greatly appreciative of the presenters for their time and effort in producing excellent technical presentations and the delegates for supporting the conference by their attendance and active participation.

Special thanks are given to the fifteen (15) sponsoring organizations and fourteen (14) exhibitors without whose financial support we would not have been able to convene the 18th International Conference:

Diamond Sponsor:

- LuminUltra Technologies Ltd.

Premium Sponsors:

- AD Systems
- CFR Engines
- Conidia Biosceince
- DLA Energy
- D-2, Inc.
- Energy Institute
- Falex
- PAC
- Stanhope Seta
- University of Dayton Research Inst

Classic Sponsors:

- Choice Analytical
- ECHA Microbiology
- Southwest Research Institute

Sponsors:

- Aviation Fuels Projects Consulting

Media Partner:

- F+L Asia

Respectfully submitted,

Mickaël Sicard
Conference Chair/1st Vice-Chair IASH

And

Matthew J. DeWitt
IASH Chairperson

AUTHOR INDEX

N

Danika Nicoletti
Oscar Nyman

O

Grant S. Ochoa
Mikael Orain
Patrick Oßwald
Lucy Owen [2]

P

Robert Parker
Fred Passman
Asha Patel
Eric Payne
John Pendray
Sai Pinappu
Łukasz Plis
M. Pourkashanian
Florian Pütz [2]

R

Moshe Rabaev [2]
Osman Radwan
Bruno Raepsaet
Astrid Ramirez
Zuzanna Rataj
Fausto Reyher
Mark Romanczyk [2]
Boris Roux
Oscar N. Ruiz
Oliver James Rumford-Warr
Stephan Ruoff

S

Maor Saban
Zainab Saihati
Reetu Sallinen
Kati Sandberg
Jordan Schmidt
Amanda Schroeder [2]
Frédéric Ser
Stafford Sheehan
Mickaël Sicard [2][3]
Baptiste Sirjean [2]
Yalfal Siyum [2]
Joseph L Stark [2]
Robert E. Synovec

T

Russell Tobias
Jacek Tomaszewski
Timothy J. Trinklein
Cody Tschantz
Scott Turn

U

Onome Ugono

V

Pieter Vandezande
Jesse Vilja

W

Zachary J. West [2][3][4][5]
Gareth Williams [2]
Dan Wispinski
Don Wolfe
Paul J. Wrzesinski [2][3]

Y

Harrison Yang
Zhibin Yang
Ulas Yildirim

Z

Steven Zabarnick

TABLE OF CONTENTS

SESSION 1: FUEL PRODUCTION, OPTIMIZATION AND CERTIFICATION

Zainab Al-Saihati and Krege Christison, Session Co-Chairs

SESSION 2: ANALYTICAL TECHNIQUE DEVELOPMENT

Melanie Thom and Georg Eckel, Session Co-Chairs

SESSION 3: FUEL SUPPLY CHAIN, STORAGE, HANDLING AND LOGISTICS

Jordan Metz and Joseph L. Stark, Session Co-Chairs

SESSION 4: FUEL QUALITY CONTROL

Victoria Kompanijec and Paul P. Wells, Session Co-Chairs

SESSION 5: FUEL PROPERTIES AND EFFECTS

Robert Parker and Jack Burgazli, Session Co-Chairs

SESSION 6: PANEL DISCUSSION
TOWARDS 100% SUSTAINABLE FUELS

SESSION 7: THERMAL STABILITY
Mark Romanczyk and Zach West, Session Co-Chairs

SESSION 8: FUEL CHEMISTRY, RESEARCH AND DEVELOPMENT

Amanda Mae Arts and David Evans, Session Co-Chairs

SESSION 9: MICROBIOLOGICAL CONTAMINATION / MONITORING

Danika Nicoletti and Graham Hill, Session Co-Chairs

SESSION 10: GROUND FUELS (GASOLINE, DIESEL AND BIODIESEL) AND MARINE FUELS

Nathalie Brassart and Joanna Bauldreay, Session Co-Chairs

SESSION 11: FUEL TESTING AND TEST METHODS

Ryma Benrabah and Dietmar Posselt, Session Co-Chairs

SESSION 12: POSTER SESSION

Matt Fielder, Session Chair

IASH 2024, the 18TH INTERNATIONAL CONFERENCE ON STABILITY, HANDLING AND USE OF LIQUID FUELS
Louisville, Kentucky, USA
September 8 – 12, 2024

KEYNOTE

RAPID TRANSPORTATION DECARBONIZATION WITH BIOMASS DERIVED LIQUID FUELS

John Pendray*

Senior Technical Advisor-Technology Development, Cummins Inc., OC2003850 Victoria St. N. Shoreview, MN 55126

ABSTRACT

Liquid bio-fuels can help to rapidly decarbonize transportation, though there are concerns regarding limited feedstock availability. Treating biomass as a carbon source instead of an energy source has the potential to more than double currently planned biofuels availability. Using renewable electricity and green hydrogen to increase biofuel yields allows renewable electricity to effectively integrate into the existing transportation fueling infrastructure, allowing a more rapid path to transportation decarbonization, even within legacy vehicles.

*Corresponding author: John Pendray, *John.Pendray@Cummins.com*

IASH 2024, the 18TH INTERNATIONAL CONFERENCE ON STABILITY, HANDLING AND USE OF LIQUID FUELS
Louisville, Kentucky, USA
8 – 12 September 2024

CHARACTERIZATION OF OXYGEN-CONTAINING COMPOUNDS IN STABLE AND THERMALLY UNSTABLE JET FUELS BY USING HIGH-RESOLUTION MASS SPECTROMETRY: INSIGHTS INTO COMPOSITION-THERMAL INSTABILITY RELATIONSHIPS

Mark Romanczyk

US Naval Research Laboratory, Washington, District of Columbia 20375, United States. mark.romanczyk.civ@us.navy.mil

ABSTRACT: Oxygen-containing compounds (OCCs) found naturally in jet fuels are known to participate in adverse autoxidative reactions when exposed to high heat for short durations during flight. For certain jet fuels, the adverse reactions produce sufficent quantites of deposits that negatively influence engine performance/fuel circulation systems. Such thermally unstable fuels fail ASTM thermal stability testing (*i.e.,* ASTM D3241) and significantly hinder military and commerical operations. Currently, it remains unclear which classes of OCCs have the greatest propensities to contribute to fuel failures upon thermal stressing. Hence, there is a burgeoning need to establish strong links between OCC compositions and thermally unstable fuels. If links are recognized, overall fuel quality and thermal stability predictions may be further improved. The central aims of this work were to (1) complete thorough qualitative characterization of ionized OCCs derived from four stable and four thermally unstable Jet A fuels, (2) compare and contrast the OCC compositions of the stable and unstable fuels and (3) establish new links between composition and thermally unstable fuels. To complete characterization, (-) electrospray ionization (ESI) coupled to a high-resolution orbitrap mass spectrometer (MS) was utilized. The soft ionization of ESI and excellent mass accuracy and resolution of the MS provided comprehensive characterization of OCCs not previously documented in the literature. This is significant as identification is a prerequisite for developing judicious composition-property relationships. Upon measurements of the eight fuels, ionized OCCs contained 1 – 4 oxygen atoms in stable and unstable fuels. The unstable fuels had greater abundances of OCCs with ≥ 2 or more oxygen atoms. Further, the ionized OCCs were represented by twenty-four empirical formulas. Therefore, at least twenty-four classes of OCCs were detected. Ionized OCCs with empirical formulas of $C_nH_{(2n-1)}O_2$, $C_nH_{(2n-3)}O_2$ and $C_nH_{(2n-5)}O_2$ were greater in abundance in unstable fuels. For example, the average relative abundances of ions with the empirical formula of $C_nH_{(2n-3)}O_2$ were approximately four times greater in unstable fuels. Additional qualitative information was reported for the detected

OCCs in fuels (*i.e.,* average molecular masses, carbon numbers and ring double bond equivalent (RDB) values). Further, Kendrick Mass Defect plots and additional plots were made to facilitate straightforward comparison of the OCC compositions of stable and unstable fuels. In general, important OCC compositional differences were noted when the stable and unstable fuels were directly compared. The results may have applications for the crude oil industry to improve fuel quality by targeting potentially harmful OCCs upstream. Additionally, new composition-property relationships may be expanded from this work to facilitate better assessments of fuel thermal stability prior to end users' procurement of jet fuels.

KEYWORDS: oxygen-containing compounds, thermal stability, fuel failures, high-resolution orbitrap mass spectrometer, Kendrick Mass Defect plots, composition-property relationships

1. INTRODUCTION

In high-performance aeronautical vehicles, jet fuel serves as a propellent to facilitate propulsion and a coolant to attenuate excess heat from various aircraft systems (*i.e.,* fuel/oil heat exchanger (FOHE), hydraulic systems, combustor, etc.) to maintain acceptable temperature ranges (**Figure 1**).[1,2] The increase in fuel temperature is of critical interest as the compounds present in fuel may adversely react with dissolved oxygen to form oxidized byproducts. These products may further react to form insoluble deposits. The presence of deposits is significant as they may obstruct fuel flow, decrease heat-transfer and spray efficiencies, and completely occlude fuel filters and nozzels, resulting in fuel starvation.

The propensity of jet fuels to form harmful deposits and insoluble byproducts when exposed to high heat is predicated by the thermal oxidative stability (TOS) of the fuel. In the context of this work, TOS is defined by the resistance of a fuel to degrade in quality when exposed to high heat for short durations. TOS of fuels is germane to innovative aeronautical development, such as the Integrated High-performance Turbine Engine Technology program where a central goal was to double the thrust-to-weight ratio of gas turbine engines.[3,4] To meet this demand, high heat loads were anticipated, especially when engines propelled vehicles to supersonic speeds.

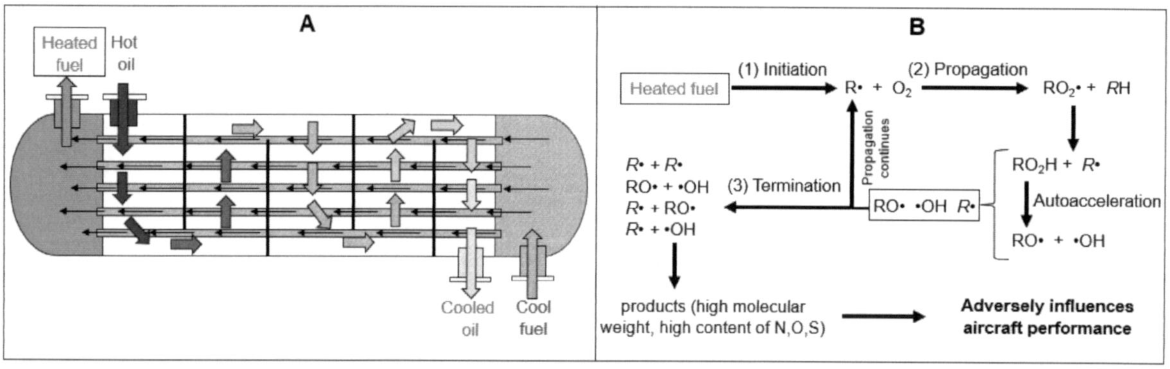

Figure 1. (**A**) Fuel/oil heat exchanger where fuel removes excess heat from lubricant oil. (**B**) Heated jet fuel experiences autoxidation reactions that contribute to thermal stability failures. R•, RO_2•, RO_2H, RO•, and •OH represent alkyl radical, alkyl peroxy radical, hydroperoxide, alkyloxy radical and hydroxy radical, respectively.

The compositions of jet fuels are important as they significantly influence important chemical and physical properties, such as seal swell, density, viscosity, freezing point, etc.[5-9] Albiet low in concentration, heteroatom-containing compounds (HCCs) were shown to significantly influence the TOS of fuels.[1,6,10-15] Although the mechanisms that contribute to thermal stability failures of fuels are complex and not fully understood, the process involves radical reactions where HCCs play an essential role (**Figure 1**).[12,15,16] Briefly, upon thermal stressing, compounds in fuel (either hydrocarbons or HCCs) form free radicals (initiation). The free radicals react with dissolved oxygen (propagation) and produce additional free radicals (propagation is continued). Upon the forming of hydroperoxides, autoxidation may also occur producing supplementary free radicals.[12]

Autoxidation continues until the free radicals react with each other (termination). Collectively, the free radical process produces byproducts that are greater in molecular weight and heteroatom content than compared to the original compounds found in the pre-thermally stressed fuel (see **Reaction scheme 1**). The byproducts are commonly referred to as soluble macromolecular oxidatively reactive species (SMORS). Upon thermal conditions, SMORS may convert into deposits that adhere to surfaces, obstructing fuel flow in circulation systems, posing a series threat to the performance of the aeronautical vehicle.

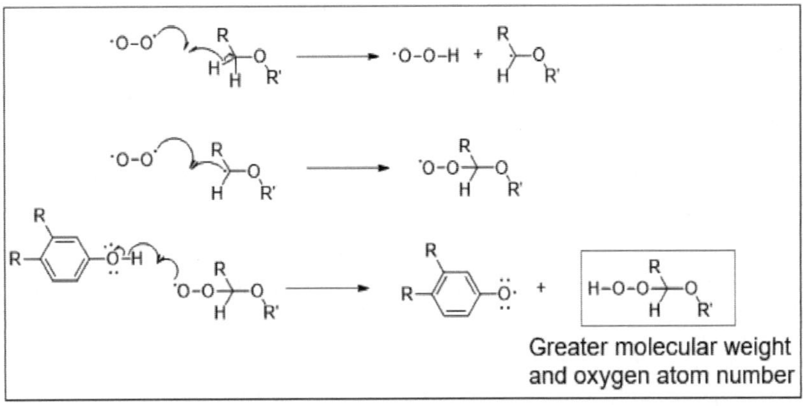

Reaction scheme 1. Free radical process that converts an ether molecule into a byproduct with greater molecular weight and oxygen atom number. The free radical phenol continues to react.

Of the HCCs present in fuels, oxygen-containing compounds (OCCs) found naturally in petroleum products contribute to thermal instability via deposit formation and play an active role in the autoxidation process.[10,13,16] Currently, it remains unclear which classes of OCCs have the greatest propensity to contribute to thermal stability failures of fuels. As observed in Balster *et al.*, 2006, the percentage of OCCs in fuels did not strongly correlate with Quartz Crystal Microbalance deposition. The absence of a trend between the total percentage of OCCs in fuels and thermal failures implies the importance of discovering how specific classes of OCCs individually influence thermal stability. Further, OCCs in the presence of other HCCs (nitrogen- and sulfur-containing compounds) likely influence the susceptibility of fuels to fail thermal stability.[13] Hence, detailed characterization of HCCs in fuels is necessary to improve critical links between fuel composition and fuel failures.

Previous analytical methods and instrumentation were developed to characterize OCCs in fuels.[10,17-21] For specific qualitative and quantitative information from these studies, see a recent review summarizing the compositional findings.[22] Newly developed high-resolution orbitrap mass spectrometer methods employing electrospray ionization (ESI) improved qualitative HCC characterization in fuels,[23,24] especially for OCCs.[24] By leveraging the high-resolution and mass accuracy of the orbitrap and soft ionization of ESI, comprehesive qualitative analysis of fuels was obtained,[24] which may aid at linking OCC fuel compositions to TOS. If such compositional differences are determined, fuel quality may be improved as industrial processes can target/remove the harmful compounds before the fuel reaches end users. Further, predictive TOS of fuels may be enhanced, providing adequate warning of imminent fuel failures.

In this study, the first aim was to characterize ionized OCCs derived from four stable and four thermally unstable fuels prior to thermal stressing. Second, the OCC compositions of stable and unstable fuels were compared to determine / evaluate differences. Third, important trends were established to link OCC compositions to thermally unstable fuels. The compositional fuel data was also presented by using Kendrick Mass Defect (KMD) plots, which provide straightforward comparison and interpretation of complex data.[23-26] Additional plots were made to draw attention to critical qualitative differences between stable and unstable fuels. For each fuel, the average molecular weight, carbon number, and ring and double bond equivalent (RDB) value (see Glossary of Terms at the end of this document for definition and calculation of RDB) were determined for all detected OCCs in fuels and for the individual homologue ion series (see Glossary of Terms). The qualitative information reported in this work was not previously documented in literature. Overall, several OCC compositional trends were noted that differentiated stable fuels from unstable fuels, which may be critical to understanding the TOS of fuels.

2. EXPERIMENTAL

2.1. Chemicals, fuel samples and materials

In this study, eight Jet A aviation fuels were received. Four Jet A fuels were thermally stable, abbreviated as S1-S4, and the remaining four were thermally unstable, abbreviated as U1-U4. HPLC/MS grade methanol was purchased from Sigma Aldrich and was used as received. Milli-Q water (18 MΩ•cm at 25 °C) was obtained in house by using a Milli-Q Direct Water Purification System equipped with Quantum TEX Polishing and LC-PAK cartridges and a Millipak Express 0.22 µm filter.

2.2 Sample preparation mass spectrometry measurements

High-resolution mass spectrometry (HRMS) measurements were completed by using a Thermo Orbitrap Exploris 480 mass spectrometer (Thermo Scientific, Waltham, MA, USA) equipped with an ESI source used in the negative ion mode. Prior to HRMS measurements, 3.5 mL of fuel was added to individual glass scintillations vials. Then, 2.0 mL of methanol was added to the individual vials and shaken by hand for 1 minute. Approximately, 100 µL of the methanol layer was removed and diluted by a factor of ten with a 50:50 mixture of Milli-Q water and methanol. The diluted samples were directly injected into the ESI probe at a flow rate of 10 µL / min by using a 500 µL Hamilton syringe and Chemyx syringe pump. A spray voltage of 3.2 kV was applied, and the ion transfer capillary temperature was set to 320 °C. To maintain a stable spray, N_2 gas was employed as sheath and auxiliary gas at flow rates of 2 and 6 arbitrary units, respectively. The m/z range and resolution were set to m/z 80 – 330 and 240,000, respectively.

2.3 Qualitative analysis of fuel samples

For HRMS analysis, ionized OCCs derived from fuel extracts with relative abundances of ca. $\geq$ 0.5 in the MS^1 spectra were processed by using the Thermo Xcalibur software. The software provided the m/z values, relative abundances, ring and double bond equivalence values (RDB), mass accuracies (in ppm units) and elemental compositions of the detected ionized OCCs. The qualitative information was exported to Microsoft Excel (Redmond, WA) and Origin Software (Irvine, CA) to compile and organize the data, and construct several different plots, respectively. KMD plots were made and facilitated compositional comparison of stable and thermally unstable fuels.

KMD plots were made by converting the m/z of $[M-H]^-$ ions to Kendrick masses (KM) by multiplying the m/z values by 14/14.01565, Eq. (**1**).[25] The KM of the ions were then converted to nominal mass and subtracted by their respective KM values to acquire the KMD values, Eq (**2**). Finally, the KMD and KM values were plotted on the y- and x-axes, respectively, as a function of relative abundances.

$$Kendrick\ mass = \frac{m}{z}\ x\ \frac{14}{14.01565} \qquad \textbf{(1)}$$

$$Kendrick\ mass\ defect = nominal\ Kendrick\ mass - Kendrick\ mass \qquad \textbf{(2)}$$

Ionized OCCs with the same relative KMD values correspond to a homologue ion series that have the same empirical formula. Hence, qualitative information may be individually assessed and assigned for each series. As a result, improved compositional information of the ionized OCCs is achieved, which may help to differentiate important compositional differences between stable and unstable fuels. For each ion series, the average molecular mass and average carbon number were determined by using Eqs. (**3**) and (**4**), respectively.

$$Average\ molecular\ mass\ (ion\ series) = \frac{\Sigma(\ m/z\ of\ ions\ in\ the\ homologous\ ion\ series\ x\ \ signal\ of\ m/z\ )}{\Sigma(signal\ of\ all\ [M-H]^-\ in\ the\ homologous\ ion\ series)} \qquad \textbf{(3)}$$

$$Average\ carbon\ number\ (ion\ series) = \frac{\Sigma(carbon\ atoms\ in\ ions\ within\ homologuous\ ion\ series\ x\ \ signal\ of\ m/z\ )}{\Sigma(signal\ of\ all\ [M-H]^-\ \ in\ the\ homologous\ ion\ series)} \qquad \textbf{(4)}$$

Further, the relative abundances of the individual ion series were determined in each fuel extract by adding the individual abundances of the $[M-H]^-$ with the same empirical formula. Then, the summed value was divided by the total relative abundances of all the ionized OCCs detected and multipled by 100 Eq. (**5**).

$$Relative\ abundance\ of\ a\ homologous\ ion\ series\ = \frac{\Sigma(Relative\ abundances\ of\ [M-H]^-\ \ in\ the\ homolgous\ ion\ series)}{\Sigma(Relative\ abundances\ of\ all\ [M-H]^-\ detected)}\ x\ 100 \qquad \textbf{(5)}$$

Further, the OCCs were grouped by total oxygen atom number in the ions from each extract. For each grouping, the relative abundances, average molecular masses, RDB values and carbon numbers were determined by using Eqs. (**6**), (**7**), (**8**) and (**9**), respectively. These steps were completed to facilitate additional comparison of fuels. Supplementary graphs plotted the empirical formulas of OCCs by total oxygen atom number (y-axis) vs carbon number (x-axis) as a function of relative abundances to compliment KMD plots.

$$Relative\ abundance\ of\ OCCs\ with\ "x"\ oxygen\ atoms = \frac{\sum(Relative\ abundances\ of\ OCCs\ with\ "x"\ oxygen\ atoms)}{\sum(Relative\ abundances\ of\ all\ OCCs\ detected)}\ x\ 100 \quad (6)$$

$$Average\ molecular\ mass\ of\ OCCs\ with\ "x"\ oxygen\ atoms = \frac{\sum(m/z\ of\ OCC\ with\ "x"\ oxygen\ atoms\ x\ \ signal\ of\ m/z)}{\sum(Relative\ abudances\ of\ OCCs\ with\ "x"\ oxygen\ atoms)} \quad (7)$$

$$Average\ RDB\ of\ OCCs\ with\ "x"\ \text{oxygen atoms} = \frac{\sum(RDB\ \ of\ OCC\ with\ "x"\ oxygen\ atoms\ x\ \ signal\ of\ m/z)}{\sum(Relative\ abudances\ of\ OCCs\ with\ "x"\ oxygen\ atoms)} \quad (8)$$

$$Average\ carbon\ number\ of\ OCCs\ with\ "x"\ \text{oxygen atoms} = \frac{\sum(carbon\ atoms\ in\ OCCs\ with\ "x"\ oxygen\ atoms\ \ x\ \ signal\ of\ m/z)}{\sum(Relative\ abudances\ of\ OCCs\ with\ "x"\ oxygen\ atoms)} \quad (9)$$

Lastly, to determine the average molecular masses, carbon numbers and RDB values for all detected OCCs in each fuel extract, Eqs. (**10**), (**11**) and (**12**) were used.

$$Average\ molecular\ mass\ (fuel) = \frac{\sum(\ m/z\ of\ ionized\ OCCs\ \ x\ \ \ signal\ of\ m/z)}{\sum(signal\ of\ ionized\ OCCs)} \quad (10)$$

$$Average\ carbon\ number\ (fuel) = \frac{\sum(carbon\ atoms\ in\ ionized\ OCCs\ \ x\ \ \ signal\ of\ m/z)}{\sum(signal\ of\ ionized\ OCCs)} \quad (11)$$

$$Average\ RDB\ value\ (fuel) = \frac{\sum(RDB\ value\ \ of\ ionized\ OCCs\ \ x\ \ \ signal\ of\ m/z)}{\sum(signal\ of\ ionized\ OCCs)} \quad (12)$$

3. RESULTS AND DISCUSSION

In this study, ionized OCCs extracted from stable and thermally unstable Jet A fuels were compositionally characterized by using (-)ESI coupled to an orbitrap mass spectrometer. As the HCC composition influences crucial fuel properties,[11-15] the ionized OCCs derived from stable and unstable fuels were compared to evaluate qualitative differences. If certain classes and/or unique abundances of OCCs are exclusive to thermally unstable fuels, correlations between OCC compositions and TOS may be further advanced. With strong correlations, more effective mitigation strategies may be developed to target/remove the adverse compounds/classes before the fuel reaches end-users.

3.1 General MS¹ qualitative data for stable and unstable fuels.

The MS^1 spectra of stable and unstable fuel extracts are shown in **Figures S1** and **S2**. The m/z values of the ionized OCCs are odd indicating that deprotonation $[M-H]^-$ was the dominant ionization reaction. An example of the raw data is provided in **Table S1**. To demonstrate the importance of high-resolution, the MS^1 spectra of fuels S4 and U4 contain an inset plot showing the separation of isobaric ions $[C_{14}H_{13}O]^-$ and $[C_{12}H_{21}O_2]^-$ (**Figures S1** and **S2**). With high-resolution attributed to the orbitrap, more effective characterization of OCCs is accomplished than compared to mass spectrometers with less resolution (*i.e.,* single quadrupole, time-of-flight mass spectrometers).

From the MS^1 spectra, thorough qualitative OCC data was acquired. To simplify the large quantities of compositional data, a comprehensive summary of the ionized OCCs detected in S1 is shown in **Table 1.** Additional summarized qualitative data of the remaining extracts are shown in **Tables S2-S8.** The tables include empirical formulas, RDB values, average KMD values, carbon number ranges, average carbon numbers, m/z ranges and average m/z values for the OCCs of each homologue ion series derived from the extracts. The detailed qualitative information reported may prove critical to expand the current understanding of TOS of fuels.

Overall, the ionized OCCs detected in stable and unstable fuels contained 1 – 4 oxygen atoms. Based on relative abundances, the OCCs with one oxygen atom (C_xH_yO) were the most abundant in both stable and unstable fuels followed by OCCs with two oxygen atoms ($C_xH_yO_2$) (**Figure 2**). The unstable fuels generally had greater relative abundances of OCCs with two oxygens than stable fuels; on average, 16.6 % and 8.0 % were found in the unstable and stable fuels, respectively.

Table 1. Summarized qualitative OCC data acquired for S1.

Oxygen atoms in OCCs	Empirical formula of $[M-H]^-$	RDB value	Avg. KMD	Relative abundance	Carbon number range and avg.	m/z range and avg.
C_xH_yO	$C_nH_{(2n-3)}O$	2.5	0.0429	0.2	5 – 8 (7)	83 – 125 (98)
	$C_nH_{(2n-5)}O$	3.5	0.0563	0.3	5 – 9 (7)	81 – 137 (111)
	$C_nH_{(2n-7)}O$	4.5	0.0700	60.1	6 – 18 (10)	93 – 261 (150)
	$C_nH_{(2n-9)}O$	5.5	0.0834	15.8	8 – 18 (12)	119 – 259 (177)
	$C_nH_{(2n-11)}O$	6.5	0.0967	1.7	9 – 17 (13)	131 – 243 (193)
	$C_nH_{(2n-13)}O$	7.5	0.1101	1.3	10 – 14 (12)	143 – 199 (164)
	$C_nH_{(2n-15)}O$	8.5	0.1236	1.7	12 – 18 (15)	169 – 253 (205)
$C_xH_yO_2$	$C_nH_{(2n-1)}O_2$	1.5	0.0527	2.1	4 – 18 (10)	87 – 283 (170)
	$C_nH_{(2n-3)}O_2$	2.5	0.0661	2.1	5 – 17 (10)	99 – 267 (165)
	$C_nH_{(2n-5)}O_2$	3.5	0.0795	1.7	5 – 16 (9)	97 – 251 (158)
	$C_nH_{(2n-7)}O_2$	4.5	0.0928	2.9	5 – 15 (9)	95 – 235 (153)
	$C_nH_{(2n-9)}O_2$	5.5	0.1063	2.4	7 – 16 (10)	121 – 247 (163)
	$C_nH_{(2n-11)}O_2$	6.5	0.1197	0.6	9 – 15 (12)	147 – 231 (182)
	$C_nH_{(2n-13)}O_2$	7.5	0.1330	0.4	9 – 13 (11)	145 – 201 (174)
$C_xH_yO_3$	$C_nH_{(2n-3)}O_3$	2.5	0.0889	0.8	4 – 12 (7)	101 – 213 (142)
	$C_nH_{(2n-5)}O_3$	3.5	0.1023	1.4	4 – 12 (7)	99 – 211 (147)
	$C_nH_{(2n-7)}O_3$	4.5	0.1158	1.8	5 – 13 (9)	111 – 223 (163)
	$C_nH_{(2n-9)}O_3$	5.5	0.1292	1.0	6 – 15 (10)	123 – 249 (175)
	$C_nH_{(2n-11)}O_3$	6.5	0.1426	0.4	8 – 14 (11)	149 – 233 (190)
$C_xH_yO_4$	$C_nH_{(2n-5)}O_4$	3.5	0.1253	0.4	5 – 11 (9)	129 – 213 (178)
	$C_nH_{(2n-7)}O_4$	4.5	0.1388	0.6	6 – 13 (9)	141 – 239 (188)
	$C_nH_{(2n-9)}O_4$	5.5	0.1522	0.3	7 – 13 (10)	153 – 237 (195)

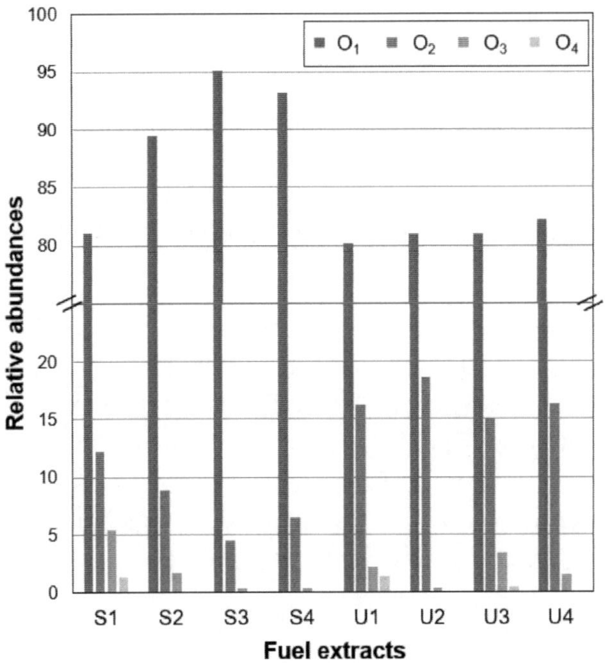

Figure 2. Relative abundances of ionized OCCs in fuel extracts by total oxygen atom number in the OCC. Generally, the unstable fuels had greater abundances of OCCs with two oxygen atoms.

The average molecular masses, carbon numbers and RDB values of all detected OCCs in each fuel extract are shown in **Table 2.** Of note, the average RDB values for OCCs derived from unstable fuels were slightly less than compared to the OCCs in stable fuels, 4.3 vs 4.6. This finding indicated that the average OCC derived from unstable fuels had greater degrees of saturation (fewer double bonds). The average molecular masses and carbon numbers did not show any relationship for stable or unstable fuels. Additional measurements on fuels are required to determine if the correlations described above have important influences on the TOS of stressed fuels.

Table 2. Summary of the average molecular masses, carbon numbers and RDB values of all detected ionized OCCs derived from the individual fuel samples.

Fuel	Avg. molecular mass (Da)	Avg. carbon number	Avg. RDB value
S1	161.1	10.6	4.7
S2	157.5	10.5	4.6
S3	155.3	10.4	4.6
S4	156.0	10.4	4.6
U1	161.4	10.5	4.3
U2	161.0	10.6	4.3
U3	149.8	9.8	4.2
U4	157.8	10.4	4.4

3.2 Kendrick Mass Defect Plots and advanced homologue ion series data of fuels.

Although the data recorded in **Figure 2** and **Table 2** is important to establish simple correlations between OCCs and TOS, using the highly detailed OCC ion series may be more informative for investigating composition-thermal stability relationships. Therefore, KMD plots for stable and unstable fuels were made and are shown in **Figures 3** and **4**, respectively. As stated previously, the plots provide a straightforward comparison of the OCC compositions of fuels than compared to traditional tables where compositional details are difficult to extrapolate.

As indicated in the Experimental section, each horizontal row in the individual KMD plots corresponds to a collection of ions with the same empirical formula with increasing masses of 14 Da (a CH_2 unit). Each row is represented by a specific class or classes (*i.e.,* isomers) of OCCs. For example, OCCs categorized by the letter identifier "e" in **Figures 3** and **4** are ionized phenols with the empirical formula $C_nH_{(2n-7)}O$. For S1 (**Figure 3**), the ionized phenols with a KM of 93 Da have a relative abundance between 2 – 6 %. The elemental composition of this ionized phenol is $[C_6H_5O]^-$ (empirical formula $C_nH_{(2n-7)}O$ where n = 6; 2 x 6 = 12 – 7 = 5 hydrogens).

Up to twenty-four differentiated ion series were identified in the fuel samples. Therefore, at least twenty-four different classes of OCCs were detected. This finding was not documented previously in literature.[17-20,24] Fuels S1 and U3 contained the greatest amount of ion series with 21 and 22, respectively. Conversely, fuels S3 and U1 contained the least with 11. Hence, the number of OCC ion series in the jet fuels did not correlate strongly with the thermal stability failures of fuels. Regarding the twenty-four classes, further structural elucidation studies are needed to assign the exact OCCs classes found in the fuel extracts and is area of ongoing research.

Upon comparison of the KMD plots, the ionized phenols are the most abundant in both stable and unstable fuels with relative abundances ranging from 60.1 – 83.9 % and 63.0 – 66.5 %, respectively (**Tables 1, S2-S8**). Furthermore, ions with the empirical formula of $C_nH_{(2n-9)}O$ (letter identifier "h", likely hydroxytetralins / tetrahydronaphthols) are the second most abundant ion series in stable and unstable fuels ranging from 9.6 – 15.8 % and 12.7 – 14.1 % relative abundances, respectively.

When comparing the OCC compositions of unstable fuels to stable fuels, important differences were noted for OCCs with two oxygen atoms categorized with letter identifiers "b" ($C_nH_{(2n-1)}O_2$), "d" ($C_nH_{(2n-3)}O_2$) and "g" ($C_nH_{(2n-5)}O_2$) (compare **Figures 3** and **4**); the three homologue ion series are consistently more abundant in unstable fuels (**Table 3**). The sum of the relative abundances of ions in "b", "d" and "g" classes are also larger in unstable fuels However, for OCCs with letter identifiers "j", "m", "q" and "u" (also with two oxygens) the trend is not observed (**Figure 5**).

Table 3. Relative abundances and averages of specific OCC with two oxygen atoms in fuels.

Class	Stable fuels					Unstable fuels				
	S1	S2	S3	S4	**Avg.**	U1	U2	U3	U4	**Avg.**
$C_nH_{(2n-1)}O_2$	2.1	3.6	2.2	3.3	2.8	8.0	4.9	5.8	7.5	6.6

Distribution Statement A. Approved for public release: distribution is unlimited.

$C_nH_{(2n-3)}O_2$	2.1	1.9	0.5	1.3	1.5	4.9	9.5	2.9	3.9	5.3
$C_nH_{(2n-5)}O_2$	1.7	1.2	0.2	0.8	1.0	1.9	3.1	1.8	2.9	2.4
Total	5.9	6.7	2.9	5.4	-	14.8	17.5	10.5	14.3	-

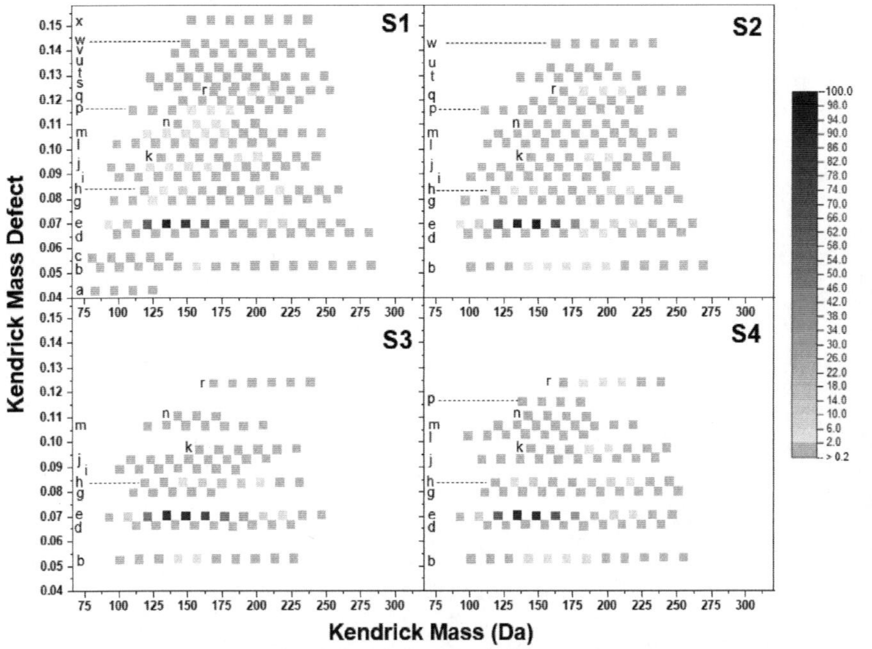

Letter Identifier	Empirical formula	RDB value
a	$C_nH_{(2n-3)}O$	2.5
b	$C_nH_{(2n-1)}O_2$	1.5
c	$C_nH_{(2n-5)}O$	3.5
d	$C_nH_{(2n-3)}O_2$	2.5
e	$C_nH_{(2n-7)}O$	4.5
f	$C_nH_{(2n-1)}O_3$	1.5
g	$C_nH_{(2n-5)}O_2$	3.5
h	$C_nH_{(2n-9)}O$	5.5
i	$C_nH_{(2n-3)}O_3$	2.5
j	$C_nH_{(2n-7)}O_2$	4.5
k	$C_nH_{(2n-11)}O$	6.5
l	$C_nH_{(2n-5)}O_3$	3.5
m	$C_nH_{(2n-9)}O_2$	5.5
n	$C_nH_{(2n-13)}O$	7.5
o	$C_nH_{(2n-3)}O_4$	2.5
p	$CnH_{(2n-7)}O_3$	4.5
q	$C_nH_{(2n-11)}O_2$	6.5
r	$C_nH_{(2n-15)}O$	8.5
s	$C_nH_{(2n-5)}O_4$	3.5
t	$C_nH_{(2n-9)}O_3$	5.5
u	$C_nH_{(2n-13)}O_2$	7.5
v	$C_nH_{(2n-7)}O_4$	4.5
w	$C_nH_{(2n-11)}O_3$	6.5
x	$C_nH_{(2n-9)}O_4$	5.5

Figure 3. Kendrick Mass Defect plots of stable fuels. The relative abundance scale, and letter identifiers for the ionized OCCs in each homologous ion series are shown to the right.

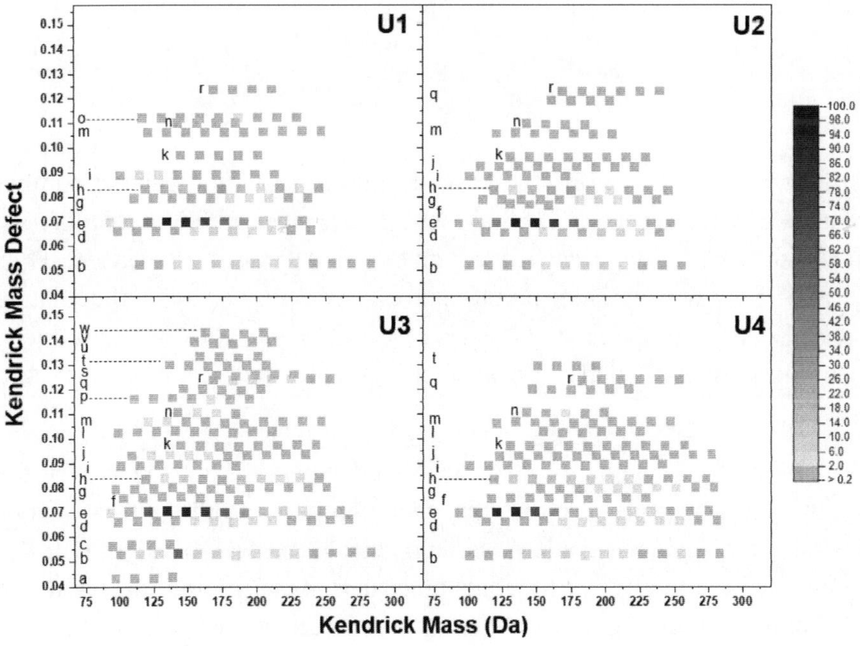

Letter Identifier	Empirical formula	RDB value
a	$C_nH_{(2n-3)}O$	2.5
b	$C_nH_{(2n-1)}O_2$	1.5
c	$C_nH_{(2n-5)}O$	3.5
d	$C_nH_{(2n-3)}O_2$	2.5
e	$C_nH_{(2n-7)}O$	4.5
f	$C_nH_{(2n-1)}O_3$	1.5
g	$C_nH_{(2n-5)}O_2$	3.5
h	$C_nH_{(2n-9)}O$	5.5
i	$C_nH_{(2n-3)}O_3$	2.5
j	$C_nH_{(2n-7)}O_2$	4.5
k	$C_nH_{(2n-11)}O$	6.5
l	$C_nH_{(2n-5)}O_3$	3.5
m	$C_nH_{(2n-9)}O_2$	5.5
n	$C_nH_{(2n-13)}O$	7.5
o	$C_nH_{(2n-3)}O_4$	2.5
p	$CnH_{(2n-7)}O_3$	4.5
q	$C_nH_{(2n-11)}O_2$	6.5
r	$C_nH_{(2n-15)}O$	8.5
s	$C_nH_{(2n-5)}O_4$	3.5
t	$C_nH_{(2n-9)}O_3$	5.5
u	$C_nH_{(2n-13)}O_2$	7.5
v	$C_nH_{(2n-7)}O_4$	4.5
w	$C_nH_{(2n-11)}O_3$	6.5
x	$C_nH_{(2n-9)}O_4$	5.5

Figure 4. Kendrick Mass Defect plots of unstable fuels. The relative abundance scale, and letter identifiers for the ionized OCCs in each homologous ion series are shown to the right.

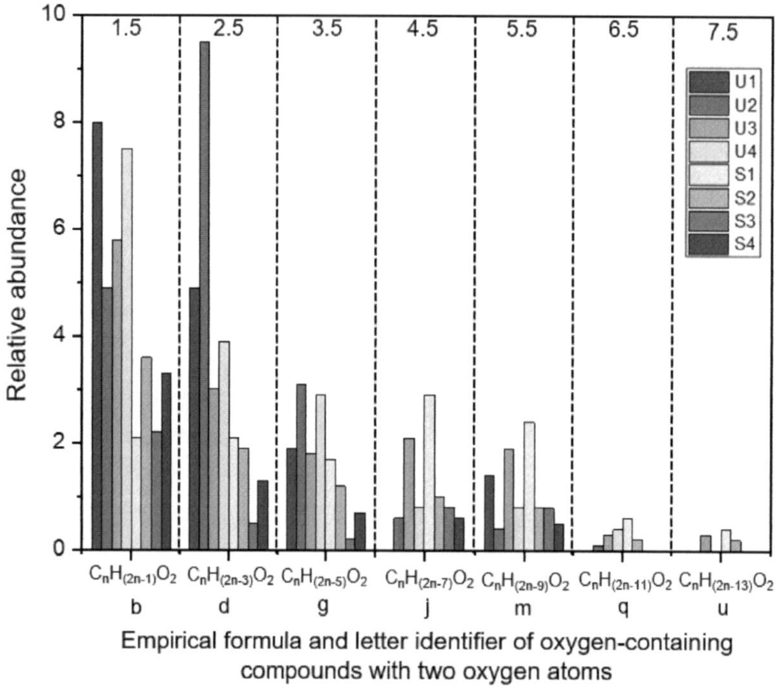

Figure 5. Relative abundances of OCCs with two oxygen atoms in stable and unstable fuels. Ions with empirical formulas of $C_nH_{(2n-1)}O_2$, $C_nH_{(2n-3)}O_2$ and $C_nH_{(2n-5)}O_2$ are greater in relative abundances in the unstable fuels. For reference, the letter identifiers used in Figures 3 and 4 are shown on the x-axis. The RDB values are provided in the upper portion of the bar chart.

Additional plots were made that categorized ions by oxygen number and total carbon number (**Figures 6** and 7). As can be observed, ions classified into the $C_xH_yO_2$ class are more abundant in the unstable fuels, specifically ions with numerical identifiers of "8", "9" and "10" (*i.e.,* $C_nH_{(2n-1)}O_2$, $C_nH_{(2n-3)}O_2$ and $C_nH_{(2n-5)}O_2$). When considering OCCs with two oxygen atoms, the relative abundances in stable and unstable fuels range from 4.4 – 12.2 % and 15.1 – 18.6 %, respectively. For a detailed qualitative summary of OCCs derived from fuels categorized by total oxygen atom number, see **Table 4.**

For the outlier S1 fuel, OCCs with two oxygen atoms had an abundance of 12.2 %. However, the composition of OCCs with two oxygens in S1 are different than compared to the unstable fuels. Specifically, the OCCs in S1 are smaller in molecular mass and have greater RDB values (*i.e.,* "j" and "m", **Figure 5**) than compared to the OCCs in the unstable fuels. This finding indicated that the OCCs have more double bonds / exhibit aromaticity. Currently, it is unclear why these chemical attributes would influence the TOS of fuel but provides important justification to examine the OCCs with two oxygen atoms more intently.

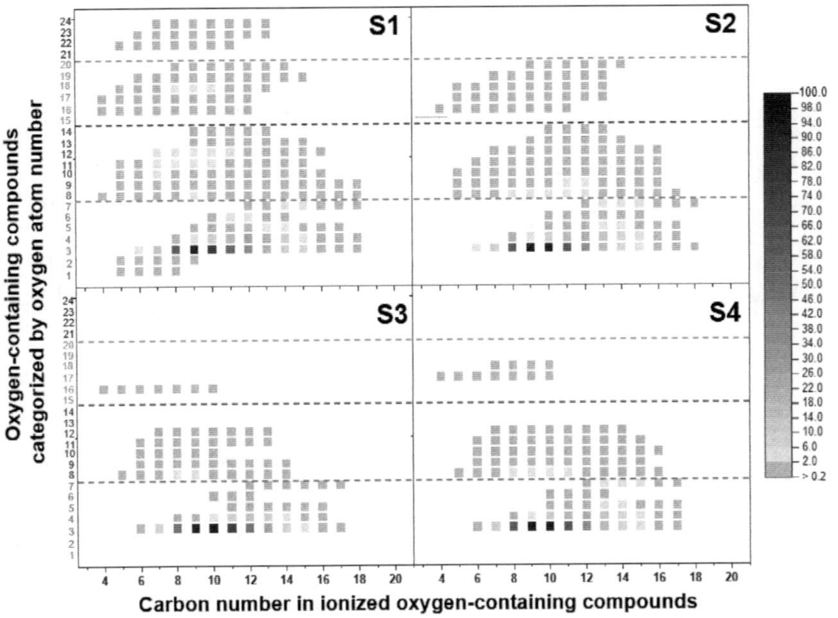

Classified by oxygen atom number	Numerical identifier	Empirical formula
$C_xH_yO_4$	24	$C_nH_{(2n-9)}O_4$
	23	$C_nH_{(2n-7)}O_4$
	22	$C_nH_{(2n-5)}O_4$
	21	$C_nH_{(2n-3)}O_4$
$C_xH_yO_3$	20	$C_nH_{(2n-11)}O_3$
	19	$C_nH_{(2n-9)}O_3$
	18	$CnH_{(2n-7)}O_3$
	17	$C_nH_{(2n-5)}O_3$
	16	$C_nH_{(2n-3)}O_3$
	15	$C_nH_{(2n-1)}O_3$
$C_xH_yO_2$	14	$C_nH_{(2n-13)}O_2$
	13	$C_nH_{(2n-11)}O_2$
	12	$C_nH_{(2n-9)}O_2$
	11	$C_nH_{(2n-7)}O_2$
	10	$C_nH_{(2n-5)}O_2$
	9	$C_nH_{(2n-3)}O_2$
	8	$C_nH_{(2n-1)}O_2$
$C_xH_yO_1$	7	$C_nH_{(2n-15)}O$
	6	$C_nH_{(2n-13)}O$
	5	$C_nH_{(2n-11)}O$
	4	$C_nH_{(2n-9)}O$
	3	$C_nH_{(2n-7)}O$
	2	$C_nH_{(2n-5)}O$
	1	$C_nH_{(2n-3)}O$

Figure 6. Relative abundances of ionized OCCs derived from the four stable fuels categorized by oxygen atom number as a function of carbon number. The relative abundance scale, and numerical identifiers for the ionized OCCs by empirical formula are shown at right.

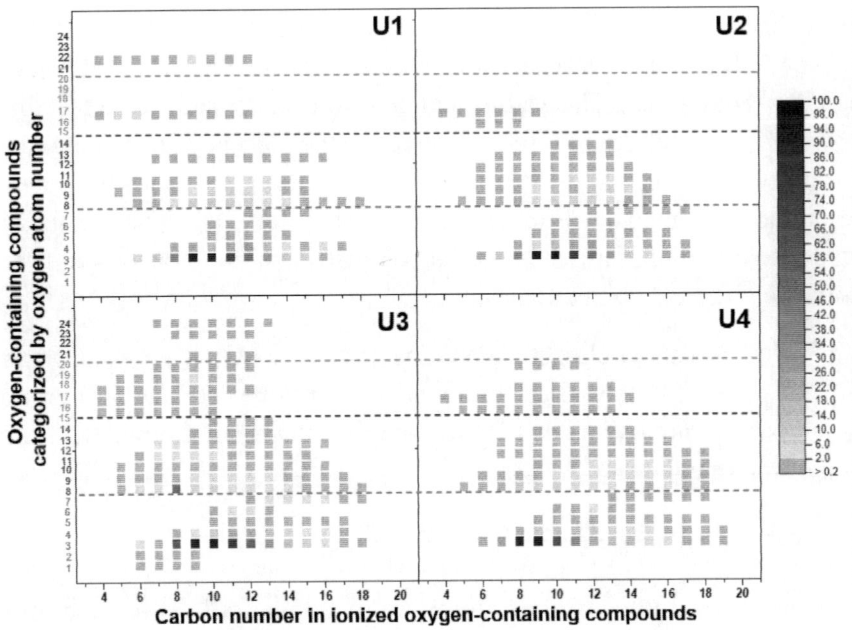

Classified by oxygen atom number	Numerical identifier	Empirical formula
$C_xH_yO_4$	24	$C_nH_{(2n-9)}O_4$
	23	$C_nH_{(2n-7)}O_4$
	22	$C_nH_{(2n-5)}O_4$
	21	$C_nH_{(2n-3)}O_4$
$C_xH_yO_3$	20	$C_nH_{(2n-11)}O_3$
	19	$C_nH_{(2n-9)}O_3$
	18	$CnH_{(2n-7)}O_3$
	17	$C_nH_{(2n-5)}O_3$
	16	$C_nH_{(2n-3)}O_3$
	15	$C_nH_{(2n-1)}O_3$
$C_xH_yO_2$	14	$C_nH_{(2n-13)}O_2$
	13	$C_nH_{(2n-11)}O_2$
	12	$C_nH_{(2n-9)}O_2$
	11	$C_nH_{(2n-7)}O_2$
	10	$C_nH_{(2n-5)}O_2$
	9	$C_nH_{(2n-3)}O_2$
	8	$C_nH_{(2n-1)}O_2$
$C_xH_yO_1$	7	$C_nH_{(2n-15)}O$
	6	$C_nH_{(2n-13)}O$
	5	$C_nH_{(2n-11)}O$
	4	$C_nH_{(2n-9)}O$
	3	$C_nH_{(2n-7)}O$
	2	$C_nH_{(2n-5)}O$
	1	$C_nH_{(2n-3)}O$

Figure 7. Relative abundances of ionized OCCs derived from the four unstable fuels categorized by oxygen atom number as a function of carbon number. The relative abundance scale and numerical identifiers for the ionized OCCs by empirical formula are shown at right.

Table 4. Summarized qualitative information for the ionized OCCs derived from the fuel extracts arranged by the number of oxygen atoms in the OCCs. The chemical property is listed where MM and RA are abbreviations for molecular mass and relative abundance, respectively.

Class by oxygen number	Chemical property	Stable fuels				Unstable fuels			
		S1	S2	S3	S4	U1	U2	U3	U4
$C_xH_yO_1$	MM (Da)	158.1	155.2	155.5	154.7	157.6	155	157.9	150.1
	C number	10.7	10.5	10.5	10.4	10.6	10.5	10.7	10.1
	RDB	4.9	4.8	4.7	4.8	4.7	4.7	4.8	4.8
	RA	81.1	89.5	95.2	93.1	80.2	81.0	80.6	82.2
$C_xH_yO_2$	MM (Da)	162.8	179.4	152.9	175.7	181.9	185.4	169.1	195.6
	C number	9.7	10.8	8.9	10.5	10.9	11.2	10.1	11.9
	RDB	3.9	2.9	2.9	2.5	2.4	2.5	3.1	2.6
	RA	12.2	8.8	4.4	6.5	16.2	18.6	15.1	16.3
$C_xH_yO_3$	MM (Da)	160.1	162.7	136	159.8	142.6	141.2	156.4	167.7
	C number	8.4	8.7	6.5	8.4	7	6.8	8.1	8.7
	RDB	4.3	4.2	2.5	4.1	2.2	1.8	3.8	2.2
	RA	5.4	1.7	0.4	0.4	2.2	0.4	3.6	1.5
$C_xH_yO_4$	MM (Da)	186.3	-	-	-	163.3	-	192.2	-
	C number	9.2	-	-	-	7.3	-	9.6	-
	RDB	4.3	-	-	-	2.5	-	4.1	-
	RA	1.3	-	-	-	1.4	-	0.7	-

Ions with the empirical formulas of $C_nH_{(2n-1)}O_2$, $C_nH_{(2n-3)}O_2$ and $C_nH_{(2n-5)}O_2$ may be organic acids and/or hydroperoxides. The latter is less likely as anions of hydroperoxides are energetically unfavorable and do not readily form in (-)ESI. To further rule out the presence of hydroperoxides, ASTM D3703 (Standard Test Method for Hydroperoxide Number of Aviation Turbine Fuels, Gasoline and Diesel Fuels) was completed. The stable and unstable fuels had no detectable quantities of hydroperoxides. Hence, the ASTM results support the absence of a peroxide functionality for the OCCs with two oxygen atoms. It is well known that organic acids are easily formed in (-)ESI and were previously detected in jet fuels.[10,20,24] Therefore, the ions are likely carboxylic acids. As indicated in Section 3.1, further structural elucidation of these ions is required to confirm the specific chemical classes of ionized OCCs.

Despite the absolute certainty of the chemical classes / identities of $C_nH_{(2n-1)}O_2$, $C_nH_{(2n-3)}O_2$ and $C_nH_{(2n-5)}O_2$, these classes of ions are relatively larger in abundance unstable fuels. Considering previous kinetic studies and autoxidative reactions,[16,27-30] the rate of formation of thermal deposits is depended on concentration of intermediates, Eqs. **12**.

$$Rate = -\frac{\Delta\,[intermidates]}{\Delta\,t} = \frac{\Delta\,[deposits]}{\Delta\,t} \qquad \textbf{(12)}$$

If a critical mass of the adverse OCCs with two oxygen atoms are found naturally in fuels, it may generate more harmful deposits and facilitate greater propensities for the fuel to fail thermal stability upon stress. Future study should evaluate how carboxylic acids influence the thermal stability of fuels by utilizing traditional ASTM testing (*i.e.,* JFTOT, Quartz Crystal Microbalance measurements).

3.3 Compositional trends of OCCs in stable and unstable fuels

The qualitative information derived from MS[1] measurements was further evaluated to develop trends between composition and TOS of fuels. When the relative abundances of OCCs with one oxygen atom was divided by OCC with ≥ 2 oxygen atoms, the ratios were generally greater for stable fuels (**Figure 8**). S1 was the outlier as the fuel had a ratio comparable to that of unstable fuels. This finding was attributed to greater abundances of OCCs with 2 oxygen atoms. However, the OCCs with two oxygen atoms in S1 were shown to have greater RDB values and smaller molecular masses than compared to the OCCs in the unstable fuels. These characteristics may explain why S1 was stable despite having less OCCs with one oxygen atom (Section 3.2).

The presence of more oxygen atoms in the structures of OCCs increases the overall polarity of the molecule. Upon adverse oxidative reactions, the polyatomic oxygen-containing species may react and gain additional molecular weight and more oxygen atoms, adversely affecting the solubility of the byproducts in a non-polar hydrocarbon jet fuel matrix (SMORS formation, **Reaction scheme 1**).[31] Upon thermal stressing, the newly formed OCCs may precipitate out of the fuel contributing to pressure failure and/or deposit formation on the surfaces of fuel circulation systems and/or engine components.

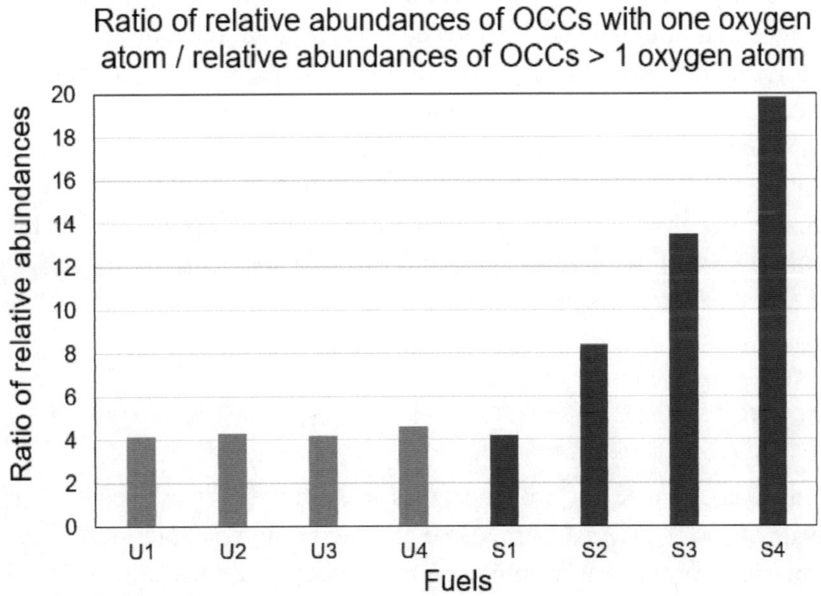

Figure 8. The ratio of relative abundances of OCCs with one oxygen atom divided by the relative abundances of OCCs with ≥ 2. Generally, stable fuels contained greater abundances of OCCs with one oxygen atom.

When the relative abundances of ionized OCCs with two oxygen atoms were plotted against their relative abundances divided by their average RDB values an important trend was observed (**Figure**

9); OCCs in the unstable fuels had smaller RDB values than compared to the stable fuels. This finding supports that OCCs with greater saturation (*i.e.,* less double bonds and or cyclic hydrocarbons) pose greater risk for adversely influencing TOS.

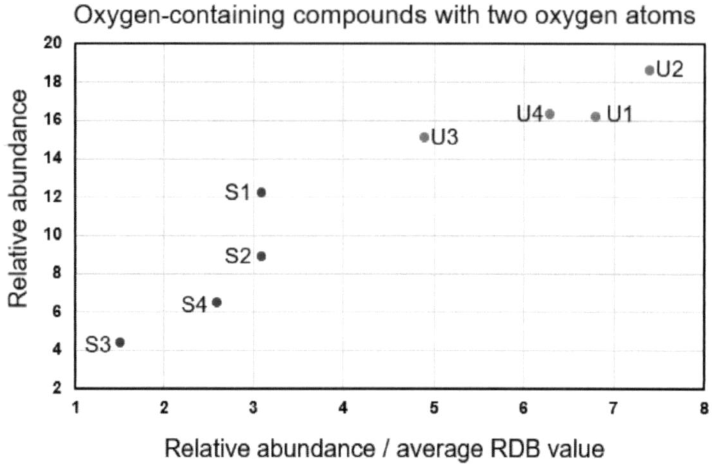

Figure 9. OCCs with two oxygen atoms in the unstable fuels had smaller RDB values than compared to same class of compounds derived from stable fuels.

With high-fidelity MS[1] data, future work implementing statistical programs / chemometrics will be initiated to establish additional links between OCCs and the unstable fuels. Additional HCCs (nitrogen- and sulfur-containing compounds) will also be detected in the stable and unstable fuels as their presence likely influences thermal stability. Indeed, multivariant interactions of HCCs likely play an imperative role on the TOS of fuels. The initial work completed on OCCs in jet fuels provides a starting point to continuously build stronger links between HCC compositions and TOS. Indeed, with greater processing power of chemometrics and comprehensive MS[1] data, exciting new composition-property links will be investigated with applications to TOS predictions and improvement of fuel quality.

4. CONCLUSIONS

In this work, (-)ESI coupled to an orbitrap mass spectrometer was used to perform thorough qualitative analysis of OCCs derived from stable and thermally unstable Jet A fuels. Up to twenty-four different empirical formulas were reported for the detected OCCs in the Jet A fuels. Hence, the number of classes of OCCs was significantly expanded compared to previous studies. The number of classes of OCCs did not strongly correlate with thermal stability. However, unstable fuels generally had greater abundances of OCCs with ≥ 2 oxygen atoms than compared to stable fuels. Furthermore, ions with the empirical formulas of $C_nH_{(2n-1)}O_2$, $C_nH_{(2n-3)}O_2$ and $C_nH_{(2n-5)}O_2$ were greater in abundances for the unstable fuels. The ions are likely representative of carboxylic acids. Additional structural analysis is required to confirm the chemical class of the compounds.

Overall, ESI coupled to the orbitrap facilitated comprehensive characterization of OCCs in Jet A fuels not documented previously in the literature. The important qualitative differences derived from the analytical method may further improve our understanding of the TOS of fuels upon thermal stressing. With greater elucidation of TOS, mitigation strategies may be developed to remove harmful compounds from fuels and more efficient thermal stability predictions of fuels may be determined prior to procurement and usage.

5. REFERENCES

1. Maurice LQ, Lander H, Edwards T, Harrison WE. Advanced Aviation Fuels: A Look Ahead Via a Historical Perspective. Fuel 2001;80(5):747-56.

2. Sobel DR, Spadaccini LJ. Hydrocarbon Fuel Cooling Technologies for Advanced Propulsion. J. Eng. Gas Turbines Power 1997;119:344-51.

3. Simoneau RJ. Hudson DA. CFD in the Context of IHPTET – The Integrated High Performance Turbine Technology Program. 25th AIAA, SME, SAE and ASEE Joint Propulsion Conference, July 10-12, 1989.

4. Edwards T. Liquid Fuels and Propellants for Aerospace Propulsion: 1903 – 2003. J Propuls. Power 2003;(19):1089-107.

5. Romanczyk M, Ramirez Velasco JH, Xu L, Vozka P, Dissanayake P, Wehde KE, et al. The Capability of Organic Compounds to Swell Acrylonitrile Butadiene O-rings and Their Effects on Mechanical Properties. Fuel 2019;238:483-92.

6. Corporan E, Edwards T, Shafer L, DeWitt MJ, Klingshirn C, Zabarnick S, et al. Chemical, Thermal Stability, Seal Swell and Emissions Studies of Alternative Fuels. Energy Fuels 2011;25:955-66.

7. Berrier KL, Freye CE, Billingsley MC, Synovec RE. Predictive Modeling of Aerospace Fuel Properties Using Comprehensive Two-Dimensional Gas Chromatography with Time-off-flight Mass Spectrometry and Partial Least Squares Analysis. Energy Fuels 2020;34:4084-94.

8. Shi X, Li H, Song Z, Zhang X, Liu G. Quantitative Composition-property Relationship of Aviation Hydrocarbon Fuel Based on Comprehensive Two-dimensional Gas Chromatography with Mass Spectrometry and Flame Ionization Detector. Fuel 2017;100:395-406.

9. Heyne J, Bell D, Feldhausen J, Yang Z, Boehm R. Towards Fuel Composition and Properties from Two-dimensional Gas Chromatography with Flame Ionization and Vacuum Ultraviolet Spectroscopy. Fuel 2022;312:122709.

10. Balster LM, Zabarnick S, Striebich RC, Shafer LM, West ZJ. Analysis of Polar Species in Jet Fuel and Determination of Their Role in Autoxidative Deposit Formation. Energy Fuel 2006;20:2561-71.

11. Beaver B, Gao L, Burgress-Clifford C, Sobkowiak M. On the Mechanism of Formation of Thermal Oxidative Deposits in Jet Fuels. Are Unified Mechanisms Possible for Both Storage and Thermal Oxidative Deposit Formation for Middle Distillate Fuels? Energy Fuels 2005;19:1574-79.

12. Heneghan SP, Zabarnick S. Oxidation of Jet Fuels and the Formation of Deposit. Fuel 1994;739:35-43.

13. Zabarnick S, West ZJ, Shafer LM, Mueller SS, Striebich RC, Wrzesinski PJ. Studies of the Role of Heteroatomic Species in Jet Fuel Thermal Stability: Model Fuel Mixtures and Real Fuels. Energy Fuels 2019;33(9):8557-65.

14. Liu G, Cao Y, Jiang R, Wang L, Zhang X, Mi Z. Oxidative Desulfurization of Jet Fuels and Its Impacts on Thermal-oxidative Stability. Energy Fuels 2009;23:5978-85.

15. Christison, KM, Lorenz RM, Xue L, Sparkman OD. Exploring the Molecular Origin of Jet Fuel Thermal Oxidative Instability Through Statistical Analysis of Mass Spectral Data. Energy Fuels 2019;33:830-36.

16. Kuprowicz NJ, Zabarnick S, West ZJ, Ervin JS. Use of Measured Species Class Concentrations with Chemical Kinetic Modeling for the Prediction of Autoxidation and Deposition of Jet Fuels. Energy Fuels 2007;21(2):530-44.

17. Striebich RC, Contreras J, Balster LM, West Z, Shafer LM, Zabarnick S. Identification of Polar Species in Aviation Fuels Using Multidimensional Gas Chromatography Time-of-flight Mass Spectrometry. Energy Fuels 2009;23:5474-82.

18. Yuan SY, Li H, Liu Z, Wang Y, Wang L, Zhang X. Measurement of Non-hindered and Hindered Phenolic Species in Aviation Fuels via Tandem-SPE with Comprehensive GCxGC-MS/FID. Fuel 2021;287:119561.

19. Adams RK, Zabarnick S, West ZJ, Striebich RC, Johnson DW. Chemical Analysis of Jet Fuel Polar, Heteroatomic Species via High-Performance Liquid Chromatography with Electrospray Ionization Mass Spectrometric Detection. Energy Fuels 2013;27:2390-98.

20. Link DD, Baltrus JP, Zandhuis Pm Hreha DC. Extraction, Separation and Identification of Polar Oxygen Species in Jet Fuel. Energy Fuels 2005;19:1693-98.

21. Guthrie JD, Romanczyk M, Kilaz G, Peretich ME, Kenttämaa, HI. Chemical Characterization and Quantitation of Phenols in Fuel Extracts by Using Gas Chromatography/Methane Chemical Ionization Triple Quadrupole Mass Spectrometry. Energy Fuels 2023;36:11962-71.

22. Romanczyk, M. Chemical Compositional Analysis of Jet Fuels: Contributions of Mass Spectrometry in the 21st Century. Mass Spec Rev 2022:1-24. https://doi.org/10.1002/mas.21825.

23. Romanczyk M, Loegel TN, Meyers KM. Qualitative Characterization of Nitrogen-containing Compounds in Jet and Diesel Fuel Extracts by Using (+) Electrospray Ionization Coupled to a Linear Quadrupole Ion Trap/Orbitrap Mass Spectrometer. Fuel 2023;344:128062.

24. Romanczyk M, Loegel TN. Qualitative Characterization of Oxygen-containing Compounds in Fuel Extracts by Using (-) Electrospray Ionization Coupled to a Linear Quadrupole Ion Trap/Orbitrap Mass Spectrometer. Fuel 2024;359:130502.

25. Kendrick E. A mass scale based on CH_2 = 14.00000 for High Resolution Mass Spectrometry of Organic Compounds. Anal Chem 1963;35:2146-54.

26. Hughey CA, Hendrickson CL, Rodgers RP, Marshall AG. Kendrick Mass Defect Spectrum: A Compact Visual Analysis of Ultrahigh Resolution Broadband Mass Spectra. Anal Chem 2001;73:4678-81.

27. Zabarnick S. Psuedo-detailed Chemical Kinetic Modeling of Antioxidant Chemistry for Jet Fuel Applications. Energy Fuels 1998;12:547-53.

28. Bacha K, Ben-Amara A, Vannier A, Alves-Fortunato M, Nardin M. Oxidation Stability of Diesel/biodiesel Fuels measured by a PetroOxy Device and Characterization of Oxidation Products. Energy Fuels 2015;29:4345-55.

29. Jia T, Gong S, Pan, L, Deng C, Zou JJ, Zhang X. Impact of Deep Hydrogenation on Jet Fuel Oxidation and Deposition. Fuel 2020;264:116843.

30. Jia T, Zhang X, Liu Y, Gong S, Deng C, Pan L, Zou JJ. A Comprehensive Review of the Thermal Oxidation Stability of Jet Fuels. Chem Eng Sci 2021;229:116157.

31. Kabana CG, Botha S, Schmucker C, Woolard C, Beaver B. Oxidative Stability of Middle Distillate Fuels. Part 1: Exploring the Soluble Macromolecular Oxidatively Reactive Species (SMORS) Mechanism with Jet Fuels. Energy Fuels 2011;25(11):5145-57.

Supporting Information

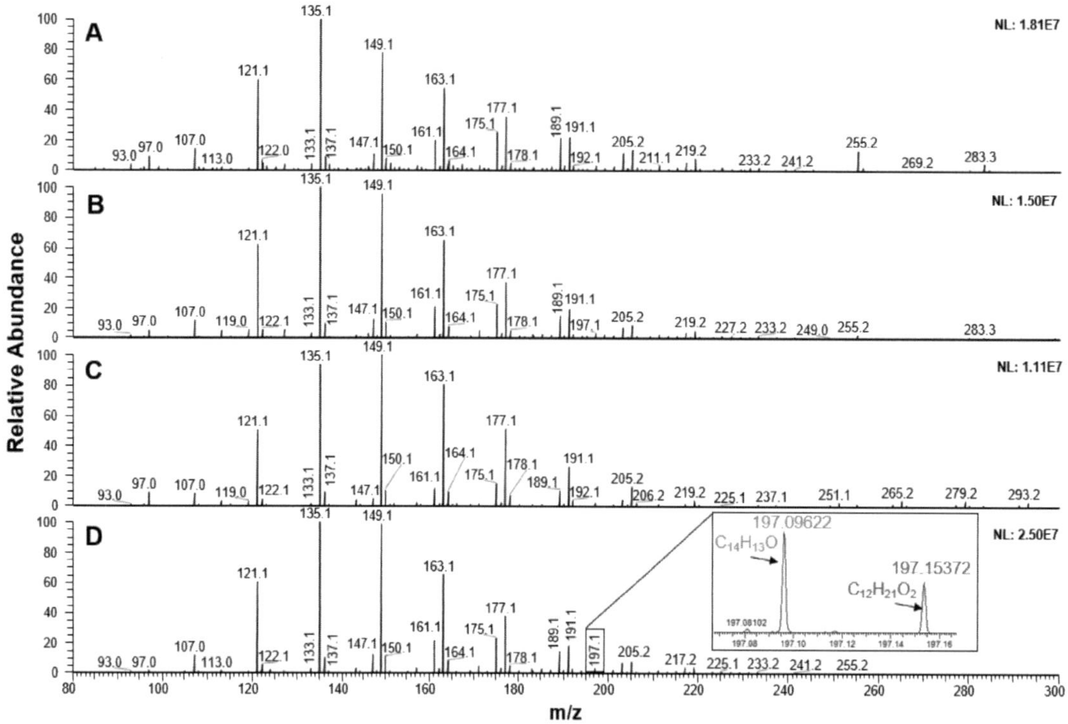

Figure S1. MS[1] spectra of stable fuels (A) S1, (B) S2, (C) S3, and (D) S4.

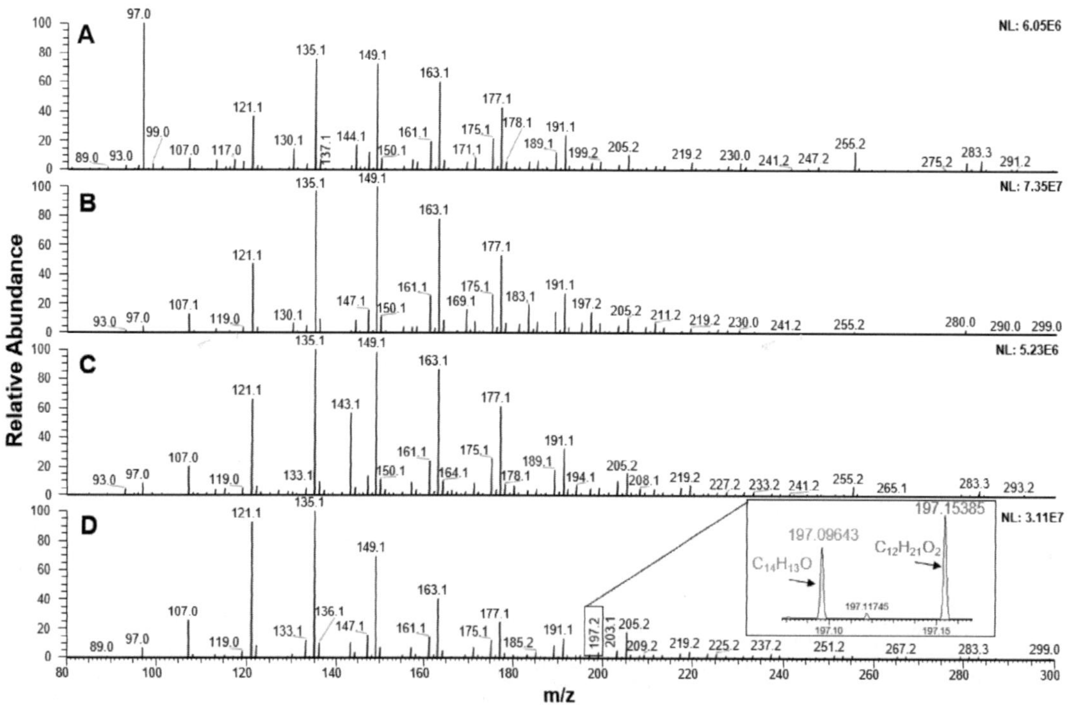

Figure S2. MS[1] spectra of unstable fuels (A) U1, (B) U2, (C) U3, and (D) U4.

Distribution Statement A. Approved for public release: distribution is unlimited.

Table S1. Example of raw data of ionized OCCs derived from U3 by using (-)ESI orbitrap MS.

m/z	Relative abundance	Mass accuracy (ppm)	RDB value	Elemental composition
119.04940	0.8	2.2	5.5	$C_8H_7O^-$
133.06464	4.7	-1.2	5.5	$C_9H_9O^-$
147.08044	13.2	0.0	5.5	$C_{10}H_{11}O^-$
161.09631	23.1	1.4	5.5	$C_{11}H_{13}O^-$
175.11224	25.4	2.9	5.5	$C_{12}H_{15}O^-$
189.12804	17.4	3.4	5.5	$C_{13}H_{17}O^-$
203.14313	9.5	0.4	5.5	$C_{14}H_{19}O^-$
217.15869	5.5	0.0	5.5	$C_{15}H_{21}O^-$
231.17426	2.3	-0.4	5.5	$C_{16}H_{23}O^-$
245.18983	0.8	-0.6	5.5	$C_{17}H_{25}O^-$

Table S2. Qualitative OCC data for S2 Jet A.

Number of oxygen atoms	Empirical formula	RDB value	Avg. KMD	Relative abundance	Carbon number range and avg.	m/z range and avg.
C_xH_yO	$C_nH_{(2n-7)}O$	4.5	0.0698	70.8	6 – 18 (10)	93 – 261 (149)
	$C_nH_{(2n-9)}O$	5.5	0.0832	14.9	8 – 17 (12)	119 – 245 (174)
	$C_nH_{(2n-11)}O$	6.5	0.0966	1.4	10 – 17 (14)	145 – 243 (196)
	$C_nH_{(2n-13)}O$	7.5	0.1100	0.5	10 – 15 (12)	143 – 213 (165)
	$C_nH_{(2n-15)}O$	8.5	0.1235	1.9	12 – 18 (14)	169 – 253 (199)
$C_xH_yO_2$	$C_nH_{(2n-1)}O_2$	1.5	0.0526	3.6	5 – 17 (11)	101 – 269 (180)
	$C_nH_{(2n-3)}O_2$	2.5	0.0660	1.9	5 – 16 (11)	99 – 253 (187)
	$C_nH_{(2n-5)}O_2$	3.5	0.0794	1.2	5 – 16 (12)	97 – 251 (191)
	$C_nH_{(2n-7)}O_2$	4.5	0.0928	1.0	6 – 16 (10)	109 – 249 (160)
	$C_nH_{(2n-9)}O_2$	5.5	0.1061	0.8	7 – 16 (10)	121 – 247 (169)
	$C_nH_{(2n-11)}O_2$	6.5	0.1195	0.2	9 – 14 (12)	147 – 217 (181)
	$C_nH_{(2n-13)}O_2$	7.5	0.1329	0.2	10 – 13 (11)	159 – 201 (177)
$C_xH_yO_3$	$C_nH_{(2n-3)}O_3$	2.5	0.0888	0.4	4 – 11 (7)	101 – 199 (142)
	$C_nH_{(2n-5)}O_3$	3.5	0.1023	0.4	5 – 13 (9)	113 – 225 (163)
	$C_nH_{(2n-7)}O_3$	4.5	0.1157	0.3	5 – 13 (9)	111 – 223 (161)
	$C_nH_{(2n-9)}O_3$	5.5	0.1291	0.3	7 – 13 (10)	137 – 221 (172)
	$C_nH_{(2n-11)}O_3$	6.5	0.1426	0.2	9 – 14 (11)	163 – 233 (193)

Table S3. Qualitative OCC data for S3 Jet A.

Number of oxygen atoms	Empirical formula	RDB value	Avg. KMD	Relative abundance	Carbon number range and avg.	m/z range and avg.
C_xH_yO	$C_nH_{(2n-7)}O$	4.5	0.0701	83.9	6 – 17 (10)	93 – 247 (152)
	$C_nH_{(2n-9)}O$	5.5	0.0835	9.6	8 – 16 (12)	119 – 231 (175)
	$C_nH_{(2n-11)}O$	6.5	0.0969	0.4	11 – 16 (14)	159 – 229 (195)
	$C_nH_{(2n-13)}O$	7.5	0.1102	0.2	10 – 12 (11)	143 – 171 (160)
	$C_nH_{(2n-15)}O$	8.5	0.1238	1.0	12 – 17 (14)	169 – 239 (200)
$C_xH_yO_2$	$C_nH_{(2n-1)}O_2$	1.5	0.0528	2.2	5 – 14 (9)	101 – 227 (154)
	$C_nH_{(2n-3)}O_2$	2.5	0.0663	0.5	6 – 14 (10)	113 – 225 (172)
	$C_nH_{(2n-5)}O_2$	3.5	0.0796	0.2	6 – 10 (8)	111 – 167 (137)
	$C_nH_{(2n-7)}O_2$	4.5	0.0930	0.8	6 – 13 (9)	109 – 207 (150)
	$C_nH_{(2n-9)}O_2$	5.5	0.1064	0.8	7 – 13 (9)	121 – 205 (146)
$C_xH_yO_3$	$C_nH_{(2n-3)}O_3$	2.5	0.0891	0.4	4 – 10 (7)	101 – 185 (136)

Table S4. Qualitative OCC data for S4 Jet A.

Number of oxygen atoms	Empirical formula	RDB value	Avg. KMD	Relative abundance	Carbon number range and avg.	m/z range and avg.
C_xH_yO	$C_nH_{(2n-7)}O$	4.5	0.0702	73.8	6 – 17 (10)	93 – 247 (149)
	$C_nH_{(2n-9)}O$	5.5	0.0837	15.7	8 – 17 (12)	119 – 245 (173)
	$C_nH_{(2n-11)}O$	6.5	0.0971	1.5	10 – 17 (14)	145 – 243 (196)
	$C_nH_{(2n-13)}O$	7.5	0.1102	0.4	10 – 13 (11)	143 – 185 (160)
	$C_nH_{(2n-15)}O$	8.5	0.1239	1.8	12 – 17 (14)	169 – 239 (197)
$C_xH_yO_2$	$C_nH_{(2n-1)}O_2$	1.5	0.0530	3.2	5 – 16 (10)	101 – 255 (172)
	$C_nH_{(2n-3)}O_2$	2.5	0.0664	1.3	6 – 15 (11)	113 – 239 (189)
	$C_nH_{(2n-5)}O_2$	3.5	0.0798	0.8	6 – 16 (12)	111 – 251 (193)
	$C_nH_{(2n-7)}O_2$	4.5	0.0932	0.6	6 – 15 (10)	109 – 235 (160)
	$C_nH_{(2n-9)}O_2$	5.5	0.1065	0.5	7 – 14 (9)	121 – 219 (153)
$C_xH_yO_3$	$C_nH_{(2n-5)}O_3$	3.5	0.1026	0.2	4 – 10 (8)	99 – 183 (149)
	$C_nH_{(2n-7)}O_3$	4.5	0.1160	0.2	7 – 10 (9)	139 – 181 (168)

Table S5. Qualitative OCC data for U1 Jet A.

Number of oxygen atoms	Empirical formula	RDB value	Avg. KMD	Relative abundance	Carbon number range and avg.	m/z range and avg.
C_xH_yO	$C_nH_{(2n-7)}O$	4.5	0.0699	64.5	6 – 16 (10)	93 – 233 (153)
	$C_nH_{(2n-9)}O$	5.5	0.0834	13.8	8 – 17 (12)	119 – 245 (173)
	$C_nH_{(2n-11)}O$	6.5	0.0967	0.5	10 – 14 (12)	145 – 201 (176)
	$C_nH_{(2n-13)}O$	7.5	0.1101	0.6	10 – 13 (11)	143 – 185 (162)
	$C_nH_{(2n-15)}O$	8.5	0.1235	0.8	12 – 15 (14)	169 – 211 (190)
$C_xH_yO_2$	$C_nH_{(2n-1)}O_2$	1.5	0.0523	8.0	6 – 18 (11)	115 – 283 (184)
	$C_nH_{(2n-3)}O_2$	2.5	0.0661	4.9	5 – 15 (11)	99 – 239 (179)
	$C_nH_{(2n-5)}O_2$	3.5	0.0795	1.9	6 – 15 (11)	111 – 237 (187)
	$C_nH_{(2n-9)}O_2$	5.5	0.1063	1.4	7 – 16 (11)	121 – 247 (178)
$C_xH_yO_3$	$C_nH_{(2n-3)}O_3$	2.5	0.0890	2.2	4 – 12 (7)	101 – 213 (143)
$C_xH_yO_4$	$C_nH_{(2n-3)}O_4$	2.5	0.1120	1.4	4 – 12 (7)	117 – 229 (164)

Table S6. Qualitative OCC data for U2 Jet A.

Number of oxygen atoms	Empirical formula	RDB value	Avg. KMD	Relative abundance	Carbon number range and avg.	m/z range and avg.
C_xH_yO	$C_nH_{(2n-7)}O$	4.5	0.0692	65.0	6 – 17 (10)	93 – 247 (152)
	$C_nH_{(2n-9)}O$	5.5	0.0826	14.1	8 – 17 (12)	119 – 245 (169)
	$C_nH_{(2n-11)}O$	6.5	0.0960	0.5	9 – 16 (13)	131 – 229 (180)
	$C_nH_{(2n-13)}O$	7.5	0.1094	0.6	10 – 13 (11)	143 – 185 (159)
	$C_nH_{(2n-15)}O$	8.5	0.1228	0.8	12 – 17 (14)	169 – 239 (196)
$C_xH_yO_2$	$C_nH_{(2n-1)}O_2$	1.5	0.0520	4.9	5 – 16 (11)	101 - 255 (186)
	$C_nH_{(2n-3)}O_2$	2.5	0.0654	9.5	6 – 15 (11)	113 – 239 (185)
	$C_nH_{(2n-5)}O_2$	3.5	0.0788	3.1	6 – 15 (12)	111 – 237 (192)
	$C_nH_{(2n-7)}O_2$	4.5	0.0922	0.6	6 – 14 (10)	109 – 221 (166)
	$C_nH_{(2n-9)}O_2$	5.5	0.1056	0.4	7 – 13 (10)	121 – 205 (162)
	$C_nH_{(2n-11)}O_2$	6.5	0.1188	0.1	10 – 13 (11)	161 – 203 (179)
$C_xH_yO_3$	$C_nH_{(2n-1)}O_3$	1.5	0.0768	0.3	6 – 8 (7)	131 – 159 (144)
	$C_nH_{(2n-3)}O_3$	2.5	0.0884	0.1	4 – 9 (6)	101 – 171 (134)

Table S7. Qualitative OCC data for U3 Jet A.

Number of oxygen atoms	Empirical formula	RDB value	Avg. KMD	Relative abundance	Carbon number range and avg.	m/z range and avg.
C_xH_yO	$C_nH_{(2n-3)}O$	2.5	0.0433	0.2	6 – 9 (7)	97 – 139 (112)
	$C_nH_{(2n-5)}O$	3.5	0.0567	0.3	6 – 9 (7)	95 – 137 (113)
	$C_nH_{(2n-7)}O$	4.5	0.0702	63.0	6 – 18 (10)	93 – 261 (152)
	$C_nH_{(2n-9)}O$	5.5	0.0837	13.2	8 – 17 (12)	119 – 245 (175)
	$C_nH_{(2n-11)}O$	6.5	0.0970	1.1	10 – 17 (13)	145 – 243 (193)
	$C_nH_{(2n-13)}O$	7.5	0.1102	1.1	10 – 13 (11)	143 – 185 (163)
	$C_nH_{(2n-15)}O$	8.5	0.1239	2.1	12 – 18 (14)	169 – 253 (203)
$C_xH_yO_2$	$C_nH_{(2n-1)}O_2$	1.5	0.0531	5.8	5 – 18 (9)	101 – 283 (156)
	$C_nH_{(2n-3)}O_2$	2.5	0.0665	2.9	5 – 17 (11)	99 – 267 (181)
	$C_nH_{(2n-5)}O_2$	3.5	0.0798	1.8	5 – 16 (11)	97 – 251 (177)
	$C_nH_{(2n-7)}O_2$	4.5	0.0931	2.1	6 – 15 (9)	109 – 235 (156)
	$C_nH_{(2n-9)}O_2$	5.5	0.1066	1.9	7 – 16 (9)	121 – 247 (154)
	$C_nH_{(2n-11)}O_2$	6.5	0.1198	0.3	9 – 13 (11)	147 – 203 (174)
	$C_nH_{(2n-13)}O_2$	7.5	0.1331	0.3	10 – 13 (11)	159 – 201 (176)
$C_xH_yO_3$	$C_nH_{(2n-1)}O_3$	1.5	0.0758	0.5	4 – 10 (7)	103 – 187 (147)
	$C_nH_{(2n-3)}O_3$	2.5	0.0892	0.6	4 – 10 (7)	101 – 185 (141)
	$C_nH_{(2n-5)}O_3$	3.5	0.1026	0.8	4 – 12 (8)	99 – 211 (152)
	$C_nH_{(2n-7)}O_3$	4.5	0.1160	0.8	5 – 11 (9)	111 – 195 (160)
	$C_nH_{(2n-9)}O_3$	5.5	0.1294	0.5	7 – 12 (9)	137 – 207 (170)
	$C_nH_{(2n-11)}O_3$	6.5	0.1427	0.2	9 – 12 (11)	163 – 205 (187)
$C_xH_yO_4$	$C_nH_{(2n-5)}O_4$	3.5	0.1256	0.2	8 – 12 (10)	171 – 227 (194)
	$C_nH_{(2n-7)}O_4$	4.5	0.1390	0.3	7 – 13 (10)	155 – 239 (191)

Table S8. Qualitative OCC data for U4 Jet A.

Number of oxygen atoms	Empirical formula	RDB value	Avg. KMD	Relative abundance	Carbon number range and avg.	m/z range and avg.
C_xH_yO	$C_nH_{(2n-7)}O$	4.5	0.0701	66.5	6 – 19 (10)	93 – 275 (145)
	$C_nH_{(2n-9)}O$	5.5	0.0835	12.7	8 – 19 (12)	119 – 273 (167)
	$C_nH_{(2n-11)}O$	6.5	0.0969	0.9	9 – 18 (14)	131 – 257 (202)
	$C_nH_{(2n-13)}O$	7.5	0.1101	0.8	10 – 14 (12)	143 – 199 (165)
	$C_nH_{(2n-15)}O$	8.5	0.1237	1.1	13 – 18 (15)	183 – 253 (209)
$C_xH_yO_2$	$C_nH_{(2n-1)}O_2$	1.5	0.0529	7.5	5 – 18 (11)	101 – 283 (178)
	$C_nH_{(2n-3)}O_2$	2.5	0.0663	3.9	6 – 18 (13)	113 – 281 (213)
	$C_nH_{(2n-5)}O_2$	3.5	0.0798	2.9	9 – 18 (14)	153 – 279 (227)
	$C_nH_{(2n-7)}O_2$	4.5	0.0931	0.8	7 – 18 (13)	123 – 277 (213)
	$C_nH_{(2n-9)}O_2$	5.5	0.1064	0.8	7 – 16 (10)	121 – 247 (156)
	$C_nH_{(2n-11)}O_2$	6.5	0.1197	0.4	9 – 14 (11)	147 – 217 (172)
$C_xH_yO_3$	$C_nH_{(2n-1)}O_3$	1.5	0.0756	0.6	5 – 13 (8)	117 – 229 (164)
	$C_nH_{(2n-3)}O_3$	2.5	0.0891	0.7	4 – 14 (9)	101 – 241 (165)
	$C_nH_{(2n-5)}O_3$	3.5	0.1026	0.2	8 – 13 (11)	155 – 225 (199)
	$C_nH_{(2n-9)}O_3$	5.5	0.1292	0.2	8 – 11 (9)	151 – 193 (167)

Glossary of terms

(1) Ring and double bond equivalent value (**RDB**) – the degree of unsaturation of an organic molecule. The measurement considers double bonds and rings.

$$RDB = X - \frac{Y}{2} + \frac{Z}{2} + 1$$

Where X is the number of carbons, Y is the number of hydrogens and halogen atoms, and Z is the number of nitrogen and phosphorus atoms in the molecule. Please note the number of oxygen atoms is excluded from the formula.

Example: Neutral phenol – C_6H_6O vs. deprotonated phenol – $[C_6H_5O]^-$

RDB Neutral phenol = 4.0

RDB Deprotonated phenol = 4.5

(2) Homologue ion series – ionized oxygen-containing compounds derived from fuel extracts that have the same empirical formula, but differ by 14 Da (a CH_2 unit).

An empirical formula is the simplest whole number ratio of the atoms present in a compound. It does not indicate the total number of atoms in the molecule.

For example, ionized phenols have the same empirical formula, $C_nH_{2n-7}O$. The ions may include m/z of 93 (C_6H_5O), 107 (C_7H_7O), 121 (C_8H_9O), 135 ($C_9H_{11}O$), 149 ($C_{10}H_{13}O$), 163 ($C_{11}H_{15}O$), etc. The ions all have the same empirical formula of $C_nH_{2n-7}O$, but differ by a CH_2. The KMD plots of the stable fuels are shown below. The red boxes highlight the homologue ion series of phenols.

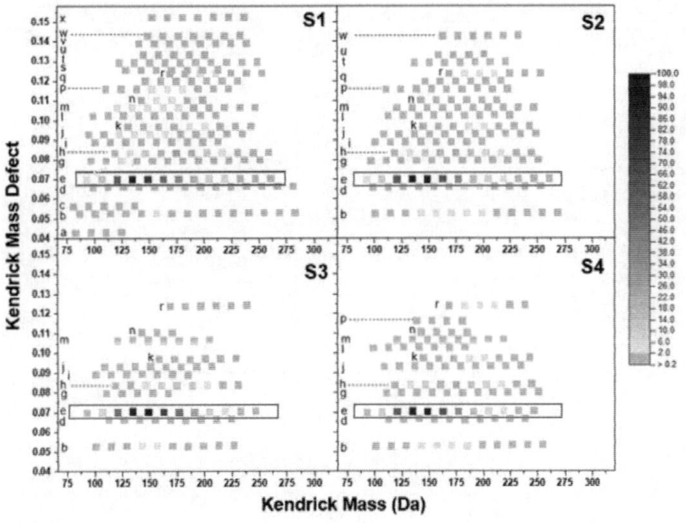

Letter Identifier	Empirical formula	RDB value
a	$C_nH_{(2n-3)}O$	2.5
b	$C_nH_{(2n-1)}O_2$	1.5
c	$C_nH_{(2n-5)}O$	3.5
d	$C_nH_{(2n-3)}O_2$	2.5
e	$C_nH_{(2n-7)}O$	4.5
f	$C_nH_{(2n-1)}O_3$	1.5
g	$C_nH_{(2n-5)}O_2$	3.5
h	$C_nH_{(2n-9)}O$	5.5
i	$C_nH_{(2n-3)}O_3$	2.5
j	$C_nH_{(2n-7)}O_2$	4.5
k	$C_nH_{(2n-11)}O$	6.5
l	$C_nH_{(2n-5)}O_3$	3.5
m	$C_nH_{(2n-9)}O_2$	5.5
n	$C_nH_{(2n-13)}O$	7.5
o	$C_nH_{(2n-3)}O_4$	2.5
p	$CnH_{(2n-7)}O_3$	4.5
q	$C_nH_{(2n-11)}O_2$	6.5
r	$C_nH_{(2n-15)}O$	8.5
s	$C_nH_{(2n-5)}O_4$	3.5
t	$C_nH_{(2n-9)}O_3$	5.5
u	$C_nH_{(2n-13)}O_2$	7.5
v	$C_nH_{(2n-7)}O_4$	4.5
w	$C_nH_{(2n-11)}O_3$	6.5
x	$C_nH_{(2n-9)}O_4$	5.5

IASH 2024, the 18TH INTERNATIONAL CONFERENCE ON STABILITY, HANDLING AND USE OF LIQUID FUELS
Louisville, Kentucky USA
8—12 September 2024

UTILIZING HYDROISOMERIZATION OF NORMAL PARAFFINS TO IMPROVE COLD FLOW PROPERTIES AND ENHANCE STORAGE AND TRANSPORT LOGISTICS.

Victoria Kompanijec[1], Cheryl Leong[1], Mihir Shah[1], Mahlet Garedew[1], Chi Chen[1], Stafford W. Sheehan[1].

[1]*Air Company, 407 Johnson Avenue, Brooklyn, NY 11206 USA.*
victoria.kompanijec@aircompany.com

ABSTRACT

Sustainable aviation fuel (SAF) is a revolutionary synthetic alternative to jet fuel, with an aim to significantly reduce the environmental impacts of air travel. But SAF needs to provide operational performance that is equal to or better than conventional jet fuels made from petroleum. This includes performing safely over a wide range of environmental conditions, such as operating in temperatures as low as –47 °C. The composition of conventional jet fuel produced from petroleum consists of four major hydrocarbon components: isoparaffins, normal paraffins (n-paraffins), cycloparaffins and aromatics. Each of these play an important role in achieving the desired performance and safety of the fuel. Isoparaffins are branched-chain molecules that contribute to improved cold-flow performance: a critical property when operating at high altitudes and for storage at low temperatures. In our CO2-to-SAF process, the CO2 hydrogenation reaction produces primarily linear hydrocarbons, so hydroisomerization must be done to form branched isomers before it can be used as a jet fuel or jet fuel blending component. This study investigates the n-paraffin to isoparaffin conversion efficiency of precious metal lab-scale catalyst using a silica-alumina support material in a model fuel. This study showcases the potential of a CO2-to-paraffin conversion process to formulate a jet fuel for use and storage at lower temperatures, given the efficiency and simplicity of the hydroisomerization process.

Keywords: Sustainable aviation fuel, SAF, hydroisomerization, hydrotreatment, isoparaffins.

1. INTRODUCTION :

Sustainable aviation fuel (SAF) holds the potential to dramatically reduce the environmental footprint of air travel, and it is an area of heightened interest due to the increasing demand for greener alternatives to fossil fuels. However, formulating high-performance SAF that has the same properties as conventional jet fuel presents a challenge, as all the components must be tightly controlled to ensure that the fuel falls within regulatory and safety standards. One crucial aspect of this is meeting demanding cold flow property requirements. This is typically accomplished by incorporating isoparaffins into the fuel blend. These branched-chain hydrocarbons offer several advantages for SAF.[1] Compared to linear paraffins, the lower freezing point of isoparaffins provides improved cold flow properties, which is necessary at the low temperatures experienced at flight altitudes. Second, the energy density of isoparaffins is slightly higher than that of *n*-paraffins, leading to improved SAF fuel efficiency. Finally, the addition of isoparaffins makes SAF more compositionally similar to conventional jet fuel, thereby helping to meet other critical jet fuel properties required by aircraft, engines and the existing aircraft fuel infrastructure. For these reasons, SAF producers may need to include processing steps such as isomerization if their conversion process yields high concentrations of n-paraffins and low concentrations of isoparaffins in their synthetic jet fuel compositions. However, some reaction pathways from CO_2 to paraffins do not yield iso-alkanes in the desired ratio, so additional processing is required before it can be used in a fuel blend.

The most prevalent method of isomerizing paraffins is with a hydroisomerization catalyst containing a molecular sieve loaded with a noble metal such as palladium or platinum (Pt). The functionality of catalysts is driven by parameters such as pore size, along with the loading, or proportions of catalytic material type (metal or acid) and the material's dispersion pattern on the catalyst surface. Hydroisomerization catalysts typically exhibit medium pore size, high metal dispersion on the catalyst surface, low acidity and an equal mixture of acid and metal sites. The type of molecular sieve can be tuned for specific purposes such as loading a certain metal, but the commonly used classes are zeolites such as ZSM-5, ZSM-22, ZSM-48, and silicoaluminophosphates (SAPOs) like SAPO-11 and SAPO-31.[2] For hydroisomerization of long chain n-paraffins, SAPO has been shown to be more effective than other zeolites due to its lower acidity.[3] A 2004 study demonstrated high yields of isomers (up to 90% conversion) from a *n*-tetradecane feed using 0.4% Pt loaded onto in-house synthesized SAPO-11 powder. Another recent experiment showed an 81.8% conversion of *n*-hexadecane to isomers with a selectivity of 86.5%. Based on these results, we selected to use a Pt-impregnated SAPO-11 catalyst for our investigation.

In SAF, there must be a mixture of both isoparaffins and paraffins, though the exact ratio can vary depending on the attributes of the rest of the fuel components. Because of this, full isomerization of *n*-paraffins is not desirable. Instead, we target approximately 70% isoparaffins after reaction for use in SAF to better align with the composition of conventional jet fuel.

Additionally, a concern for hydroisomerization is hydrocracking of long-chain paraffins to lower carbon number species, which occurs following hydroisomerization.[4,5] These lighter hydrocarbons would be unusable for SAF, a disadvantage for commercial applications. For these reasons, we hypothesize that a lower Pt loading than is typically found in lab-scale studies may be more appropriate for processing larger volumes of SAF due to less activity for cracking. An added benefit of this would be a reduction in the cost of materials needed for this process. In this work, we investigate the effects of different Pt loading on the isomerization activity of SAPO-11 and the impact of varying the calcination temperature. Our model fuel blend contains C_9-C_{16} *n*-paraffins, designed to mimic the composition of a CO_2 derived stream, to improve upon the existing body of literature that examines one model compound alone. Additionally, we explore the pathway for deactivation of the Pt/SAPO-11 catalytic material by oxygenates, which is vital to the processing of fuels that may contain these species.

2. MATERIALS AND METHODS:

Materials and Equipment

The model fuel was comprised of Nonane, decane, undecane, dodecane, tridecane, tetradecane, pentadecane and hexadecane that were purchased from Fisher Scientific and used without further purification. In this experiment, chloroplatinic acid hydrate (≥99.9%, trace metals basis) was obtained from Sigma-Aldrich and was used as the source of catalytic Pt. Extrudates of the SAPO-11 catalytic base material that were ready for loading were purchased from a commercial source. Ultra-high purity hydrogen (100%) from Airgas was used during hydroisomerization. All catalysis took place in a fixed bed reactor scheme with temperature control provided by an internal thermocouple and an MTI single-zone tube furnace. Gas flow was controlled using a Brooks mass flow controller (MFC) and liquid flow was controlled with an Eldex Optos Series pump.

Experimental Conditions

Catalyst preparation: Application, or loading, of the Pt metal component of the catalyst onto the SAPO-11 extrudates was accomplished using incipient wetness impregnation (IWI) of the chloroplatinic acid carrier solution on to the SAPO-11 extrudate structure. IWI is a technique for loading a catalyst support with a precursor solution by drawing it in with capillary action, using just enough liquid to fill the support's pores. Chloroplatinic acid was weighed out such that the mass was 0.0.095% or 2% by weight of the SAPO-11 extrudates to be impregnated via IWI, allowing the extrudates to be evenly coated by the chloroplatinic acid carrier solution. The catalyst was calcined at 700 C, unless otherwise stated, before pretreatment began. Calcination is used to heat the catalyst in the presence of oxygen to remove the organic template and expose active sites within the framework. The reactor used was a stainless-steel 1" tube reactor and the loading was: quartz wool, SiC (inert), quartz wool, catalyst, quartz wool, SiC (inert), quartz wool.

Catalyst Pretreatment: Hydrogen was flushed through the reactor for 10 minutes to remove any possible oxygen from the system. At that point, the pressure on the reactor was increased to 100 psig and, after pressure was stable, the furnace temperature was adjusted to ramp to 350 °C over 30 minutes. This temperature was maintained for four hours.

Reaction: The reactor temperature was set to 280 C, and the pressure was adjusted up to 350 psig with a constant H_2 flow of 300 sccm. Once the conditions were stable, the liquid feed of a surrogate fuel mixture was introduced through the top of the reactor at a flow rate of 0.16 mL/min. The weight hourly space velocity (WHSV), which is a metric of the weight of sample fed into the reactor per weight of the catalyst per unit time, was 1 hr^{-1}. The pump was primed directly before beginning feed to ensure that the flow had no interruptions, and the level of liquid in the feed container was monitored periodically to check for proper flow.

Data collection and analysis: Liquid samples were collected and analyzed using GC-MS (Agilent 5973 Mass Selective Detector) and GC-FID (Trace 1310). To determine the full selectivity of hydroisomerization, gaseous products of the reaction were also periodically monitored using GC. The gas outlet of the reactor was connected to an INFICON Micro GC Fusion® Gas Analyzer.

The accuracy of the collected data was verified by calculating the carbon and mass balance for each reaction. All experimental results were within $\pm$ 3% of 100% for both figures of merit.

The isomer conversion was calculated using the normalized weight percent for each *n*-paraffin and the isomers of each carbon number, using equation 1.

$$(1)\quad Total\ isoparaffin\ composition = \sum_{C_9}^{C_{16}} \left(\frac{C_n\ isomer\ wt.\%}{total\ C_n\ wt.\%} \cdot C_n\ wt.\% \right)$$

3. RESULTS AND DISCUSSION:

3.1 Isomerization of n*-paraffins*

Upon isomerization under standard conditions, the overall conversion of *n*-paraffins to isoparaffins was 70%. It was observed that longer hydrocarbon chains underwent a higher degree of isomerization, which aligns with expected results based on well-established knowledge of isomerization using other Pt/SAPO-11 catalysts.[6] Table 1 reports the conversion for each *n*-paraffin along with the isoparaffinic percent and weight percent for each carbon number.

Table 1. Conversion data of *n*-paraffins to isoparaffins under typical reaction conditions (280 C, 350 psig, WHSV = 1 hr^{-1}, 0.095 wt.% Pt/SAPO-11 calcined at 700 C).

Carbon number	Isomer composition (%)	Total weight percent of fuel (%)
C_7 and below	--	2.35
C_8	--	1.09
C_9	28.3	10.7
C_{10}	45.9	12.3
C_{11}	61.0	15.0
C_{12}	75.2	15.5
C_{13}	81.5	14.2
C_{14}	85.3	13.0
C_{15}	93.2	10.9
C_{16}	92.6	4.86

Notably, the same distribution was not observed when using Pt/ZSM-22; Claude et al. reported no statistically significant difference in the total isomerization conversion of C_{10}-C_{24} alkanes, although there was a clear increase in multi-branched products for the higher carbon number species.[7] This implies a difference in the isomerization mechanism dependent on the molecular sieve used. The resulting chromatograph of the model fuel blend before and after isomerization is shown in Figure 1.

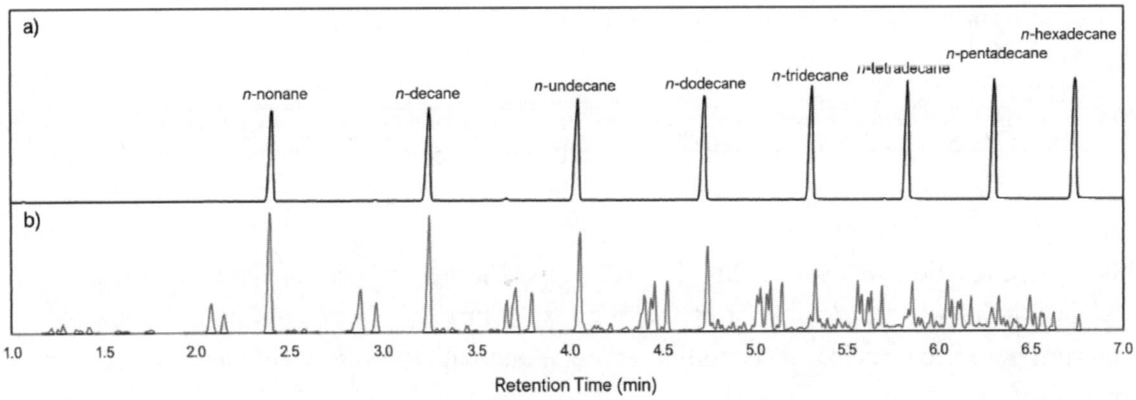

Figure 1. Gas chromatograms depicting model fuel blend (a) before and (b) after hydroisomerization at 280 C, 350 psig, WHSV = 1 hr^{-1} with 0.095 wt.% Pt/SAPO-11 calcined at 700 °C.

After hydroisomerization, the model fuel contained 70% isoparaffins by weight percent, which met the target set for the reaction. The density of the fuel was also decreased, with a density of 0.7727 g/mL prior to hydroisomerization and an average density of 0.7273 g/mL after. The amount of cracking was determined through the presence of any compound with a carbon number below C_9, detected both as a liquid or gas product. Of the reaction products, 98.97% were liquid phase, while 1.03% were gas phase. Of the liquid products, 3.44% were C_8 or lighter, meaning that the overall cracking for the reaction was 4.43%. Previous work done on hydroisomerization of *n*-paraffins lists 4% cracking selectivity as "almost negligible" and falls below the standard target for many similar systems.[8,9]

3.2 Effect of calcination temperature on isomerization

To gain a more complete understanding of the catalyst framework and how the activation process interplays with isomerization activity, we conducted an experiment with a lower calcination temperature. Calcination at 550 °C is a commonly reported procedure in the literature, but 700 °C can lead to a more thermally stable SAPO-11 structure, increasing the mesoporous pore volume and diameter.[10] It also is postulated that the higher calcination temperature may remove more moisture and/or residual organic material, which could be useful when working with bulk quantities of SAPO-11 extrudates. However, it is possible that this high temperature can lead to the sintering of the metal components of Pt/SAPO-11, which could lead to lower activity and degradation of the material over time.[11] In our test, all conditions were held identical other than the temperature of calcination. Overall, there was not much change in the reactivity for hydroisomerization. A comparison between results of the two calcination conditions can be found in table 2.

Table 2. Conversion results for Pt/SAPO-11 tested after being calcined at 550 °C and 700 °C. Reaction conditions: 280 °C, 350 psig, WHSV = 1 hr^{-1}.

Calcination Temperature (C)	**Total isomer composition (%)**	**Cracking (%)**
550	68	5.00
700	70	4.39

Although the reaction yields for both experiments were largely the same, there was a slight difference in the product distribution of the isomers obtained. Figure 2 outlines the isomer composition for each calcination condition tested, along with the total weight percent of each carbon number.

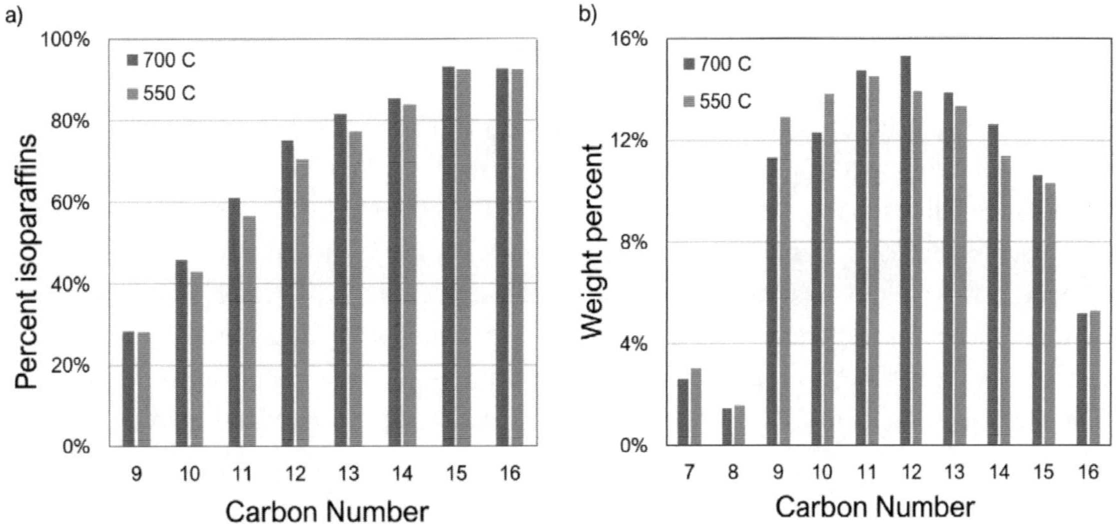

Figure 2. Comparison of the impact of different calcination temperatures on the model fuel blend distribution post-isomerization. Reaction conditions: 280 °C, 350 psig, WHSV = 1 hr^{-1}. Panel (a) shows the relative amount of isoparaffins for each carbon number, while panel (b) shows the weight percent of each carbon number found in the fuel mixture.

Looking at the products of isomerization closely, it is observed that the 700 °C calcination conditions lead to slightly higher isoparaffinic content in C_{10}-C_{15}, which may be attributed to a slight variation in the pore dimensions of each catalyst. Additionally, the weight distribution of the 550 °C calcined Pt/SAPO-11 lies marginally lower than that of the 700 °C calcined Pt/SAPO-11. This is a finding with potentially useful ramifications for fuel processing, as the calcination temperature may be used to fine-tune product selectivity. This would be useful to allow for slight adjustments to the catalytic system without the need to change the type of catalyst or reaction conditions. Solids characterization of the molecular sieve immediately following calcination may provide more insight into the specific changes to the SAPO-11 structure occurring at different temperatures.

3.3 Effect of Pt loading on SAPO-11 isomerization activity

It is well known that Pt loading is necessary for SAPO-11 to be active for hydroisomerization of *n*-paraffins, as it is responsible for the dehydrogenation of the hydrocarbons, enabling the molecular rearrangements.[12] To confirm that this pathway is taking place, we examined a reaction with SAPO-11 without any loaded Pt. In addition, we hypothesized that our catalytic system would lead to less cracking due to a lower Pt loading relative to other catalysts reported in the literature. A representative of higher loading was chosen to be 2 wt.% Pt/SAPO-11, as it falls within typical ranges for this application.[13] The comparison between the results of 0%, 0.095% and 2% Pt loading are found in Figure 3.

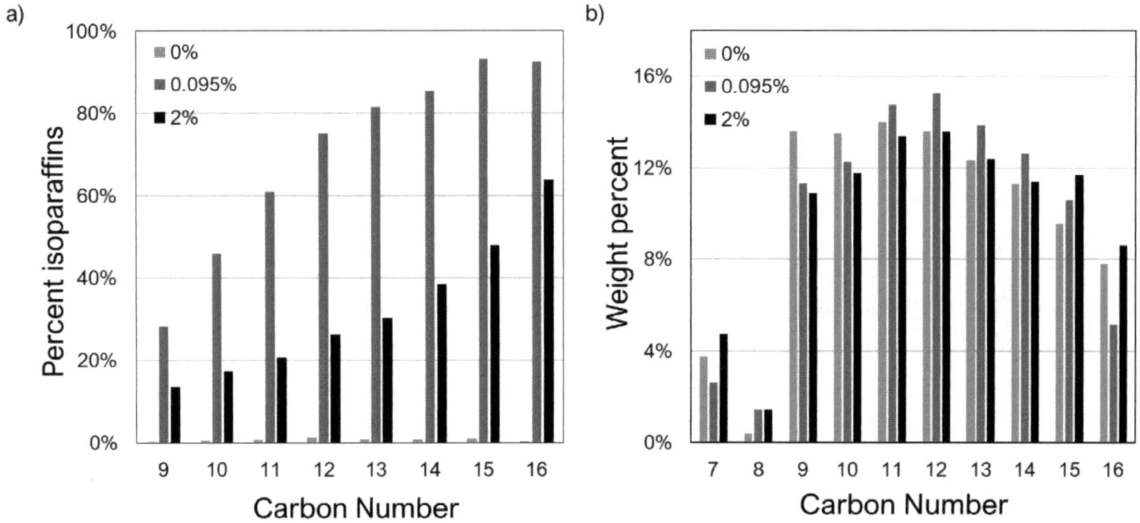

Figure 3. Comparison of SAPO-11 with 0, 0.095 and 2 wt.% Pt loading effects on (a) the isoparaffin content and (b) weight percent distribution in model fuel. Reaction conditions: 280 C, 350 psig, WHSV = 1 hr^{-1}.

The results of this study aligned with the expected outcomes: SAPO-11 with no Pt loading did not achieve meaningful isomerization and 2 wt.% Pt/SAPO-11 had some isomerization, but the cracking was more pronounced than the 0.095 wt.%. SAPO-11 alone had trace isomerization (5%), but it remains unclear whether any amount took place or if the isomers can be attributed to residue from a previous reaction within the liquid receiver. What is unambiguous, however, is that there was cracking observed over the SAPO-11, indicating that the molecular sieve alone can break apart higher carbon number species. The cracking was calculated to be 5.66%, which is higher than 0.095 wt.% but without any significant isomerization. This confirms that SAPO-11 is not a suitable catalyst for hydroisomerization of paraffins without metal loading. The 2% Pt/SAPO-11 resulted in 35% isomers after reaction, which is half of the reactivity seen by the 0.095 wt.%. This could be explained by an overabundance of cracking sites on the catalyst structure, as a high density of Pt can contribute to heightened cracking of hydrocarbon species rather than maintaining isomeric structures.[14] Another possible explanation for this could be that the higher Pt loading is not compatible with 700 °C calcination – Pt clusters can block access to the acidic sites on the SAPO-11 framework, and the high Pt density may lead to heightened sintering effects that may not be prevalent at lower loading.[15] Also, higher Pt loading may lead to the formation of Pt atom clusters, which are not favorable for isomerization compared to well-dispersed Pt atoms.[16,17] The results show that the 0.095% Pt loading is the most appropriate for applications in SAF processing, though additional studies could be performed to identify if there is a more suitable level of loading to balance dehydrogenation metal sites with acid sites.

3.3 Investigation into deactivation threshold of Pt/SAPO-11 for oxygenates.

One major consideration for hydroisomerization is the deactivation of the catalyst. At elevated temperatures, the structure of SAPO-11 becomes intolerant of moisture, experiencing framework degradation.[18] This has two major impacts on the use of Pt/SAPO-11: first, it becomes impossible to return to ambient conditions after calcination, due to the adsorption of water present in the air, which will result in structural damage upon heating to reaction conditions again. Second, it means that the catalyst is sensitive to impurities found in liquid feeds. In SAF, there are often small molecules present that may exist as byproducts of the reaction pathway or can be formed during other hydrotreating processes such as HDO. These molecules can include alcohols, carbonyls, and carboxylic acids, among others.[19] Fuel processing may also result in the formation of water when these species are present, such as from decarboxylation.[20] To assess the Pt/SAPO-11 catalyst under realistic feed conditions, we conducted a study to determine the threshold of deactivation under the addition of a representative acid and water. We selected hexanoic acid for this study because it is the mean carbon number acid comprising our typical CO_2-to-paraffin stream. The gas chromatograms before and after hydrogenation are shown for 0 wt.% hexanoic acid to 1 wt.% hexanoic acid in Figure 4.

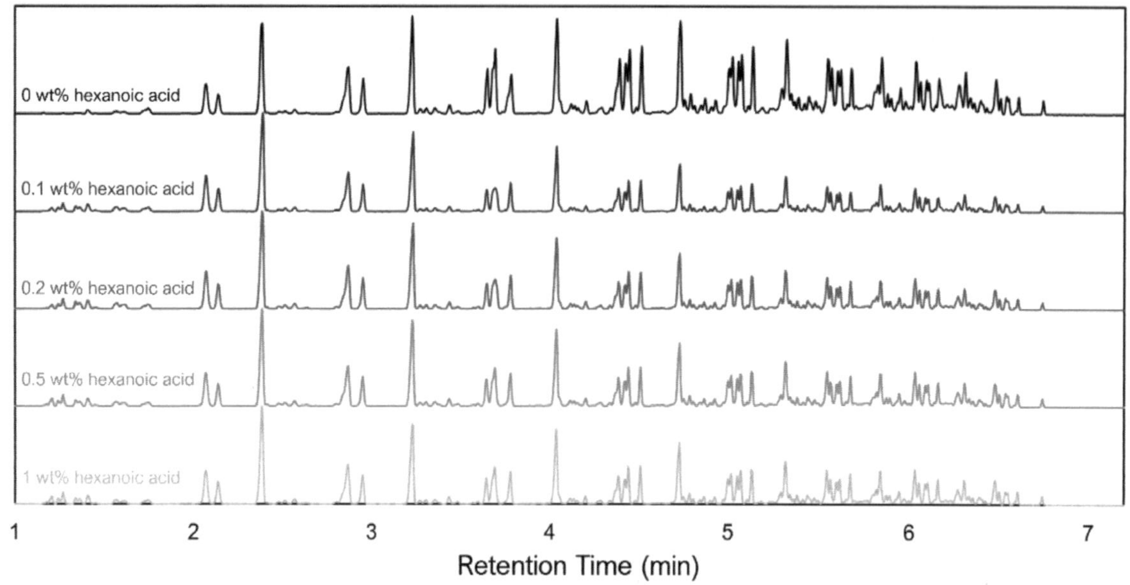

Figure 4. Gas chromatograms of model fuel with increasing concentration of hexanoic acid after hydroisomerization. Reaction conditions: 280 °C, 350 psig, WHSV = 1 hr^{-1}.

Looking at the activity of Pt/SAPO-11 under increasing concentrations of hexanoic acids, it is apparent that there is no meaningful decrease in the hydroisomerization conversion (verified by peak area). This leads to the conclusion that carboxylic acids are not species which will lead to catalyst deactivation at this set of reaction conditions.

In addition to hexanoic acid, we studied water the addition of water at different concentrations. This study's purpose was to determine if a different type of functional group could deactivate the catalyst, which would give insight to what classes of compounds would be deleterious to the process. As water is not fully miscible in the model fuel, sonication was performed on the liquid feed immediately before addition to ensure a homogenous suspension. The resultant chromatograms are pictured in Figure 5.

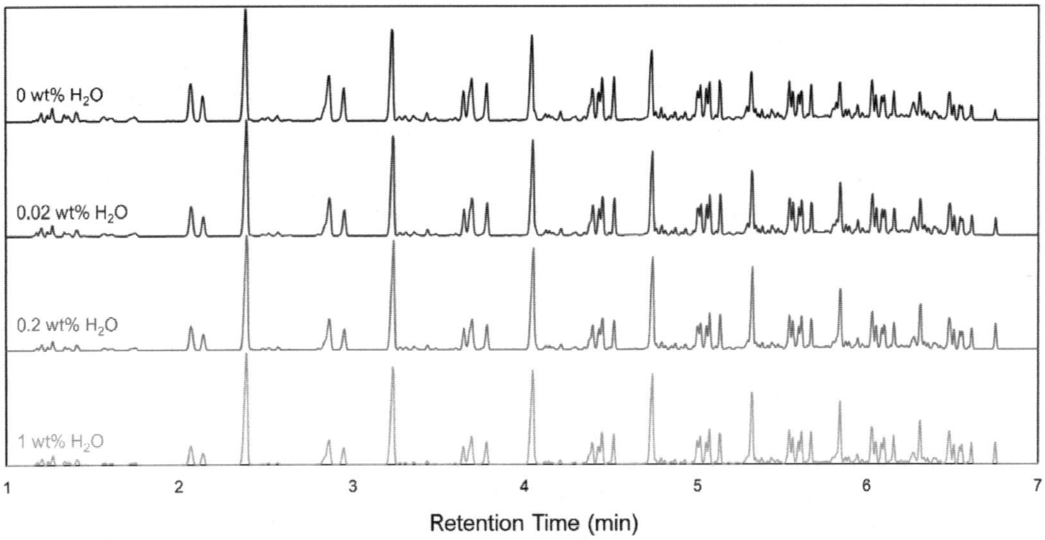

Figure 5. Gas chromatograms of model fuel with increasing water concentration post-hydroisomerization. Reaction conditions: 280 °C, 350 psig, WHSV = 1 hr^{-1}.

In this case, the increasing addition of water to the model fuel led to a substantial decrease in the conversion of *n*-paraffins to isoparaffins. The ratio of paraffins to isoparaffins for C_9 compounds decreased from 0.61 to 0.27 upon the addition of 1 wt.% water, while the ratio remained unchanged when feeding 1 wt.% hexanoic acid. There is reason to believe, based on literature reports, that SAPO-11 can tolerate water at elevated temperatures such as 280 °C. For example, one study used SAPO-11 as a catalyst for ethanol dehydration from 250 °C – 330 °C and reported no deactivation, considering that water was one of the major products of the reaction.[21] The conclusion of this is that the SAPO-11 framework was unaffected by the water, but the Pt metal sites were deactivated, which resulted in lower activity. The poisoning of Pt is a well-understood process, in which the metal site of the catalyst is irreversibly bonded to a heteroatom species.[22] We propose that a similar mechanism to Pt poisoning is taking place, in which the –OH group of water is binding to Pt sites, rendering them inactive for dehydrogenation of *n*-paraffins. To further probe this, a representative alcohol could be fed into the system and assessed for similar results. This demonstrates that, though Pt/SAPO-11 is a highly effective catalyst for hydroisomerization, the composition of the feed must be carefully considered.

4. CONCLUSION:

In this work, we have shown that a low loading of Pt on SAPO-11 (0.095 wt.%) can facilitate the production of 70% isoparaffins from a feed of varying length *n*-paraffins, with minimal cracking. We also demonstrated that the final composition of the model fuel can be tuned using simple parameters such as calcination temperature, which would be advantageous for large-scale processing. The presence of a metal site is necessary to result in isomerization, but excess Pt results in lower yield and higher cracking of hydrocarbons. Finally, the deactivation of the catalyst was investigated, showing that certain impurities like water will be detrimental to catalysis. However, acids, a common byproduct when producing SAF, do not impact the reactivity of 0.095 wt.% Pt/SAPO-11 up to 1 wt.% of the total fuel. In all, this catalytic system has shown that an n-paraffin stream can be converted to a target iso-paraffin composition in one pass, which is desirable for SAF production. Iso-paraffins are critical components of fuel blends, so having a simple, effective route to achieving the correct ratio has positive ramifications for the feasibility of SAF in the aviation industry.

5. REFERENCES:

[1] "Sustainable Aviation Fuel." *Alternative Fuels Data Center*, https://afdc.energy.gov/fuels/sustainable_aviation_fuel.html Accessed 13 Feb. 2024.

[2] Deldari, H. Suitable catalysts for hydroisomerization of long-chain normal paraffins. *Appl. Catal. A Gen.* **2005**, *293*, 1-10.

[3] Khouzami, R.; Coudurier, G.; Mentzen, B. F.; Vedrine, J. C. Structural, Acidic and Catalytic Properties of SAPO-11 Molecular Sieves. *Stud. Surf. Sci.* **1988**, *37*, 355-363.

[4] Calemma, V.; Peratello, S.; Giardina, R.; Perego, C. Hydrocracking and Hydroisomerization of Long-Chain *n*-Paraffins. Reactivity and Reaction Pathway for Base Oil Formation. *Ind. Eng. Chem. Res.* **2004**, *43*, 934-940.

[5] Calemma, V.; Peratello, S.; Perego, C. Hydroisomerization and hydrocracking of long chain *n*-alkanes on Pt/amorphous SiO_2—Al_2O_3 catalyst. *Appl. Catal. A Gen.* **1999**, *190*, 207-218.

[6] Geng, C.; Zhang, F.; Gao, Z.; Zhao, L.; Zhou, J. Hydroisomerization of *n*-tetradecane over Pt/SAPO-11 catalyst. *Catal. Today* **2004**, *93-95, 485-491.*

[7] Claude, M. C.; Vanbutsele, G.; Martens, J. A. Dimethyl Branching of Long *n*-Alkanes in the Range from Decane to Tetracosane on Pt/H-ZSM-22 Bifunctional Catalyst. *J. Catal.* **2001**, *203*, 213-231.

[8] Oloye, F. F.; Aliyev, R.; Anderson, J. A. Hydroisomerization of *n*-heptane over Pt/sulfated zirconia catalyst at atmospheric pressure. *Fuel* **2018**, *222*, 569-573.

[9] Misra, P.; Alvarez-Majmutov, A.; Chen, J. Isomerization catalysts and technologies for biorefining: Opportunities for producing sustainable aviation fuels. *Fuel* **2023**, *351*, 128994.

[10] Hu, M.; Yuan, H.; Liu, W. Decarboxylation of Oleic Acid over Ordered Mesoporous Pt/SAPO-11. *Energy Fuels* **2019**, *33*, 9956-9964.

[11] Lv, G.; Wang, C.; Chi, K.; Liu, H. Wang, P.; Ma, H.; Qu, W.; Tian, Z. Effects of Pt site distributions on the catalytic performance of Pt/SAPO-11 for *n*-dodecane hydroisomerization. *Catal. Today* **2018**, *316*, 43-50.

[12] Liu, Y.; Liu, C.; Liu, C.; Tian, Z.; Lin, L. Sn-modified Pt/SAPO-11 catalysts for selective hydroisomerization of n-paraffins. *Energy & Fuels* **2004**, *18*, 1266-1271.

[13] Frey, S. J.; Wang, H.; Bozzano, A. G. Process for Producing Jet Fuel from Isomerizing a Biorenewable Feed. US 2023/0193143 A1, June 22, 2023.

[14] van der Wal, L. I.; Oenema, J.; Smulders, L. C.; Samplonius, N. J.; Nanpersad, K. R.; Zecevic, J.; de Jong, K. P. Control and Impact of Metal Loading Heterogeneities at the Nanoscale on the Performance of Pt/Zeolite Y Catalysts for Alkane Hydroconversion. *ACS Catal.* **2021**, *11*, 3842-3855.

[15] Yang, S.; Liu, X.; Zhang, X.; Sun, W.; Ma, Q.; Song, N.; Lu, M.; Hu, J. Isomerization Properties of Pt/SAPO-11 Catalysts for the Production of Bio-Aviation Kerosene. *Catalysts* **2023**, *13*, 1100.

[16] Chen, N.; Gong, S.; Shirai, H.; Watanabe, T.; Qian, E. W. Effects of Si/Al ratio and Pt loading on Pt/SAPO-11 catalysts in hydroconversion of Jatropha oil. *Appl. Catal. A Gen.* **2013**, *466*, 105-115.

[17] Al-Rawi, U. A.; Sher, F.; Hazafa, A.; Rasheed, T.; Al-Shara, N. K.; Lima, E. C.; Shanshool, J. Catalystic Activity of Pt Loaded Zeolites for Hydroisomerization of *n*-Hexane Using Supercritical CO_2. *Ind. Eng. Chem. Res.* **2020**, *59*, 22092-22106.

[18] Lutz, W.; Kurzhals, R.; Sauerbeck, S.; Toufar, H.; Buhl, J. C.; Gesing, T.; Altenburg, W.; Jager, C. Hydrothermal stability of zeolite SAPO-11. *Microporous Mesoporous Mater.* **2010**, *132*, 31-36.

[19] United States. Dept. Of Energy. *Sustainable Aviation Fuel Review of Technical Pathways*. EE-2041, 2020.

[20] Ni, W.; Zhu, G.; Liu, F.; Xie, C.; Han, Y. Carboxylic Acids in Petroleum: Separation, Analysis, and Geochemical Significance. *Energy Fuels* **2021**, *35*, 12828-12844.

[21] Effects of the SAPO-11 synthetic process on dehydration of ethanol to ethylene. *Front. Chem. Sci. Eng.* **2011**, *5*, 60-66.

[22] Chang, J. R.; Chang, S. L.; Lin, T. B. γ-Alumina-Supported Pt Catalysts for Aromatics Reduction: A Structural Investigation of Sulfur Poisoning Catalyst Deactivation. *J. Catal.* **1997**, *169*, 338-346.

***IASH 2024, the 18TH INTERNATIONAL CONFERENCE ON STABILITY, HANDLING AND USE OF LIQUID FUELS
Louisville, Kentucky, USA
September 8 – 12, 2024***

IMPROVING THE QUALITY OF ADVANCED RECYCLED OILS

Joseph L Stark*, Paul Biggerstaff, Randy Garver

Innospec Fuel Specialties LLC, 200 Executive Dr, Newark, DE 19702

ABSTRACT

The demand for plastics has steadily increased since production began in the early 1950s due its durable yet light weight properties and flexibility to be used in a variety of forms to improve our daily lives. Plastics are utilized in nearly all aspects of our daily lives in, from healthcare, food packaging, clothing, and building materials just to name a few. Over 400 million tons of plastic are produced globally per year, and this volume is expected to triple by 2050. The majority of plastic ends up in landfills as waste. Currently only around 9% of plastic is recycled through traditional mechanical methods and those plastics suitable for mechanical recycling can only go through the process once or twice before the plastic degrades beyond additional use. Recent innovations in advanced recycling, such as pyrolyzation, allows for plastic waste to be broken down into its original building blocks. The oil generated through pyrolysis can be recycled infinitely in petrochemical facilities to remake plastic in a circular process. Pyrolysis oil can also be sent to refineries as a feedstock to process units or fractionated directly into fuels. Waste plastic pyrolysis oil (WPPO) generated from advanced recycling tends to be highly olefinic, contain a high concentration of long chain carbon waxes and may be corrosive due to high levels of acidic compounds. In this paper, the cold flow properties, and thermal and oxidative stability properties, of WPPO will be reviewed. Potential solutions for mitigating instability, reducing the potential for fouling issues during process, storage, and transport, and improving the flow properties of these oils will be discussed.

*Corresponding author: Joseph L Stark, *Joseph.Stark@innospecinc.com*

IASH 2024, the 18TH INTERNATIONAL CONFERENCE ON STABILITY, HANDLING AND USE OF LIQUID FUELS
Louisville, Kentucky, USA
September 8 – 12, 2024

UNIQUE ORGANIC CARBONATE BASED SOLVENTS DERIVED FROM CO2 FOR SULFUR REMOVAL TECHNOLOGY

Sankaran Murugesan*, Onome Ugono, Sai Pinappu

Baker Hughes, Inc., 12645 West Airport Blvd, Sugarland,Texas 77478 USA

ABSTRACT

The presence of unreactive organic sulfur species in hydrocarbon stream cause both operational and environmental problems. As such regulatory bodies set limits to the allowable amounts of these contaminants present in finished fuels. These serve as the required specifications for those finished fuel streams. In order to meet these specifications several approaches are used to remove the sulfur compounds. Harsh chemicals like caustic are widely employed to remove these sulfur bearing impurities. However, in many circumstances, caustics and other chemicals are unable to effectively reduce the amount of these organosulfur impurities to the required levels. This paper presents the development of an environmentally responsible solvent medium to remove sulfur compounds. This class of solvents (organic carbonates) is derived from CO2. The efficiency of the extraction of the sulfur compounds was studied for different hydrocarbon stream and also across the two process phases i.e., gas-liquid and liquid-liquid. The sulfur species carrying capacity and number of cycles required for efficient extraction were also analyzed. Furthermore, the regeneration and reuse of this solvent system was also explored. The observations from this study offered an improved process for organosulfur removal comparing to commercially available alternatives.

*Corresponding author: Sankaran Murugesan, sankaran.murugesan@bakerhughes.com

IASH 2024, the 18TH INTERNATIONAL CONFERENCE ON STABILITY, HANDLING AND USE OF LIQUID FUELS
Louisville, Kentucky, USA
September 8 – 12, 2024

STORAGE, HANDLING AND COMBUSTION OF A NOVEL HEATING FUEL: EL

Ryan Kerr[1]*, Thomas Butcher[1], Neehad Islam[1], Robert O'Brien[1], Stephen Fitzpatrick[2] and Michael Cassata[2]

[1]*National Oilheat Research Alliance, 131 East Ames Court, Plainview, NY, 11757 USA.* *rkerr@noraweb.org*
[2]*Biofine Developments Northeast, 77 Exchange St, Bangor, ME 04401 USA.*

Keywords: biofuels, heating, boiler, furnace, combustion, storage

ABSTRACT

Ethyl Levulinate (EL) is being developed as a cellulosic, renewable fuel for use in boilers and furnaces. In this paper, the technical aspects of using this fuel in this market are discussed. Positive attributes of EL include very low net greenhouse gas emissions, excellent cold flow properties, acceptable flash point, and low viscosity. This candidate fuel is not generally miscible with petroleum fuels. It is miscible with biodiesel but in such blends the cold flow advantages are reduced. Use of EL alone is seen as a practical deployment pathway. EL has not been found to be compatible with elastomers commonly used in petroleum designed equipment and alternative seal materials have been identified and integrated into burner systems for testing. With an oxygen content of 33% the heating value of EL, on a volumetric basis, is 30% lower than for petroleum heating fuel. At an equal energy input rate, the NOx emissions with EL have been found to be 40% lower.

INTRODUCTION

Renewable biofuels, derived from non-food and waste sources, offer the potential to achieve net reductions in greenhouse gas emissions (GHG) by displacing fossil fuel use from a range of applications. This study is focused on the use of biomass-derived ethyl levulinate (EL) as a fuel for use in stationary boiler and furnace applications. Relative to transportation applications, these have a lower global fuel consumption but are also seen as less technically challenging for biofuel conversion.

Ethyl levulinate (EL) fuel production is based here on the Biofine Process [1, 2] which starts with the mild acid hydrolysis of cellulosic feedstock. Primary products of these steps include levulinic acid (LA), formic acid (FA), and furfural. The levulinic acid is then esterified to produce ethyl levulinate. Figure 1 provides a simplified scheme representing the multi-stage chemical cascade occurring in a typical dilute acid hydrolysis reaction resulting in LA. Under typical conditions of aqueous dilute mineral acid (3%) and elevated temperature (200 °C.) cellulose is broken down by hydrolysis to form monomeric glucose. This isomerizes to fructose followed by dehydration to the unstable intermediate hydroxymethyfurfural (HMF). HMF then hydrolyzes to form the stable end-products LA and FA.

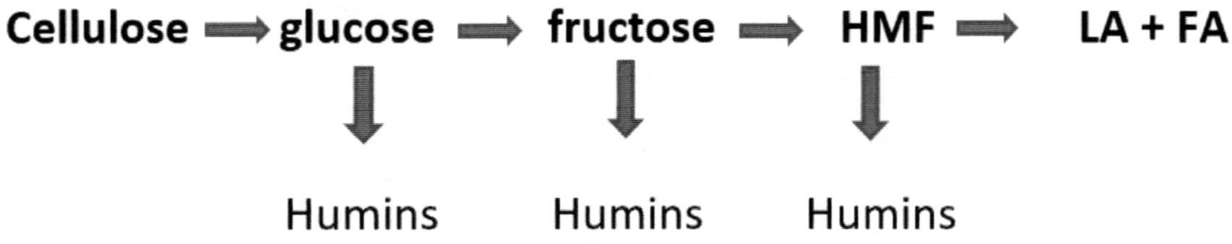

Figure 1. Simplified chemistry scheme of Biofine process

At each stage of the reaction, decomposition products collectively known as humins are formed via kinetically favored higher order side-reactions and represent local free-energy minima. In the production of LA formation of humins represents significant yield loss when the reaction is conducted in simple batch mode. The low yield has, in the past, rendered production of levulinic acid via this route commercially unattractive despite the abundant availability of low-cost cellulosic feedstock [3-5]

The Biofine process addresses the problem of excessive humin production by carrying out the conversion reaction in a continuous back-mixed mode. The mixing regime in the reactor causes the average concentrations of reactants to be diluted thus reducing the dominance of the higher order humin-forming reactions. This increases the forward flow of carbon resulting in increased levulinic and formic acid yields. With this reactor configuration, for a wide range of cellulosic feedstocks tested, the molar yield of levulinic acid from the cellulose fraction is consistently found to be at least 70%. The Biofine process thus allows a potentially economic route to produce low cost levulinic acid from biomass. This opens up the possibility of economic production of a range of chemicals and bio-fuels using levulinic acid as the intermediate platform feedstock [1, 2]

Given the near-term availability, at reasonable cost, of roughly 12 quads energy equivalent of biomass material, the Biofine process therefore has the potential to contribute significantly in combatting global carbon dioxide emissions via displacement of fossil-derived heating fuel. The Biofine process for production of biofuels and chemicals is shown schematically below:

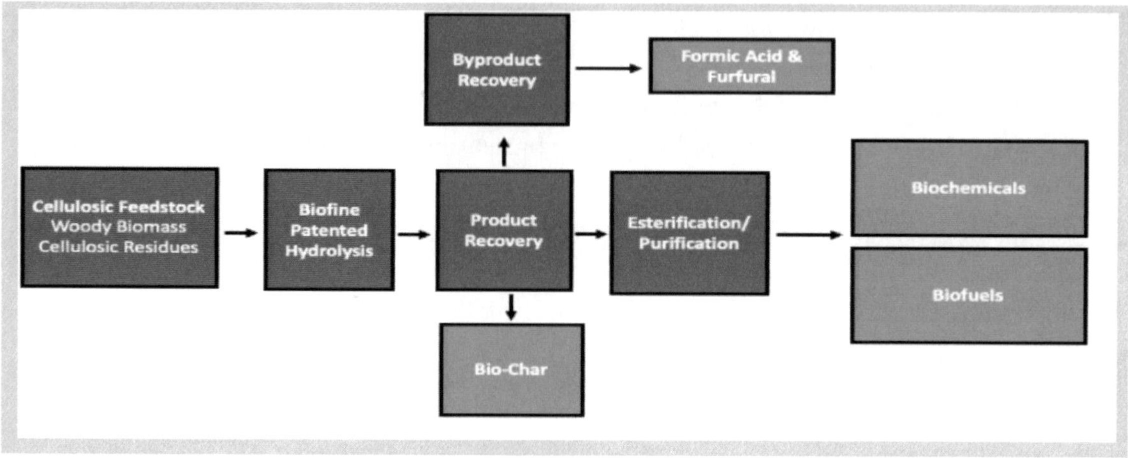

Figure 2. Block diagram of the Biofine process

The Biofine conversion process is presently operated at semi-commercial scale in a pilot demonstration plant owned and operated by the University of Maine Forest Bioproducts Research Institute (UMO FBRI) in Old Town, Maine. This plant is nominally capable of processing one dry ton per day of wood or other biomass feedstock producing around one dry ton per week of levulinic acid or ethyl levulinate. The plant has been used to produce tonnage quantities of ethyl levulinate for testing purposes.

Important Co-products from Cellulose Conversion

As indicated in the above diagram, LA production via the Biofine route also produces important co-products: Formic acid, produced in the main reaction, is a commodity chemical with a present global market of over 500,000 metric tons per year. It is produced commercially at present using fossil-derived fuel oil. In addition to its commodity market uses, formic acid shows promise as a convenient method of hydrogen storage as it can be easily split using metallic catalysts such as ruthenium into gaseous hydrogen and carbon dioxide.

Furfural is produced in the acid hydrolysis reaction from pentosan polymer occurring in many biomass feedstocks. Furfural is also a commodity chemical with a similar sized market to formic acid. As an alternative to commercial sale, furfural can be converted to levulinic acid via furfuryl alcohol using well-known technology. This would represent a significant additional output of levulinic acid.

The process also produces bio-char which is a combination of lignin in the feedstock and humins. This is produced as a black hydrophobic powder with an energy content of around 28 MJ per Kg (12,000 BTU per lb). Biofine's bio-char can be dewatered, pelletized and sold as boiler or stove fuel or can be burned in the process to produce electric power for internal use and for export to the grid. Ahmad et al. [6] and Badgujar et al.[7] recently published broad reviews of methods of producing EL from biomass which provides insight into other production technologies.

Table 1 provides a comparison of the basic fuel property data for No. 2 petroleum-sourced heating oil, biodiesel, and ethyl levulinate.

Table 1. Comparison of Basic Fuel Properties

		No. 2 heating oil	Propane	Biodiesel	Ethyl Levulinate (EL) $C_7H_{12}O_3$
Composition	C wt%	86	89	75.5	58.3
	H wt%	14	11	12.6	8.3
	O wt%	0	0	11.9	33.3
Density	g/cm3	.85	0.49	0.88	1.02
Higher Heating Value	MJ/kg	45.6	48.1	40.7	26.3
	MJ/L	38.7	25.1	36.0	26.8
Flash Point	°C	38 min	-155	93 min	91
Viscosity	mm^2/s @ 40°C	1.9 – 4.1	4.98 x10^{-6}	1.9-6.0	1.5

Due to its high in oxygen content, the heating value of EL is considerably lower than No. 2 heating oil and biodiesel, but roughly equivalent to that of propane. Studies published to date have focused on the use of EL as a blend stock or additive in diesel fuel, No. 2 heating oil, or blends also involving biodiesel.

Christensen et al. explored the use of EL and butyl levulinate (BL) in blends with diesel fuel in engines[8]. They report that saturated EL contains 8.5% water by weight. This group also explored separation of EL/diesel blends at reduced temperatures. They found that for a 10% blend of EL in diesel, fuel separation begins at 5°C.

Joshi et al. [9] studied the use of EL as an additive to improve the cold flow properties of biodiesel fuel. Reductions in cloud point as high as 5°C were found with 20% EL in the blend.

The use of EL as a blend component in commercial heating fuels, including petroleum No. 2 and blends of petroleum No. 2 and biodiesel, offers some logistical advantages. With blends, a heating fuel marketer with a large number of end customers could distribute just one fuel, the EL blend, to all customers, avoiding the need for segregation of their storage and delivery systems. In an early, informal field test, the Dead River Company in Maine distributed a blend of 10% EL by volume in petroleum to 11 home sites for two years. In the present study, blends of EL, at different levels, with petroleum No. 2 heating fuel and biodiesel were explored. The primary focus in this work was on simple separation over a range of temperatures. These blend tests also included an evaluation of one candidate co-solvent reported to be at least somewhat effective with these types of blends.

Based on the results of the blend testing with No. 2 heating oil as described in the following sections, a decision was made to focus on 100% EL, unblended, as the target fuel for this application. Combustion studies, also described in the following sections, were done with the unblended fuel.

The liquid fuels heating industry provides heating oil, biodiesel, and kerosene to over 5 million homes and business in the United States. Comparatively, utility gas supplies gas to 58,702,794, and electricity 48,468,860. In total, heating oil supplies approximately 5 percent of all homes in the United States [10]. However, that modest percentage changes as we focus on the populated and colder areas of the country. In New York, there are 1,393,560 oil heated housing units, or 19 percent of all units in the state [11]. In Maine, there are 344,869 housing units using heating oil, or 60 percent [12].

The northeastern states where oilheat is prominent have also adopted greenhouse gas reduction strategies. Most of them are seeking a 30-40 percent reduction in emissions by 2030 and 80 percent or higher by 2050. The thermal load's contribution to greenhouse gases is characterized as one-third by NYSERDA. Thus, reductions of the magnitude sought cannot be achieved without efforts in the thermal sector.

The heating oil industry has attempted to provide low carbon alternatives through the adoption and use of biodiesel. However, concerns with supply, operability in cold climates, and greenhouse gas scores have prompted the industry to continuously examine other biofuels. One of these is ethyl levulinate.

A key aspect of ethyl levulinate is how its use would impact emissions of greenhouse gases from liquid fuels. Earthshift Global, a provider of life cycle analysis and sustainability services, completed an analysis of the fuel and process of Biofine in early 2019. In this analysis they examined two streams for potential conversion into ethyl levulinate, forest residue and post-consumer wastepaper. They did the analysis based on the process flow prepared by Biofine, and the co-products that would be produced in the production of EL. The study concluded that as a heat source, EL could reduce emissions from ultra-low sulfur diesel (ULSD) used as heating fuel between 110 and 120 percent [13]. Thus, this fuel could be very attractive to the heating industry as it can essentially provide a zero carbon fuel for heating oil consumers.

EXPERIMENTAL

Blend Tests

A total volume of 100 ml was prepared for each target blend as listed in Table 2.

Initially, blends were observed for visual separation for 24 hours at room temperature. Blends which did not separate at room temperature were moved to a glass-front, controlled temperature chamber to observe visual separation at lower temperatures. Chamber setpoint temperature was lowered in 2.8 °C (5 °F) intervals until separation was observed or a temperature of -9.4 °C (15 °F) was reached. At each temperature setpoint the sample was observed for 30 minutes.

Five of the sample blends did not show separation at -9.4 °C (15 °F). The Cloud Point of these samples was then measured at Iowa Central Fuel Testing Laboratory following ASTM D 2500.

Table 2. List of Blends of EL, No. 2 oil, Biodiesel, and Isoamyl Alcohol[1].

Blend Number	Blend Composition (volume %)			
	EL	Petroleum No. 2 Heating Fuel	Biodiesel	Isoamyl Alcohol[1]
1	10	90	0	0
2	20	80	0	0
3	30	70	0	0
4	50	50	0	0
5	5	0	95	0
6	10	0	90	0
7	20	0	80	0
8	50	0	50	0
9	33.3	33.3	33.3	0
10	45	10	45	0
11	20	75	0	5
12	30	65	0	5

[1]Isoamyl Alcohol used as a candidate cosolvent

Elastomers and Filter Elements

Testing was done to determine EL's compatibility with some of the elastic materials found in an oil heat system. The two materials evaluated were Nitrile and Viton. Swell testing was done following UL 157 Section 11.4.2. In this standard, an elastomer exposed to fuel must have swell within the range -1% to 25% in the fuel to pass the test. For both materials, two different sized O-rings were evaluated. Each material was immersed in the desired fuel for 70 hrs. The O-rings were measured with a micrometer before and after this swell period and the change in volume was calculated. Testing was done with multiple different fuels, including No. 2 oil, biodiesel, EL, and multiple blends of the fuels include 95/5 No. 2 oil:EL, 90/10 No. 2 oil: EL, 78/22 No. 2 oil:EL, 50/50 Bio:EL, and 1/1/1 No. 2 oil:Biodiesel:EL.

After testing was done on the materials common to the oilheat industry, other potential materials were evaluated. Many different types of materials were evaluated for their compatibility with EL, but the two that will be focused on here are silicone and Teflon. In the same manner as discussed above, silicone and Teflon O-rings were evaluated following UL 157 11.4.2. These o-rings were immersed in No. 2 oil, Biodiesel and EL. After 70 hrs of immersion, the change in volume of each o-ring was again determined.

Based on reports gathered from the field, the compatibility of EL with various fuel filter materials was also evaluated. The filter types evaluated were a cellulose "microfiber" filter, a felt filter, a wool string filter, and a spin-on filter cartridge. Each of these filters was completely immersed in EL. They were then left to be observed to see if any visible changes occurred over time.

Combustion Testing

Combustion testing was done in two heating boilers, one typical of those used for the production of heat and domestic hot water in single family home applications and one typical of those used for larger commercial buildings.

The residential-scale testing was done in the NORA lab in Plainview, N.Y. using a 3-section cast iron boiler with an integral tankless coil heat exchanger for domestic hot water (Model Peerless WBV-03, PBHeat LLC). Water on the heating side of the boiler was continually circulated without active heat extraction during these tests. The desired thermal load on the boiler was controlled through adjustment of cold water flow through the internal domestic hot water coil. The boiler temperature for all tests was steady and in the range of 70 to 75°C. A swirl, pressure-atomized, fixed firing rate burner was used in all tests (Model EZ-1, Carlin Combustion Technologies). The fuel pump with integral pressure regulator used was modified to have all elastomer static seals compatible with EL. This included some silicon and some Teflon seals. The pump shaft Nitrile lip-style seal was replaced with a Teflon-graphite composite seal. This pump was used with both petroleum heating oil and EL.

To directly measure fuel flow in the residential boiler tests, the fuel supply container was mounted on a scale with a resolution of 0.023 kg (0.05 pounds). Fuel mass flow measurement was integrated over a 30 minute steady period for each fuel. Flue gas oxygen was measured in all tests using a wet electrochemical cell analyzer (Model DC 710, Test Products International, Inc.). Flue gas NO_x and CO concentration was measured using an NDIR analyzer with external NO/NO_2 converter. (Model ZRE, California Analytical Inc.). Boiler temperature, flue gas temperature, gas composition and other basic operating parameters were recorded using a LabView-based data acquisition system.

Based on the results of blend testing, described in the next section, combustion tests focused on EL not blended with petroleum or biodiesel. Unblended EL is seen as the most practical commercial fuel. Testing of the residential boiler was done using No. 2 petroleum heating oil, 100% EL, and a blend of 91.5% EL and 8.5% water by volume. The water/EL blend represents EL which has been saturated by EL and this is seen as an extreme, limit case. It is highly unlikely that this would occur in the field.

Firing No. 2 petroleum heating oil, the heat input rate, based on measured fuel mass flow was 28.7 kW. Nozzle nominal size was 0.65 gph[1] and the burner operating pressure was adjusted to 150 psi. Firing EL the fuel mass flow was raised using a larger nozzle size (1.0 gph nominal) to produce

---

[1] Burner fuel nozzles of this type are commonly rated in flow of petroleum No. 2 heating oil at 100 psi in gallons-per-hour (gph). At higher pressure, volume flow rate can be closely estimated using the square root of the pressure ratio.

the same heat input rate. This same nozzle and pressure condition was used with the water/EL blend, producing a heat input rate 8.5% lower than with EL.

The larger commercial boiler tests were done tests in the development lab of Carlin Combustion Technologies, a major burner manufacturer in North Haven, Connecticut. Tests were conducted in a Weil-McLain 688 Cast Iron Section boiler configured with direct cold water at the boiler inlet. In the field, boilers of this type operate with much higher temperatures at the boiler inlet but with large boilers of this type it is common practice to test with cold water in, simply to reduce the water flow rate required to enable the burner to run continuously without cycling. The burner installed in the boiler for this test was a Model 601 CRD manufactured by Carlin Combustion Technologies. This burner also uses a pressure-atomized arrangement. Flue gas CO, Oxygen, and NOx were measured using an analyzer with wet electrochemical cells (Ecom Instruments). As with the residential boiler testing, the pump used for these tests was converted to have elastomer seals compatible with EL. For petroleum No. 2 heating oil, an 8.0 gph nominal rate nozzle with an operating pressure of 150 psi was used. For EL testing, a nominal 9.5 gph nozzle with an operating pressure of 250 psi was used. Testing with a water/EL blend was not done with the commercial boiler. With both tested fuels, the input rate was 410 kW.

For all fuels in both the residential and commercial boiler tests smoke number was measured using the Shell Bacharach method (ASTM D2156). In all cases reported here the measured smoke number was zero.

Steel Corrosion

Questions have been raised about the affects EL or EL contaminated with water could have on steel material. This is particularly important as many of the fuel storage tanks found throughout the industry are made of steel. To test this, small 2 inch by 1 inch sized coupons of steel were immersed in EL samples with various amounts of water. The EL samples evaluated here were EL with no water, EL with 1% water, and EL with 8% water. These samples were evaluated daily to visually check if any corrosion was forming on the metal.

RESULTS

Blend Tests

The results of mixing the various blends and leaving them at room temperature are available in Table 3. As you can see from the table, most samples did not have any issues with separation. The only samples that had issues were blends number 3, 4, and 12, as well as blends of No. 2 oil and EL at the 70/30 and 50/50 blend level. Separation was immediately apparent after shaking of the samples. Blend 12, with the addition of isoamyl as a cosolvent, also showed separation. This testing points to a compatibility issue with blending No. 2 oil and EL. Because of the separation at room temperature, these sample were removed from further testing. No other samples showed any separation at room temperature after 24 hrs.

Table 3. Blend Separation at room temperature.

Blend Number	Separation (Y/N)
1	N
2	N
3	Y
4	Y
5	N
6	N
7	N
8	N
9	N
10	N
11	N
12	Y

Next, testing in the freezer at various temperatures was done. Observations of any separation and the temperature this occurred at are shown in Table 4. From this testing, blends 1, 2, 10 and 11 showed separation below room temperature. With the separation of blends 1 and 2, it becomes clear that blending No. 2 oil and EL on their own can have serious issues. The addition of isoamyl alcohol as a cosolvent also seemed to have little to no effect. Blends of biodiesel and EL appear to have no issues with separation, though many of the samples appeared to be clouding at the lower temperatures of the test. It should be noted, though, that Blend 9, a blend of equal parts No. 2 oil, EL and biodiesel, did not have any issues with separation in the temperatures tested. Before moving to the next round of testing, all samples that showed separation were removed.

Finally, the cloud point of the various blends that did not separate were assessed using ASTM D2500. In addition, the cloud point of the No. 2 oil and biodiesel used to make the blends was tested. The cloud points of each of the fuels are listed in Table 5. As seen in the table, the cloud point of the blend with EL improved. However, these changes are small and the measured cloud points of these blends are much greater than that of unblended EL.

Table 4. Separation of blends in freezer.

Blend Number	Separation (Y/N)	Freezer Temperature at Separation (°F)
1	Y	15
2	Y	55
5	N	-
6	N	-
7	N	-
8	N	-
9	N	-
10	Y	20
11	Y	55

Table 5. Cloud point of various blends.

Blend Number	Cloud Point (°F)
5	30.2
6	28.4
7	28.4
8	26.6
9	23.0
Base No. 2 oil	12.2
Base Biodiesel	33.8

Elastomers and Filter Elements

The change in volume of each of the Nitrile and Viton o-rings can be found in Table 6. For both Nitrile and Viton o-rings, a large amount of swell was observed with any fuel blend containing EL. Only the Nitrile O-ring 2 immersed in 95/5 No.2 oil:EL was below the 25% swell limit. Interestingly, it appears that Viton swells more in EL than Nitrile, though both changes in volume are unacceptably high.

Table 6. Volume Change of Nitrile and Viton O-rings in Various Fuels After 70 Hours of Exposure

Fuel	Nitrile O-ring 1 Volume Change (%)	Nitrile O-ring 2 Volume Change (%)	Viton O-ring 1 Volume Change (%)	Viton O-ring 2 Volume Change (%)
No. 2 Oil	-1.5	7.7	-8.7	8.1
B100	6.0	20.5	-3.3	8.2
EL	94.7	148.2	181.1	195.8
95/5 No. 2 oil:EL	37.8	13.0	57.5	50.8
90/10 No. 2 oil:EL	51.4	65.4	67.9	68.4
78/22 No. 2 oil:EL	104.0	131.1	189.3	143.4
50/50 Bio:EL	58.6	101.8	114.5	100
1/1/1 No. 2 oil:Biodiesel:EL	46.1	89.5	74.3	69.4

The change in volume for both the Silicone and Teflon o-rings is listed in Table 7. The Teflon o-ring performed very well and saw almost no change when immersed in EL. The main issue with this Teflon o-ring is stiffness. This material may not be usable in all parts of a heating oil system. The Silicone O-ring was performed well with EL and is flexible enough to be used in all parts of a fuel system. It should be noted, however, that the Silicone O-ring showed a 44.9% swell in No. 2 oil. Additional fluorocarbon elastomer types were tested and found to be compatible with EL.

Table 7. Volume Change of Nitrile and Viton O-rings in Various Fuels.

Fuel	Silicone O-ring Volume Change (%)	Teflon O-ring Volume Change (%)
No. 2 oil	44.9	0.9
Biodiesel	20.1	4.0
EL	6.0	0.2

Of the various filter materials tested, only one showed a noticeable change in the time period tested. The felt filter, the wool string filter, and the spin-on, pleated paper filter cartridge all showed no obvious change after being immersed in EL for two weeks. The one filter that did show a very rapid degradation is described by the manufacturer as a microfiber filter element. Only a few minutes after being placed in the fuel, the filter began to change, almost appearing to melt. After 30 minutes, the filter had completely dissolved. Pictures of the filter before immersion and after 30 minutes of immersion can be seen in Figure 3. This common type of filter is clearly incompatible with EL. The specific element tested is a Westwood "PureOil Element" – other manufacturers offer similar products.

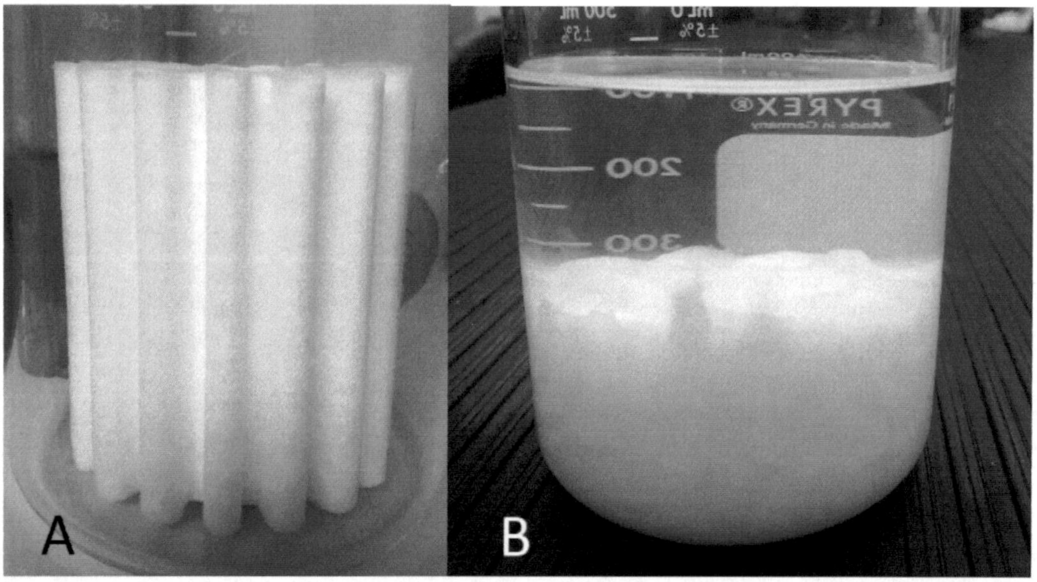

Figure 3. A) Microfiber filter element before being immersed B) filter after being immersed in EL for 30 minutes.

Combustion and Emissions

With all fuels tested in the residential and commercial boilers, ignition was smooth, and the burners operated over a wide excess air range with a stable flame. In the case of the residential boiler, flue gas CO was measured to be less than 10 ppm under all conditions and this level is considered negligible. Figure 4 provides a comparison of the measured NOx emission factors for all fuels in the residential boiler.

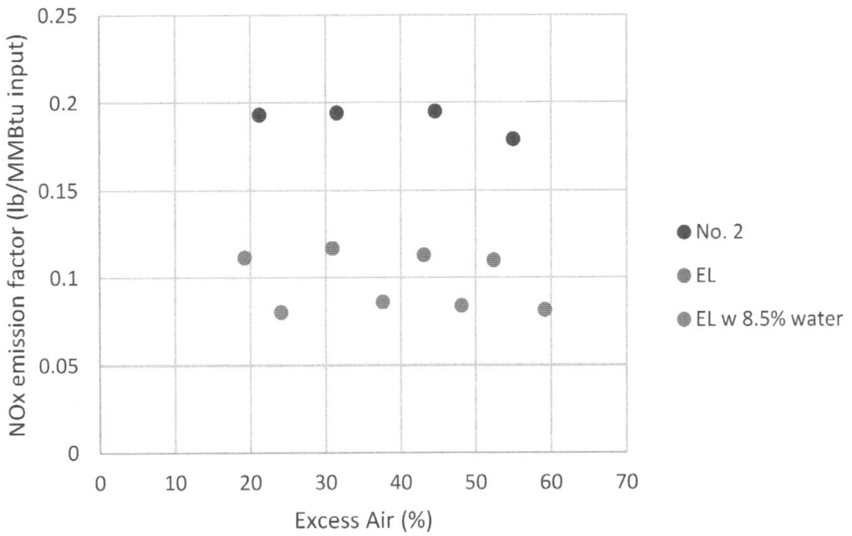

Figure 4. Residential boiler - NOx emission factor, all fuels tested

In the commercial boiler tests, flue gas CO was found to be higher for EL and these results are presented in Figure 5. NOx emission factor results for the commercial boiler are shown in Figure 6.

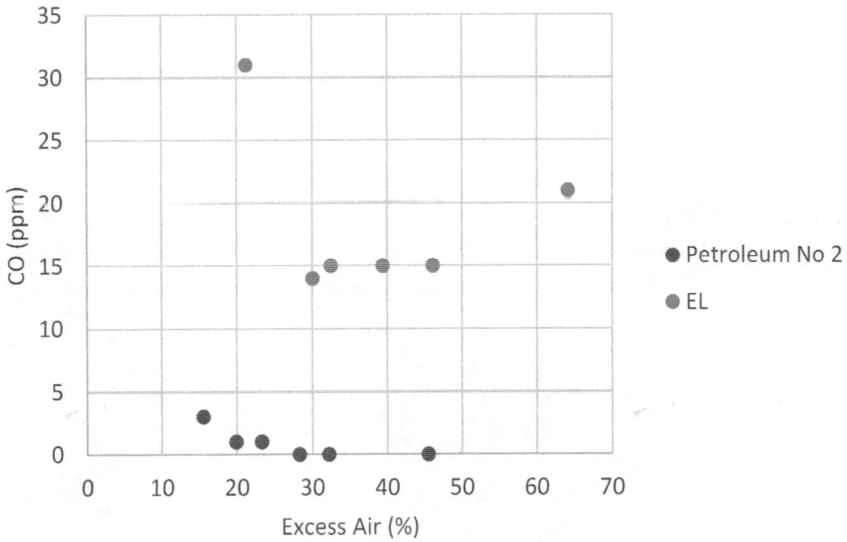

Figure 5. Commercial boiler- flue gas CO, all fuels tested

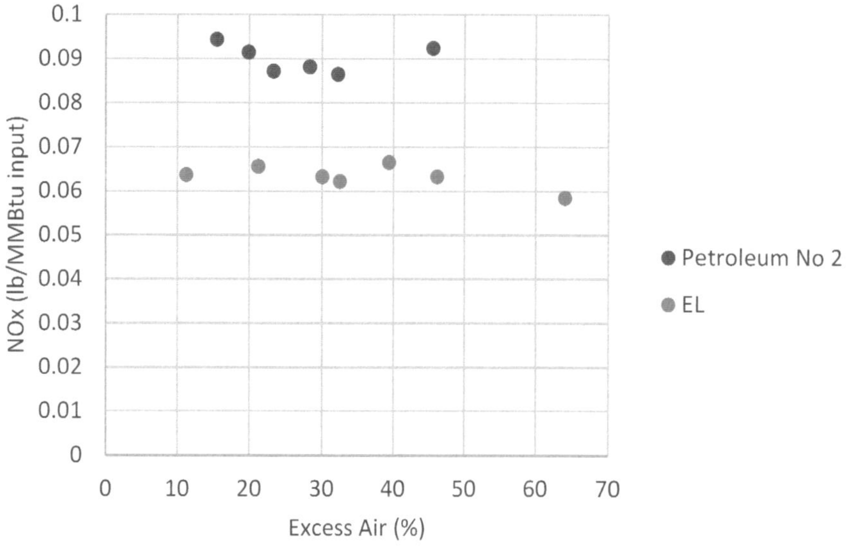

Figure 6. Commercial boiler- NOx emission factor, all fuels tested

The results in Figure 5 show higher CO for EL than for petroleum No. 2 heating fuel. However, these levels of CO are considered well within acceptable levels, particularly between 30 and 50% excess air, the common operating range. The smoke number was also found to be zero across this testing. Figures 4 and 6 show that the NOx emission factor will be significantly lower with EL, falling approximately 40% lower than for petroleum No. 2 in the residential boiler case and 30% lower in the commercial boiler case. The NOx emission factor with water saturation is 24% lower than for EL alone in the residential case.

The configuration of the commercial boiler allows direct viewing of the flame from the rear surface, opposite from the burner. Figure 7 provides these images for both petroleum No. 2 and EL at the same input rate and excess air level.

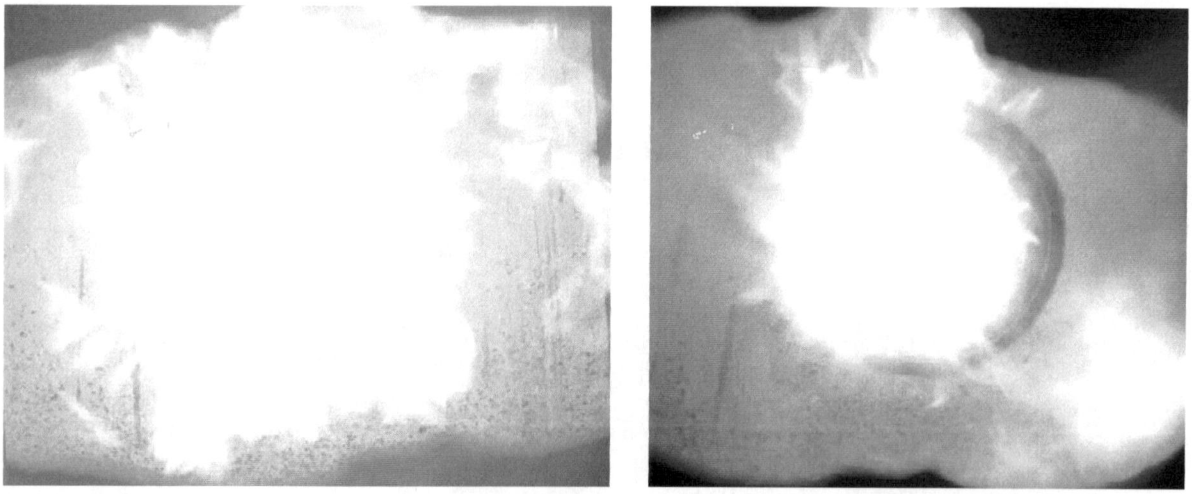

Figure 7. Comparison of the appearance of the petroleum No. 2 flame (left) and the EL flame (right). Both at ~ 6.5% flue gas oxygen and a nominal input rate of 410 kW.

Steel Corrosion

Visual inspection of the steel coupon did yield results. The sample of EL with no water and the sample of EL with 1% water show no signs of corrosion after 2 months of observation.

The sample of EL with 8% water, however, did begin to show signs of corrosion after 13 days. This can be seen in Figure 8. Additional testing over a range of water levels and using corrosion inhibitors is planned. It should be noted that 8% is not a typical operating condition for the heating fuels.

Figure 8. Steel coupon immersed in EL blended with 8% water after 13 days.

CONCLUSIONS

In this work, Ethyl Levulinate has been shown to hold significant promise as a renewable fuel for stationary boilers and furnaces. Life Cycle Analysis studies which have been done show that the calculated greenhouse gas emissions with this fuel can be less than zero. Because this fuel can be made from waste wood products the resource size is exceptionally large.

At low blend levels the fuel can be used in combination with petroleum No. 2 oil although with increasing blend levels and reducing temperatures separation becomes a significant concern. EL can be blended with No. 2 petroleum heating oil and biodiesel together with improved separation aspects. With such a blend, however, the cold flow properties are dominated by the biodiesel fraction and there remains a significant concern in colder climates. A particularly important advantage of EL in this application is the extremely good cold flow properties. For these reasons, this work focused on 100% EL as a candidate fuel for this application.

EL has been found to not be compatible with Nitrile and Viton elastomer seal materials used in current No. 2 petroleum-fired boiler and furnace fuel systems. These materials need to be replaced with silicon, Teflon, and Teflon/graphite composite materials have been found to be suitable.

The corrosion of steel found with EL mixed with high amounts of water is concerning. More work is needed to confirm the concentration of water where corrosion will most likely happen. Work is also planned to look at metal deactivators as a potential way to resolve the corrosion issue. It should be noted that the EL sample mixed with 1% water has shown no corrosion reaction so far.

Using fuel system components with acceptable elastomers, combustion testing has been done with EL and water-saturated EL. In both cases good combustion performance was observed in a residential and a commercial boiler application. NOx emissions with EL were found to be 40-60% of those with petroleum No. 2 fuel under the same conditions.

REFERENCES

1. Hayes, D.J., et al., *The Biofine Process: Production of Levulinic Acid, Furfural and Formic Acid from Lignocellulosic Feedstocks*, in *Biorefineries - Industrial Processes and Products*, B. Kamm, P.R. Gruber, and M. Kamm, Editors. 2005, Wiley-VCH Verlag GmbH&Co.

2. Fitzpatrick, S.W., *The Biofine technology: A "bio-refinery" concept based on thermochemical conversion of cellulosic biomass*, in *Feedstocks for the Future*. 2006, American Chemical Society. p. 271-287.

3. Girisuta, B., L.P.B.M. Janssen, and H.J. Heeres, *Kinetic study on the acid-catalyzed hydrolysis of cellulose to levulinic acid.* Ind. Eng. Chem. Res., 2007. **46**(6): p. 1696-1708.

4. Fachri, B.A., et al., *Experimental and kinetic modeling studies on the sulfuric acid catalyzed conversion of D-Fructose to 5- hydroxymethylfurfural and levulinic acid in water.* ACS Sustainable Chem. Eng., 2015. **3**(12): p. 3024-3034.

5. Corma, A., S. Iborra, and A. Velty, *Chemical routes for the transformation of biomass into chemicals.* Chem. Rev., 2007. **107**(6): p. 2411-2502.

6. Ahmad, E., et al., *Catalytic and mechanistic insights into the production of ethyl levulinate from biorenewable feedstocks.* Green Chemisry, 2016. **18**: p. 4804-4823.

7. Badgujar, K.C., V.C. Badgujar, and B.M. Bhanage, *A review on catalytic synthesis of energy rich fuel additive levulinate compounds from biomass derived levulinic acid.* Fuel Processing Technology, 2020. **197**.

8. Christensen, E., Williams, A., Paul, S., Burton, S., McCormick, R.L., *Properties and performance of levulinate esters as diesel blend components.* Energy and Fuels, 2011. **25**: p. 5422-5428.

9. Joshi, H., Moser, B.R., Toler, J., Smith, W.F., Walker, T., *Ethyl Levulinate: a potential bio-based diluent for biodiesel which improves cold flow properties.* Biomass and bioenergy, 2011. **35**: p. 3262-3266.

10. *Selected Housing Characteristics 2019 ACS 1-Year Estimates Data Profile.* [cited 2021 August 12]; Available from: https://data.census.gov/cedsci/table?d=ACS%201-Year%20Estimates%20Data%20Profiles&tid=ACSDP1Y2019.DP04&hidePreview=true.

11. *Selected Housing Characteristics 2019 ACS 1-Year Estimates Data Profile.* [cited 2021 August 12]; Available from: https://data.census.gov/cedsci/table?g=0400000US36&tid=ACSDP1Y2019.DP04&hidePreview=true.

12. *Selected Housing Characteristics 2019 ACS 1-Year Estimates Data Profile - Maine.* [cited 2021 August 12]; Available from: https://data.census.gov/cedsci/table?g=0400000US23,36&d=ACS%201-Year%20Estimates%20Data%20Profiles&tid=ACSDP1Y2019.DP04&hidePreview=true.

13. Earthshift, *EL Report*. 2022: Earthshift Global. www. https://earthshiftglobal.com/case-studies-reports

IASH 2024, the 18TH INTERNATIONAL CONFERENCE ON STABILITY, HANDLING AND USE OF LIQUID FUELS
Louisville, Kentucky, USA
September 8 – 12, 2024

USING GC×GC-TOFMS AND COMPARATIVE DATA ANALYSIS SOFTWARE TO GAIN A DEEPER UNDERSTANDING OF CHEMICAL COMPOSITION CHANGES IN FUEL STUDIES

Robert E. Synovec*[1], Grant S. Ochoa[1], Timothy J. Trinklein[1], Caitlin N. Cain[1], Austin D. Dobrecevich[1], Matthew C. Billingsley[2]

[1]*Department of Chemistry, University of Washington, Department of Chemistry Box 351700, Seattle, WA 98195 USA*
[2]*Air Force Research Laboratory/RQRC*

ABSTRACT

Kerosene-based jet and rocket fuels such as Jet A, JP-8, RP-1, and RP-2 are ubiquitous to aerospace propulsion. There is an ongoing need to more fully assess fuel composition to improve quantitative connections between fuel composition, properties, and system performance. To this end, we apply comprehensive two-dimensional gas chromatography with time-of-flight mass spectrometry (GC×GC-TOFMS) and multivariate "chemometric" data analysis to understand how fuel performance depends upon chemical composition, which varies due to production changes and operational handling and use scenarios. Two fuel studies will be presented. First, selective analysis of olefins using bromination chemistry was applied to gasoline, followed by GC×GC-TOFMS analysis of the original and brominated gasoline samples. The data was analyzed to pinpoint chromatographic regions distinguishing the original and brominated samples, with all of the olefins readily discovered. This study has broad implications using selective reaction chemistry to facilitate supervised discovery by targeting compound classes of interest. In the second study, solid phase extraction (SPE) is employed to target polar compounds in fuel. Treating fuel samples with SPE facilitates removal/analysis of specific compound classes, such as oxygenates, allowing for the creation of an altered "pass" sample, identical to the original "neat" sample except for the extracted polar compounds. These neat versus pass samples are analyzed to provide a global discovery of compounds affected by the SPE procedure. This second study has significant ramifications for providing an accurate assessment of the deleterious compounds in fuels, which is essential to confidently relate fuel composition to thermal stability and performance.

*Corresponding author: Robert E. Synovec, res9@uw.edu

IASH 2024, the 18TH INTERNATIONAL CONFERENCE ON STABILITY, HANDLING AND USE OF LIQUID FUELS
Louisville, Kentucky, USA
September 8 – 12, 2024

CHALLENGES AND INSIGHTS IN APPLYING AVIATION FUEL ADVANCED COMPOSITIONAL METHODS TO DIESEL FUELS

N. Scott Bergey*[1], Lena E. Flynn[1], Terrence W. Dickerson[1], Richard A. Kamin[1], Corrianna R. Boucher[2], Eric M. Johnson[2]

[1]*U.S. Navy, 22229 Elmer Rd. Bldg. 2360 Patuxent River,,Maryland 20670 USA*
[2]*Peraton, 22229 Elmer Rd. Bldg. 2360 Patuxent River,,Maryland 20670 USA*

ABSTRACT

The industry focus on understanding aviation fuel thermal stability issues has led to the development of powerful analytical techniques to characterize aviation fuel composition. Some of these methods are readily applicable to diesel fuels and require little or no additional optimization to produce similar quality results, while others require significant additional refinement to be applicable across fuel classes. This work highlights recent developments by the US Navy to expand upon and optimize the joint Air Force/Navy SPE-GCxGC-FID extractable polars methodology to be applied to diesel fuels. This expansion includes the development of a method which characterizes the LIA content in marine fuels, an analyte class that is uniquely prevalent in diesel fuels and create analytical challenges for GCxGC inlets and column chemistries.

*Corresponding author: N. Scott Bergey, scott.bergey1@navy.mil

IASH 2024, the 18TH INTERNATIONAL CONFERENCE ON STABILITY, HANDLING AND USE OF LIQUID FUELS
Louisville, Kentucky, USA
September 8 – 12, 2024

RAPID SCREENING TECHNIQUE FOR QUALITATIVE MOLECULAR CHARACTERIZATION OF POLAR ANALYTES IN FUELS USING GCxGC/(+CI)MS

Grace E. Greene[3]*, Richard J. Bell[2], N. Scott Bergey[1], Lena E. Flynn[1], Terrence W. Dickerson[1], Richard A. Kamin[1]

[1]*US Navy, 22229 Elmer Road, Patuxent River, MD, USA 20670*
[2]*Peraton, 46591 Expedition Dr, Lexington Park, MD, USA 20653*
[3]*Purdue University, 610 Purdue Mall, West Lafayette, IN 47907*

ABSTRACT

The United State Department of Defense (DoD) uses multiple fuel types for tactical vehicle, ship, and aircraft propulsion. Recent emphasis on trace heteroatomic composition and advanced algorithmic modeling has led to the US Navy developing several analytical tools for trace-level characterization of Navy tactical fuels. Extractable polar nitrogen- and oxygen-containing species have typically been measured using GCxGC analyses using FID and/or single quadrupole MS detectors, but these detectors rely on relative elution times and low-confidence mass spectral assignments to speciate analytes. The United States Navy has recently acquired a state-of-the-art GCxGC/ToF HRMS system which significantly improves the 2D resolution as well as mass accuracy of previous instrumentation, allowing for greater molecular specificity. This work highlights recent developments using the GCxGC/ToF HRMS system by demonstrating a rapid screening tool based on chemical ionization trends between different chemical moieties. This tool can be used for comparing molecular assignments between petroleum based fuels and their sustainable analogs to identify whether analytes eluting in similar regions are truly identical for petroleum versus sustainable fuels or if sustainable production pathways are introducing new trace-level oxygenated moieties.

*Corresponding author: Grace E. Greene, green497@purdue.edu

IASH 2024, the 18TH INTERNATIONAL CONFERENCE ON STABILITY, HANDLING AND USE OF LIQUID FUELS
Louisville, Kentucky, USA
September 8 – 12, 2024

WATER AND COMPOSITIONAL ANALYSIS IN LIQUID PETROLEUM GASES (LPGS) WITH THE NEW VUV LUMA GC DETECTOR

Dan Wispinski*[1], Chris Goss[2]

[1]*VUV Analytics, 1500 Arrow Point Drive Bldg 8, Suite 805Cedar Park,Texas 78613 USA*
[2] *InnoTech Alberta*

ABSTRACT

Improved test methods to measure water in LPG have long been an industry need. Propane dryness is monitored with a "functional" valve freeze test. The venting of LPG inherent to the D2713 Valve Freeze method is a safety hazard. The valve freeze test cannot be adapted to on-line analysis or process control. Methanol in the range of 50 to a maximum of 200ppmv is used to mitigate wet LPG problems that occur at sub-saturation levels.

This presentation describes a new test method using gas chromatography (GC) separations and the deconvolution (spectral) power of the VUV LUMA for water and methanol (antifreeze agent) in propane. The analysis of ethyl mercaptan (odorant) and hydrocarbon composition of propane (ASTM D2163) is also demonstrated. This single GC run on the VUV LUMA can also detect oxygen, carbon dioxide and hydrogen sulfide.

*Corresponding author: Dan Wispinski, dan.wispinski@vuvanalytics.com

IASH 2024, the 18TH INTERNATIONAL CONFERENCE ON STABILITY, HANDLING AND USE OF LIQUID FUELS
Louisville, Kentucky, USA
September 8 – 12, 2024

TECHNOLOGY DEVELOPMENT OF MOTOR GASOLINE STORAGE IN SALT CAVERNS WITH ASPECTS OF QUALITY PREDICTION, STORAGE AND END-USE.

Arkadiusz Majoch*, Jacek Tomaszewski, Łukasz Plis, Zuzanna Rataj

ORLEN S.A, ul. Chemików 7, 09-411 Płock, Poland

ABSTRACT

The growing demand for fuels obtained from crude oil is directly related to the need to increase their stocks. This translates into investments in the expansion of logistics infrastructure, including alternative storage techniques. In Poland, due to favorable geological conditions (availability of salt geological structures), salt caverns are used for long-term storage of hydrocarbons, where crude oil, natural gas and diesel fuel reserves have been successfully stored for years. There are also attempts to apply this technique to the storage of motor gasoline, especially in the face of relatively little practical experience in storage of motor gasoline in salt caverns. In 2016, ORLEN S.A. (refining and pet-chem company) launched the project "Methodology for forecasting and monitoring the quality of gasoline directed for long-term storage in salt caverns, abbreviated as "GASSTO", co-financed by the European Regional Development Fund. The aim of the Project was to develop in laboratory conditions, using the knowledge of ORLEN S.A. employees and external experts, base gasoline blends, the suitability of which for long-term storage in caverns was verified. This was done through an appropriately selected and program of research and testing. Laboratory studies and tests mimicking the conditions of real storage were the basis for developing a method for monitoring gasoline quality during storage in the cavern, as well as technology for producing final gasoline from base gasoline after the storage period. The culmination of the project was the industrial-scale production (40,000 m3) of a gasoline blend selected on the basis of the research and pumped into an real salt cavern. The presentation will cover the results of the project, including an abbreviated review of global and Polish experience in testing the feasibility of storing gasoline in salt caverns and a presentation of the method developed for predicting the properties, production technology and quality monitoring system of gasoline for long-term storage in salt caverns.

*Corresponding author: Arkadiusz Majoch, arkadiusz.majoch@orlen.pl

IASH 2024, the 18TH INTERNATIONAL CONFERENCE ON
STABILITY, HANDLING AND USE OF LIQUID FUELS
Louisville, Kentucky, USA
September 8 – 12, 2024

NAVIGATING FUEL INTEGRITY CHALLENGES IN BULK STORAGE FACILITIES

Dr. David Evans*[1], Nathan Matheson[1], Dr Ulas Yildirim[2]

[1]*Defence Science and Technology Group, 506 Lorimer Street Fishermans Bend, 3207 Australia*
[2] *Royal Australian Air Force*

ABSTRACT

Contamination incidents within bulk storage facilities pose substantial operational hurdles for companies and organisations. This paper offers an in-depth analysis of multiple contamination events within a 5 million litre diesel storage tank. It explores the ongoing struggle between procurement specifications, degradation thresholds, and the essential fit-for-purpose properties when managing in-service issues. Specifically, it investigates a case involving a 5 million litre diesel storage tank utilised as an intermediate quarantine facility, where off-specification fuel was inadvertently introduced into the system. Through this case study, the paper highlights the complexities of maintaining fuel integrity in storage facilities and offers insights into effective mitigation strategies and preventive measures.

*Corresponding author: Dr. David Evans, david.evans29@defence.gov.au

IASH 2024, the 18TH INTERNATIONAL CONFERENCE ON STABILITY, HANDLING AND USE OF LIQUID FUELS
Louisville, Kentucky, USA
September 8 – 12, 2024

RE-EVALUATION OF ASTM D5304: DETAILED COMPARISON OF STORAGE STABILITY METHODS THROUGH THE ANALYSIS OF LIQUID-PHASE CHEMISTRY

Thomas Loegel*[1], Mark Romanczyk[1], Kristina M. Myers[2], Robert E. Morris[2], Jacob Guthrie[2]

[1]*Naval Research Laboratory, 4555 Overlook Ave., S,W. Washington, DC USA*
[2]*Nova Research, Inc*

ABSTRACT

Evidence suggests that ASTM D5304 may not provide accurate assessments of storage stability for hydrorefined fuels, such as ultra low sulfur diesel (ULSD) or other hydrotreated diesel fuels. ASTM D5304 was developed with legacy high sulfur F-76 fuels, but hydrorefined or alternative fuels that do not contain natural oxidation inhibitors can react non-linearly under the accelerated conditions in ASTM D5304 with respect to ambient storage. Previous attempts to accomplish this critical work was hampered by the lack of sensitivity to track the chemical changes occurring in fuels . High-resolution mass spectrometry (HRMS) demonstrated to be more sensitive than previous GC-MS techniques and able to monitor changes in heteroatomic species within fuel.

This study is intended to develop baseline information that can facilitate the development of an improved storage stability test method appropriate for diesel fuels currently being produced focusing on ASTM D5304, ASTM D2274, ASTM D4625, and ASTM D7545. Three aspects of this issue focus on (1) liquid phase chemistry of stressed fuels using HRMS to determine the conditions under which an accelerated test crosses the regime from probable storage related chemistry to forced high temperature reactions (2) linking chemistry to stability performance, and (3) determine potential detection methods and changes that will allow for accelerated storage testing under conditions that more closely approximate ambient fuel storage.

*Corresponding author: Thomas Loegel, thomas.loegel.civ@us.navy.mil

IASH 2024, the 18TH INTERNATIONAL CONFERENCE ON STABILITY, HANDLING AND USE OF LIQUID FUELS
Louisville, USA
8-12 September 2024

SAF BLENDING AND DISTRIBUTION – SUPPLIER SHARING

Clarisse Doucet*

TotalEnergies Aviation, 24, cours Michelet - La Defense 10 Paris, 92069 France

ABSTRACT

Sustainable Aviation Fuel is a practical alternative to fossil-based jet fuel. It significantly reduces the CO2 emissions from air transportation and can be used as a drop-in fuel as of today, without any need to modify existing supply chain infrastructure, aircraft or engines.

Having started production of Sustainable Aviation fuels (SAF) since 2021, TotalEnergies would like to take the opportunity of this presentation to share feedback and good practices for blending and distribution. During this presentation, we will focus on some definitions, some key points that need to be considered when blending Jet A-1 with synthetic blending component. Incorporation of SBC is linked to the quality of fossil Jet A-1 and SBC characteristics. Understanding & Knowledge of SBC and Jet A-1 properties is needed for the blending. Potential impacts on logistic and distribution must be taken into account when considering future evolutions regarding 100% SAF formulations.

*Corresponding author: Clarisse Doucet, clarisse.doucet@totalenergies.com

IASH 2024, the 18TH INTERNATIONAL CONFERENCE ON STABILITY, HANDLING AND USE OF LIQUID FUELS
Louisville, Kentucky, USA
September 8 – 12, 2024

DEMONSTRABLE QA STANDARDS IN THE SPECIFICATION, PROCUREMENT, DELIVERY AND STORAGE OF LIQUID FUELS FOR USE IN EMERGENCY GENERATORS AND OTHER PLANT

Carl Atkinson
EDF Energy, Jaelin House Building 1420 Charlton Court, Gloucester Business Pak, Gloucester GL3 4AE UK

ABSTRACT

As end users of liquid fuels, a number of industries and other applications require the systems fuelled to be permanently available. These include essential emergency power supplies for data centres, communications, hospitals, and, together with various essential back-up plant systems, on power stations. In order to ensure that such systems perform as expected on demand the fuel quality must meet strict control standards from initial specification through to its ultimate consumption. In the case of nuclear power stations, particularly, the quality assured process must also be demonstrable.

This presentation describes the control standards and quality assured measures in place and how these standards are maintained and demonstrated from the point of procurement of liquid fuels through to their ultimate consumption. Additionally, the process includes a continual review of external factors that could positively or negatively affect the measures currently in place internal to the organisation (e.g. changes to international or British standards, information gathering via international conferences such as IASH). The presentation also details the corrective action process that is invoked when anomalies or shortfalls are identified, corrective actions monitored through to their close-out on completion, and, the process of effectiveness reviews (or equivalent) to evaluate the success (or otherwise) of measures implemented.

*Corresponding author: Carl Atkinson, carl.atkinson@edf-energy.com

IASH 2024, the 18TH INTERNATIONAL CONFERENCE ON STABILITY, HANDLING AND USE OF LIQUID FUELS
Louisville, Kentucky, USA
September 8 – 12, 2024

PARTIAL SURVEY OF AVIATION TURBINE FUEL IMPURITIES: CONTROLS AND EFFECTS

Randall C. Boehm*[1], Joshua S. Heyne[2]

[1]*Washington State University, 2710 Crimson Way, Richland, Washington 99354*
[2]*Washington State University and Pacific Northwest National Laboratory*

ABSTRACT

The standard practice for evaluation of new aviation turbine fuels and fuel additive (ASTM D4054) lists 53 materials (classes-of or specific elements or molecules) whose concentration in fuel should be less than 1%m. Additionally, 18 properties that could be impacted significantly by trace levels of certain materials in fuel are cited in that method, each with some controlling guidance. In this work we a connection between these fuel requirements and the potential issues they are intended to prevent. Potential issues include excessive deposition of solid or gelatinous material anywhere within the aircraft/engine fuel system or even systems upstream of the aircraft, material compatibility throughout the fuel system and extending into the hot section of the engine and including compatibility with additives, wear considerations, and fuel toxicity or other safety considerations. They may also include factors such as confounding influence on deposit forensics which could make it even harder to diagnose the source of contamination that may lead to an unscheduled engine removal (or worse). Our intention here is to highlight published works that address some of these issues and to document where gaps exist in the archival literature.

*Corresponding author: Randall C. Boehm, randall.boehm@wsu.edu

IASH 2024, the 18TH INTERNATIONAL CONFERENCE ON STABILITY, HANDLING AND USE OF LIQUID FUELS
Louisville, Kentucky, USA
September 8 – 12, 2024

MEASUREMENT OF JET FUEL INTERFACIAL TENSION AND ITS RELATION TO MICRO-SEPAROMETER DATA

Shane Kosir*[1], Susan Mueller[1], Zachary West[1], Paul Wrzesinski[2]

[1]*University of Dayton Research Institute, 300 College Park Dayton,Ohio 45469 USA*
[2]*Air Force Research Laboratory*

ABSTRACT

Surfactants are undesirable compounds in jet fuel since they can compromise fuel filter/separator systems, which could allow dirt and water to pass through and damage engine components. While ASTM standard methods (i.e., D3948 and D7224) are known to respond to some surfactants, there is a lack of knowledge linking these standards to fundamental fuel properties, i.e., fuel-water interfacial tension (IFT) and fuel-air surface tension (ST). The primary focus of this effort was to develop correlations between IFT, ST, and micro-separometer (MSEP) data for various surfactants. IFT, ST, and MSEP were measured for a series of F-24 and Jet A fuels to explore the relationship between the various test methods. It was found that IFT correlated moderately with MSEP for the fuels tested (R2 of 0.42) while ST did not correlate with MSEP (R2 of 0.02). Two-dimensional gas chromatography (GCxGC) hydrocarbon-type and polars data were collected and analyzed for the fuels. The GCxGC data indicated that IFT was correlated to the total polars concentration, while MSEP was more sensitive to the bulk fuel composition. The effects of corrosion inhibitor/lubricity improver (CI/LI) and firefighting foam (FFF) over a range of concentrations were studied by doping fuels with known concentrations of each additive and measuring IFT, ST, and MSEP. IFT was found to follow an exponential decay function with additive concentration, while ST was less responsive.

*Corresponding author: Shane Kosir, shane.kosir@udri.udayton.edu

IASH 2024, the 18TH INTERNATIONAL CONFERENCE ON STABILITY, HANDLING AND USE OF LIQUID FUELS
Louisville, Kentucky, USA
September 8 – 12, 2024

INSIGHT INTO SAMPLING PROCESSES OF AVIATION FUEL

Moshe Rabaev*[1], Liel Mazar[1], Yalfal Siyum[1], Haim Judah[1], Dr. Fred Passman[2]

[1]*Israeli Air Force, 8478803 Beer Sheva, 8478803 Israel*
[2]*Biodeterioration Control Associates Inc.*

ABSTRACT

For aviation to be safe and successful, it is essential to have a constant supply of high-quality fuel, which is ensured by using appropriate sampling for testing. This chapter begins with a discussion of official standards as well as other literature related to fuel sampling in different storage locations, during transport, and from aircraft fuel tanks. A description of sampling points from tanks and fuel lines, sampling tools, and containers for samples is provided for specific types of storage vessels. In some cases, problems may occur due to local defects, such as poorly positioned sampling valves, or sloping sections of piping or vessel floors. Methods for ensuring that representative or diagnostic samples are obtained from fuel lines or tanks are discussed. In the context of aircraft fuel tanks, sampling is a particular challenge due to safety concerns and spatial constraints. Aside from routine sampling procedures, accidents, and other flight incidents often necessitate sample collection for forensic purposes. Such cases may require sampling filters installed on fuel lines, dismantling various types of filters, and recovering filter residues. Lastly, a detailed discussion is provided on the important topic of sampling for microbiological tests.

*Corresponding author: Moshe Rabaev, rabaevm@gmail.com

IASH 2024, the 18TH INTERNATIONAL CONFERENCE ON STABILITY, HANDLING AND USE OF LIQUID FUELS
Louisville, Kentucky, USA
September 8 – 12, 2024

ADVANCED MEMBRANE SEPARATIONS FOR CARBON CIRCULARITY: FROM LUBE OIL PURIFICATION TO CHEMICAL RECYCLING OF PLASTICS

Pieter Vandezande

Flemish Institute for Technological Research (VITO), Materials and Chemistry Department (MATCH), Boeretang 200 Mol 2400 Belgium

ABSTRACT

For decades, mankind has been living in a ‘make, take, dispose’ economic model. Plastics, with their unrivaled properties and low cost, are an exponent of this way of living. While plastics production reached 391 million tons in 2021, it is estimated that today nearly 50% of the plastic waste in the EU is landfilled, roughly 25% incinerated and only 25% recycled. Another example are used lubricating oils (ULOs), equally derived from fossil feedstocks and essential in a wide range of industrial applications, but today recycled only to a limited extent and not in the most energy-efficient way. To mitigate greenhouse gas emissions, recent EU legislation calls for far higher recovery and recycling rates of these materials at the end of their service lifetime.

Thanks to their inherently energy-efficient, mild and clean character, as well as flexibility and scalability, membrane processes are fully aligned with current industry drivers towards electrification, carbon circularity and zero/minimal liquid discharge. Therefore, membranes are considered a powerful tool, even key enabling technology, for tackling tomorrow’s separation challenges in chemical related industries.

Recovery of base lubricating oils from solvent-free ULOs and of virgin polymers from solvent-based plastics mixtures, targeting removal of critical impurities such as lube oil and plastics additives, calls for robust membranes that can cope with high viscosities and work at elevated temperatures. To address these challenges, tubular ceramic membranes with tailored surface chemistry are being studied on various real life ULO and waste plastics cases. This research fits in the ongoing EU project CUMERI and Flemish project RENOVATE, where VITO is collaborating with several other research institutes and companies on membrane technology development for ULO purification and dissolution-based plastics recycling, respectively. By facilitating recycling and reuse, both projects aim at tackling the global challenge of climate neutrality while contributing to enhanced circularity and resilience in these industries.

*Corresponding author: Pieter Vandezande, pieter.vandezande@vito.be

IASH 2024, the 18TH INTERNATIONAL CONFERENCE ON STABILITY, HANDLING AND USE OF LIQUID FUELS
Louisville, Kentucky, USA
September 8 – 12, 2024

CHARACTERIZATION OF COMPOSITIONAL EFFECTS ON THE DIELECTRIC CONSTANT FOR BLENDS OF AVIATION FUEL SYNTHETIC BLEND COMPONENTS AND PETROLEUM-BASED AVIATION FUELS

Shane Kosir*, April Landsaw, Zachary West, Steven Zabarnick

University of Dayton Research Institute, 300 College Park Dayton, Ohio 45469 USA

ABSTRACT

The dielectric constant is used in aviation for fuel gauging, namely for the determination of fuel volume and density. Synthetic blend components (SBCs) have compositions that may be different from petroleum-based fuels, which can potentially affect the dielectric constant, its relationship with fuel density, and thus the ability to accurately gauge fuel quantity. This work explored the effect of SBC blends with each other and with a petroleum-based Jet A fuel. The dielectric constant was found to be positively correlated with density, with cycloparaffinic kerosene (CPK) blends deviating from the other SBCs. Partial least squares regression using hydrocarbon-type composition features indicates that the dielectric constant is positively correlated with aromatics (i.e., mono and cycloaromatics) and negatively correlated with aliphatics (i.e., iso- and n-alkanes). Clausius-Mossotti (CM) plots were generated using data from 0-40°C to explore the effect of fuel composition on density gauging. The CM slope was found to deviate from Jet A for the SBC/Jet A blends. Specifically, the CM slope decreased for low-aromatic SBCs such as alcohol-to-jet (ATJ) and CPK and increased for high-aromatic SBCs, namely synthetic aromatic kerosene (SAK). R2 values were high for the CM plots except for CPK, possibly due to its divergent correlation with density. Despite the variation of the CM slope with fuel composition, the average density error was only 0.15% for all the blends and temperatures tested.

*Corresponding author: Shane Kosir, shane.kosir@udri.udayton.edu

IASH 2024, the 18TH INTERNATIONAL CONFERENCE ON STABILITY, HANDLING AND USE OF LIQUID FUELS
Louisville, Kentucky, USA
September 8 – 12, 2024

LOW TEMPERATURE BEHAVIOR OF RENEWABLE DIESEL AND BLENDS WITH PETROLEUM DIESEL

Suzanne Golisz*, Jessica Grabowski

Innospec Fuel Specialties, 200 Executive Dr Newark, Delaware 19702 USA

ABSTRACT

Use of renewable diesel (a.k.a. hydrogenated vegetable oil, HVO) is growing worldwide. While it's tempting to think of renewable diesel as a drop-in replacement for petroleum diesel, we have determined that extra consideration is required for the use of renewable diesel in cold climates. This paper is an extension of our previous paper looking at blends of up to 20 volume % renewable diesel in petroleum diesel. We have now widened the range of blending and included blends of renewable diesel in petroleum kerosene. Our primary modes of investigation have been to measure the cloud point, cold filter plugging point (CFPP), and pour point of these blends. Further, we have investigated the role of cold flow improvers and pour point depressants in lowering the CFPP and pour point. We found that additives continue to provide benefits. Lastly, we have evaluated renewable diesel and blends of renewable diesel using a cold stage microscope. These experiments clearly show a difference in the wax crystallization behavior in renewable diesel versus petroleum diesel.

*Corresponding author: Suzanne Golisz, suzanne.golisz@innospecinc.com

IASH 2024, the 18TH INTERNATIONAL CONFERENCE ON STABILITY, HANDLING AND USE OF LIQUID FUELS
Louisville, Kentucky, USA
September 8 – 12, 2024

LOW TEMPERATURE DIELECTRIC CONSTANT MEASUREMENT OF JET FUEL CONSTITUENTS TO ENABLE 100% SAF

Harrison Yang*, David Bell, Conor Faulhaber, Randall Boehm, Joshua Heyne

Washington State University, 2710 Crimson Way, Richland, Washington 99354

ABSTRACT

Modern aircraft often employ capacitance-based gauging systems that rely on the dielectric constant of the fuel onboard to determine fuel quantity. Aircraft manufacturers have expressed concern over inaccuracies in fuel gauging attributed to variances in the dielectric constant between conventional jet fuels and 100% paraffinic sustainable aviation fuel. Our previous study showed that different hydrocarbon group types exert varying effects on the Clausius-Mossotti relationship. To align with the established experience range for both the dielectric constant and the Clausius-Mossotti relationship, it appears that 100% drop-in SAF will need to incorporate some aromatic compounds. The standard procedure for assessing the dielectric constant (outlined in IP 638) is limited to a temperature range of 10 to 30°C. However, this range does not account for the broader spectrum of temperatures encountered by aircraft fuel, which can range from -40 to 50°C. Currently, there is a notable gap in the literature regarding the measurement of SAF's dielectric constant at lower temperatures. This study aims to bridge this gap by investigating the dielectric constant of SAF and various jet fuel components at lower temperatures, extending the measurement range beyond the current standard.

*Corresponding author: Harrison (Zhibin) Yang, zhibin.yang@wsu.edu

IASH 2024, the 18TH INTERNATIONAL CONFERENCE ON STABILITY, HANDLING AND USE OF LIQUID FUELS
Louisville, Kentucky, USA
September 8 – 12, 2024

ENGINE PERFORMANCE AND EMISSIONS: A MODEL-BASED ASSESSMENT OF ADDED VALUES OF SUSTAINABLE AVIATION FUELS (SAF)

Stephan Ruoff*, Georg Eckel, Uwe Bauder, Patrick Le Clercq, Manfred Aigner

German Aerospace Center (DLR), Pfaffenwaldring 3840, Stuttgart 70569, Germany

ABSTRACT

The approval and production of Sustainable Aviation Fuels (SAF) has gained significant momentum in recent years. With the aviation industry's commitment to reducing its environmental and climate impact, SAF offers a promising solution for mitigating CO2 emissions. Moreover, besides the reduction in greenhouse gas emissions, SAF offers additional benefits, such as reducing non-CO2 effects or increasing the fuel's energy content. However, besides the constraints imposed by the ASTM specifications, there is also a tradeoff between certain performance and emission metrics. Therefore, a comprehensive tool which can help gaining an overview about different metrics and identifying limiting factors, is developed.

In this study, various blend ratios of representative Jet-fuels and Synthetic Blend Components (SBCs) (including ATJ, FT-SPK, and HEFA) are investigated on the added values and current constraints enforced by the ASTM limits. Machine Learning are utilized to predict fuel properties from a given GCxGC composition. Empirical correlations are then employed to predict metrics for performance and emissions, including soot formation and Lean Blow Off (LBO) behavior as well as combustion related subprocesses like atomization and evaporation.

The findings indicate that even a low percentage of SBC has a substantial impact on improving jet fuel performance. However, the potential added values depend on the jet-fuel used for blending. Moreover, the addition of SBC significantly reduces the aromatics content in jet fuels, so that the current limits for aromatics are undercut.

*Corresponding author: Stephan Ruoff, Stephan.ruoff@dlr.de

IASH 2024, the 18TH INTERNATIONAL CONFERENCE ON STABILITY, HANDLING AND USE OF LIQUID FUELS
Louisville, Kentucky, USA
September 8 – 12, 2024

EXPLORING LOW-SOOTING COMPOSITIONS FOR POTENTIAL 'DROP-IN' SUSTAINABLE AVIATION FUEL

Conor Faulhaber*, Randall Boehm, David Bell, Zhibin Yang, Joshua Heyne

Washington State University Tri-Cities, 1621 George Washington Way, Richland, WA 99354 USA

ABSTRACT

Decarbonizing commercial aviation within the timeframe of current industry goals requires significant upticks in the production and usage of sustainable aviation fuel (SAF). 'Drop-in' SAFs produced from renewable energy crop or waste feedstocks can expedite the removal of fossil carbon from aviation fuel by meeting the safety, operability, and performance needs of existing aircraft and engine designs. Current paraffinic SAF pathways often struggle to demonstrate some of these conventional fuel properties, namely dielectric constant and elastomer seal swell, as a result of low aromatic content. To regulate this, SAF blends with conventional fuels are required to contain ≥8% aromatics content per ASTM specifications. However, aromatic species in jet fuel are associated with increased sooting and contrail formation, making low aromatic, 'drop-in' SAF blends of high interest to the industry. In comparison, cycloalkanes show potential at increased concentrations to induce adequate seal swell but are less likely to meet dielectric constant requirements.

Here, an extensive property database of both jet-range hydrocarbons and real SAF products is used to minimize threshold sooting index (TSI) for potential fuel compositions while remaining within the specification or experience range for conventional fossil jet fuel properties, including predicted dielectric constant and seal swell values. These results are used to inform laboratory-producible fuel blends, allowing for experimental evaluation of the bulk property predictions. In addition to improving understanding of aromatics and cycloalkanes in aviation turbine fuel, this work advises fuel producers, regulatory bodies, and engine/airframe manufacturers on compositional pathways towards 100% 'drop-in' SAF.

*Corresponding author: Conor Faulhaber, conor.faulhaber@wsu.edu

IASH 2024, the 18TH INTERNATIONAL CONFERENCE ON STABILITY, HANDLING AND USE OF LIQUID FUELS
Louisville, Kentucky, USA
September 8 – 12, 2024

ASSESSMENT OF THE IMPACT OF ALTERNATIVE FUELS ON THE CETANE VALUES AND PROPOSED MODELS FOR CALCULATING IT

Nathalie Brassart*[1], Yanis Melliti[2], Mickaël Sicard[2], Amandine Herbaut[3], Mickaël Matrat[4]

[1]*IFPEN, 1-4 Av. du bois Preau Rueil-Malmaison, 92852 France, Metropolitan*

[2]*ONERA*, [3]*CETSEO*, [4]*IFPEN*

ABSTRACT

NATO's Single Battlefield Fuel Policy recommends the use of JP-8 kerosene as a fuel for DIESEL ground vehicles.

A minimum cetane number (CN) of 40 is required to assess jet fuel suitability for ground applications. The large chemical composition variability of sustainable aviation fuels (SBC) leads to lower or high CN compared to fossil jet fuel. Therefore, there incorporation into fossil kerosene might be a source of ignition anomalies (low CN), especially when cold or engine efficiency degradation (high CN).

Three conventional jet fuels representative of worldwide variability and four SBCs were used to establish a matrix of 36 blends. The aim was to evaluate the cetane number as a function of SBC incorporation rate and composition.

A compositional characterization and determination of the main physico-chemical properties of jet fuels, SBCs and their blends were carried out. Moreover, the CN and derived cetane number (DCN) were measured for all fossil fuels with different SBCs concentrations. The results show a linear dependence of the blend CN evolution with the rate of incorporation of the two initial constituents. Furthermore, the CI, CN and DCN values for fossil kerosenes and most SBCs are very close (□2). ATJ, on the other hand, stands out, with a difference of 40. Finally, two models based on different levels of chemical composition of the blend fuel were developed.
An original database is thus provided here. It highlights potential limitations for different SBCs while exploring incorporation rates above the current ASTM limit of 50%.

*Corresponding author: Nathalie Brassart, nathalie.brassart@ifpen.fr

IASH 2024, the 18th INTERNATIONAL CONFERENCE ON STABILITY, HANDLING AND USE OF LIQUID FUELS
Louisville, Kentucky, USA
September 8-12, 2024

AZEOTROPIC FUELS: ELUCIDATING THE INFLUENCE OF STRUCTURAL CLASSES ON AZEOTROPIC BEHAVIOR IN OXYGENATED FUELS

Jordan N. Metz[1] and David J. Abdallah[1]

[1]*ExxonMobil Technology and Engineering, 1545 Route 22 East, NJ 08801, USA,* *jordan.n.metz@exxonmobil.com*

ABSTRACT

The use of alcohols, ethers, and other oxygenated compounds as fuel blend components continues to generate substantial interest due to their renewable sourcing and positive impact on various fuel performance properties. Of unique importance, when blending these oxygenates into hydrocarbon fuels, is their characteristic effect on phase equilibrium. It is well known that certain oxygenates can promote non-ideal vapor-liquid equilibrium (VLE) effects, such as azeotropic behavior, when blended in fuel. In gasoline, this phenomenon has been well studied; however, the azeotropic behavior resulting from the individual underlying structural classes that comprise gasoline (i.e., *n*-paraffins, *iso*-paraffins, naphthenes, olefins, aromatics) has been largely unexplored. Herein, experimental ASTM D86 distillation and statistical associating fluid theory (SAFT) calculations to model phase phenomena are employed to elucidate these fundamental contributions. The use of these simple techniques will be highlighted to demonstrate the ease of access to such information and the implications of the results will provide useful insight into how these components impact fuel properties.

Introduction

Oxygenated hydrocarbons have a long history as blend components in gasoline fuels primarily observed as organic alcohols and ethers, with even the earliest prototype engines incorporating these compounds into their fuel mixtures.[1-3] The use of these compounds has not only been driven by fuel security,[4,5] economic,[6,7] and renewable interests,[8,9] but also due to noteworthy performance qualities such as high octane ratings,[10] the ability to promote complete combustion,[11] high latent heat of vaporization allowing higher in-cylinder temperatures and pressures,[12] and high volatility cold start properties.[13] On the other hand, their use can also have undesirable performance effects; for example, their lower energy density requires more fuel per energy release – introducing either a weight burden on the vehicle or a mileage range deficit. Depending on the exact molecular structure, these properties can be adjusted, promoting research efforts to evaluate the most appropriate oxygenates for fuel. As renewed interest in these compounds continues to grow, the various oxygenates under study are expanding.

Notably, ethanol is the most widely researched and implemented oxygenate, with E10, E15, and E85 (i.e., 10%, 15%, and 85% by volume) broadly applied to public automotive fueling stations.[14]

Beyond ethanol, a wide range of other alcohols and ethers (Figure 1) have been employed and studied as potential blend components, including alcohols such as methanol, propanol, *iso*-propanol, butanol, *iso*-butanol, *sec*-butanol, and *tert*-butanol and ethers such as diethyl ether, methyl *tert*-butyl ether (MTBE), ethyl *tert*-butyl ether (ETBE), *tert*-amyl methyl ether, and di-*iso*-propyl ether. In addition to common organic alcohols and ethers, other oxygenate classes have also earned appropriate attention. For example, ketone blend components have generated interest with studies of acetone, 2-butanone, 4-methyl-2-pentanone, cyclopentanone, cyclohexanone, methyl ethyl ketone, and methyl *iso*-butyl ketone discussed within the fuel community as potential future gasoline components.[15-17] Similarly, esters have recently been investigated as gasoline components, such as ethyl acetate and ethyl valerate.[18] Additive-level concentrations of organic acids also play a role in providing specific performance enhancements;[19] however, the scope of this work will focus primarily on higher-concentration blend components (>1%). With such a wide range of structural attributes, the ability to tune performance within these oxygenate species is apparent.

Likewise, the hydrocarbon blend components can also be categorized into structural classes, namely *n*-paraffins, *iso*-paraffins, olefins, naphthenes, and aromatics (Figure 1). Within these classes, a homologous series typically expresses a variation in properties such as molecular weight and alkyl branching. In a typical fuel blend, the concentrations of these classes are tuned to ensure certain properties of the overall fuel mixture are met either for specification standards or for performance outcomes.

Despite their abundance and importance, with over a century of use as blend components, these oxygenates and their interactions with individual hydrocarbon classes typically present in gasoline have significant gaps in fundamental understanding. Of particular interest is the impact on non-linear vapor-liquid equilibrium (VLE) effects, such as azeotropic behavior, observed when blends of these oxygenates and hydrocarbon classes are present. For fuel formulators, a comprehensive understanding of these VLE profiles can lead to significant improvements in critical fuel properties such as vapor pressure and distillation profiles, thereby enhancing performance aspects regarding vapor lock, cold start properties, spray and droplet evaporation characteristics, and fuel system performance in general.

Various studies have investigated this VLE effect in gasoline blends as a whole[20-26] demonstrating the differences between alcohols and ethers impact, and several reports on isolated species have been discussed[27,28] to help understand the intermolecular interactions of particular species; however, a comprehensive map encompassing the general hydrocarbon classes and their azeotropic behavior when blended with various oxygenates has, to the best of our knowledge, not been investigated. Herein, we report a simple method utilizing ASTM D86 distillation to tease out the azeotropic effect of oxygenates blended with each class of hydrocarbons. In addition to this simple method, standard statistical associating fluid theory (SAFT) models are utilized to expand the ability to investigate this phenomenon. The fundamental azeotropic behavior of these blends is explored using binary systems of each class of molecules (e.g., oxygenate:*n*-paraffin, oxygenate:aromatic). Due to the continued importance of oxygenated fuel blends in our industry, a more comprehensive analysis of their fundamental phase characteristics is required. Through understanding how various oxygenates affect common gasoline molecular classes, a metric to tune fuel properties is developed.

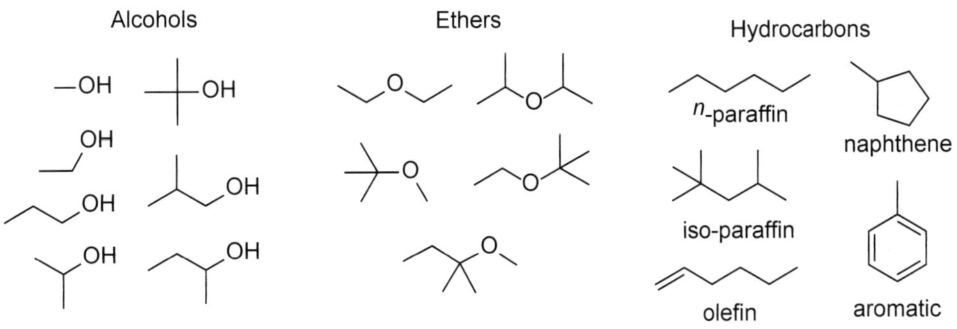

Figure 1. Common oxygenates and hydrocarbon blend components. Structural differences within each class can have varying influence on VLE profiles.

Experimental

Materials

In order to investigate the structural class effects on azeotropic behavior in oxygenated fuel blends, one hydrocarbon species to represent each hydrocarbon class (i.e., *n*-paraffins, *iso*-paraffins, naphthenes, olefins, aromatics) and the most widely used oxygenates were obtained. In particular, *n*-octane (*n*-paraffin), 2,2,4-trimethylpentane (*iso*-paraffin), 1-methylcyclohexane (naphthene), 1-octene (olefin), and toluene (aromatic) were chosen to represent the hydrocarbon classes. For oxygenates, ethanol (alcohol), *iso*-butanol (alcohol), and ETBE (ether) were chosen. All hydrocarbon species were purchased as reagent-grade chemicals, and the alcohols were purchased as anhydrous grade.

ASTM D86 Distillation

Binary systems of one part oxygenate (e.g., alcohol, ether) to one part hydrocarbon species (i.e., *n*-paraffins, *iso*-paraffins, naphthenes, olefins, aromatics) were blended (volume basis) and submitted for ASTM D86 to obtain distillation profiles for each binary system (Table 1). ASTM D86 has a long history of evaluating distillation properties in fuels, being one of the oldest ASTM methods in the industry. Nevertheless, this method provides a simple technique to study azeotropic behavior. In this method, 100 mL of material is continuously heated, with both volume and temperature monitored, thus providing a distillation profile of the liquid mixture. Azeotropic effects on distillation are well reported in the literature; however, a brief discussion on this subject follows to provide an introductory description of how this study design is implemented and how results are interpreted.

Azeotropes are mixtures that retain their composition during phase change from liquid to vapor at a constant temperature; in other words, their vapor phase composition and liquid phase composition are identical after distillation – no separation of species is afforded. This unique phenomenon is responsible for various unexpected alterations to the distillation profile, particularly: 1) higher distilled material than expected at conditions of interest and 2) the change of the initial bubble point. The bubble point is the point at which vapor is first formed in a mixture, similar to the boiling point for pure components. Due to industrial nomenclature, these terms are

Table 1. Fuel blends of binary mixtures for ASTM D86 analysis. All oxygenates were blended with hydrocarbons at 10, 20, 30, 40, 50, 60, 70, 80, and 90 percent by volume.

Oxygenate	*n*-Paraffin	*iso*-Paraffin	Olefin	Naphthene	Aromatic
ethanol *iso*-butanol ETBE	*n*-octane	2,2,4-trimethylpentane	1-octene	1-methylcyclopentane	toluene

often used interchangeably. Figure 2a describes an exemplary binary mixture containing two pure components at varying molar concentrations exhibiting a positive azeotrope (positive vapor pressure deviation from Raoult's law, negative bubble point deviation). This azeotropic behavior results in a lowering of the overall bubble point curve (lower than either pure component) to a minimum where the phase change from vapor to liquid maintains the same compositional character. This point can be viewed where the dew point curve contacts the bubble point curve resulting in the azeotrope point. Outside this point, the area between the dew point and bubble point curves is where both a liquid and vapor phase exist, and a phase change will not maintain the composition of the material distilled (lighter boiling point material will preferentially enter the gas phase and the theoretical plates and temperature rate change will determine the resultant liquid and gas phase compositions).

In the distillation curve, this phenomenon will not only result in a lowering of the bubble point (and increase in vapor pressure) but also a general trend that shifts the distillation curve to the right (figure 2b). This is due to the non-linear deviation of the bubble point across a large portion of the temperature-composition (T-X-Y) curve and, more importantly, the complete compositional phase change to vapor at the azeotrope point. A true azeotropic mixture will display this characteristic throughout the phase envelope (temperature-pressure profile), whereas zeotropic (non-azeotrope) mixtures may exhibit this characteristic at select conditions, resulting in azeotropic behavior although not classified as a true azeotrope.

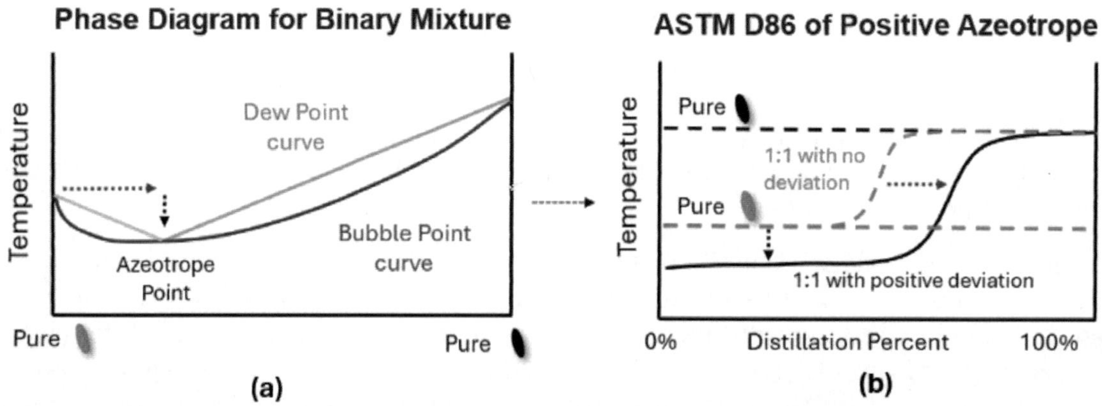

Figure 2. Effect of azeotropic behavior on: (a) T-X-Y plot, (b) distillation curve. These plots depict theoretical representations of the azeotropic effect. The azeotrope causes a decrease in distillation initial boiling point and a right-shift in the percent distilled.

The implications are that certain oxygenated blends will affect not only the initial boiling point (and vapor pressure) but also a large portion of the distillation profile at particular conditions. In this study, we will investigate binary systems of interest *via* the ASTM D86 method to explore this behavior and elucidate how these oxygenate-hydrocarbon systems can be tuned for fuel properties.

Statistical Associating Fluid Theory (SAFT)

We utilized the proprietary ExxonMobil EM-SAFT model implemented in the PRO/II steady-state process simulator (version 2022.1)[29] for the thermodynamic models. Although the EM-SAFT model is proprietary, other SAFT models are available in PRO/II and other standard engineering software. SAFT models were necessary to ensure the species' intermolecular interactions were modeled accurately.

In ideal liquid-vapor phase behavior, Raoults's and Dalton's laws can explain the vapor pressure of a mixture of components by utilizing their pure component vapor pressures. However, these idealities do not account for the crucial intermolecular interactions of the system. In oxygenated fuel blends, understanding these intermolecular interactions becomes critical for modeling phase equilibria, particularly where hydrogen bonding (H-bonding) and other strong dipole-dipole forces are involved. Azeotropic behavior is observed when significant deviations in electrostatic forces are present compared to the pure components (e.g., removal of H-bonding interactions from ethanol when in hydrocarbon-based fuel), resulting in either stronger or weaker interactions with neighboring molecules. These molecular forces either help to hold the composition into the liquid phase (stronger intermolecular interactions) or promote gaseous phase preference (weaker intermolecular interactions).

The SAFT model is an equation of state describing a hierarchy of intermolecular interactions including dispersion, dipole-dipole, and associating/H-bonding forces using statistical thermodynamics. Details of these models can be found elsewhere.[30] Inherently, an understanding of covalent and electrostatic forces are modeled that ensures the azeotropic behavior can be captured in the physical properties obtained from these models. Thermodynamic models such as the SAFT models used herein have been demonstrated to accurately predict azeotropic phenomena in gasoline fuels.[30] This tool was readily available to our group, and no additional programming was investigated.

PRO/II allows for quick access to binary VLE curves and phase envelopes and can also generate more complex information, such as distillation curves. The distillation curves are produced through a series of flash processes that mimic the distillation method. These processes were set up based on previously reported literature[30] and will be discussed in the following section.

Results and Discussion

In this section, we discuss which hydrocarbon species are responsible for azeotropic behavior in oxygenated fuels to remove generalities typically associated with gasoline as a whole and reveal the tunability of this phenomenon. Each oxygenate blend studied will start with a discussion on using ASTM D86 to obtain distillation curves, demonstrating a simple technique that can be utilized to investigate azeotropic behavior. This is followed by a discussion on the usefulness of widely available SAFT models to expand on these findings and demonstrate a useful tool that can be utilized to understand VLE phenomena and aid in experimental design.

Ethanol and Hydrocarbon Fuel Blends

ASTM D86 distillation was conducted on several blends, yielding fundamental insights into the effects of various hydrocarbon classes on azeotropic behavior in ethanol-based fuels. The results were obtained and used to reconstruct a series of distillation curves. Three features are highlighted as we discuss the distillation profiles affected by azeotropic behavior: (1) initial boiling point (IBP) suppression, (2) a right-shift in the distillation profile, and (3) final boiling point (FBP) suppression. A detailed evaluation of the *iso*-paraffin blend will review these features to facilitate understanding the differences between subsequent classes.

Figure 3 displays the distillation curves from the volumetric blends of ethanol and isooctane, revealing a clear and substantial non-ideality of the distillation profile, indicating unmistakable azeotropic behavior under these conditions. The pure component boiling points are highlighted as solid lines (black = isooctane, red = ethanol), with the binary blends represented as dashed lines to help display the boiling point suppression effect. The IBP of the 90% isooctane (10% ethanol) blend was observed to be approximately 8°C lower than ethanol's pure boiling point (more than 28°C lower than isooctane's pure boiling point). This IBP suppression is caused by the positive azeotropic deviation. Furthermore, blends from the range of 30-90% isooctane all show similar initial boiling points, with only 0-20% isooctane blends elevated – still well below the boiling point of ethanol.

In addition to the IBP suppression, there is a right-shift in the distillation curve for each isooctane-ethanol blend. Importantly, with 90% isooctane (10% ethanol), the boiling point does not reach the pure isooctane boiling point until almost 40% liquid volume is distilled instead of the 10% that might be expected (pure ethanol removal). This visibly demonstrates that the ethanol does not distill off first, followed by isooctane. There is azeotropic behavior causing both molecules (at a specific ratio) to boil off together until approximately 40% is distilled, in which typical isooctane behavior is observed, and ethanol has been largely removed. Moreover, this signifies that the azeotropic behavior in this binary mixture affects a large portion of the distillation profile. At 70% isooctane (30% ethanol), the pure isooctane boiling point is not reached until approximately 90%

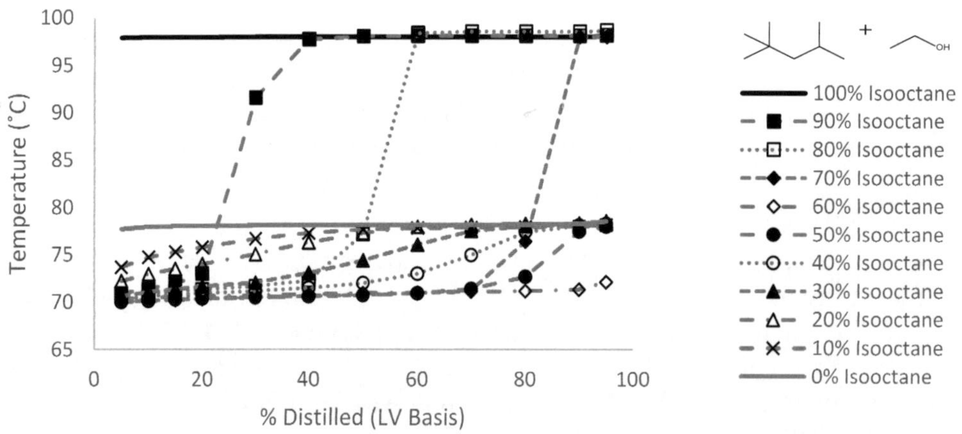

Figure 3. ASTM D86 distillation curves of ethanol:*iso*-paraffin blend. Binary blends were produced by volume percent. Barometric pressure was approximately 1 bar.

is distilled, indicating that ethanol will remain present throughout the majority of the distillation profile.

At a specific ethanol concentration (approximately 40%+), the FBP is observed only to reach that of ethanol's or lower. This can be quickly observed with only three (70%, 80%, 90% isooctane blends) blend compositions displaying distillation curves between the two pure components boiling points. Thus, a large portion of this distillation behaves much more like ethanol than isooctane.

When viewing these distillation curves, the composition of the azeotrope point can be inferred by viewing the blend concentration that is most affected. For example, in the case of isooctane-ethanol, the 60% isooctane (40% ethanol) blend can be viewed as almost a flat line (tight boiling range). This signifies that the material is boiling off together with no pure-component behavior observed. Therefore, the plot reveals a composition of the azeotrope point at approximately 60% isooctane (by volume).

Interestingly, and to highlight the importance of structural differences, the *n*-paraffin blend (Figure 4) showed little deviation from ideality. This result proves that not all five classes of hydrocarbon species create the same VLE deviations and will not produce identical performance impacts. Despite being similar in paraffinic nature, *n*-octane showed the least deviation in distillation (i.e., minimal azeotropic effect) out of all the five hydrocarbon classes. Very minimal boiling point suppression is observed. As the ethanol content increases, there is a distillation profile effect that is non-ideal, with the percent distilled (required to reach n-paraffin boiling point) not equal to the percent ethanol (slight right-shift); however, blends up to 80% *n*-paraffin still display distillation curves within the two pure components with rather steep distillation rates. Above 80% *n*-octane, the material appears to boil off with an azeotropic behavior – this will be discussed further using T-X-Y plots below. By isolating the gasoline sub-components, we demonstrate that some hydrocarbon molecules will promote this VLE effect, and others may minimize this effect. Formulations can take advantage of this unique characteristic.

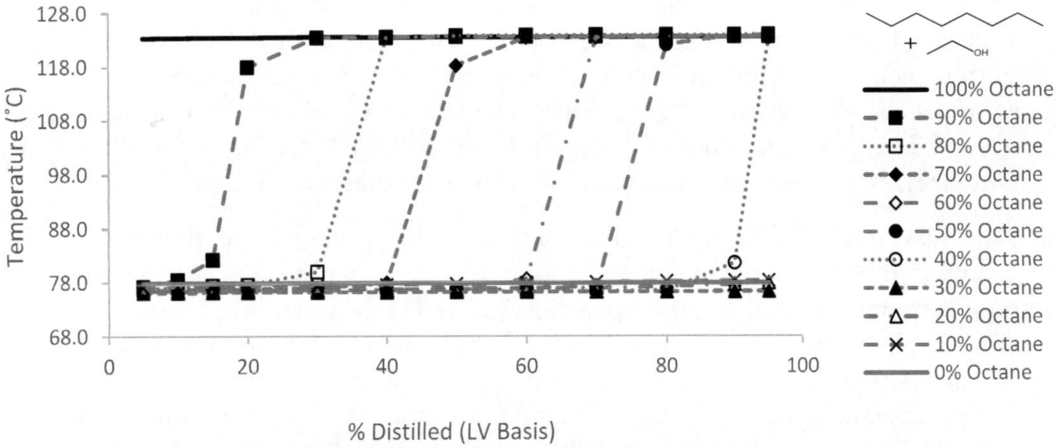

Figure 4. ASTM D86 distillation curves of ethanol:*n*-paraffin blend. Binary blends were produced by volume percent. Barometric pressure was approximately 1 bar.

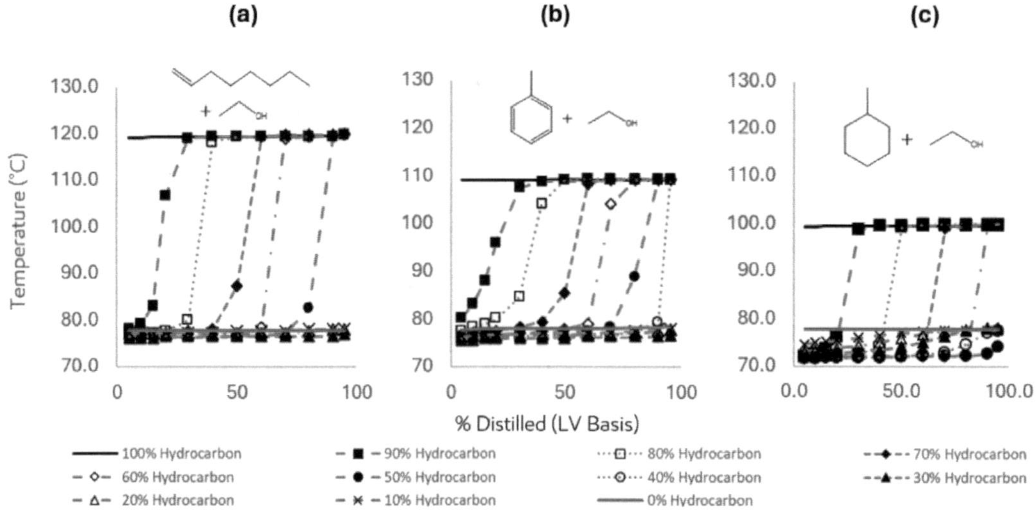

Figure 5. ASTM D86 distillation curves of: (a) ethanol:olefin blend, (b) ethanol:aromatic blend, (c) ethanol:naphthene blend. Binary blends were produced by volume percent. Barometric pressure was approximately 1 bar.

The remaining classes of hydrocarbon blends with ethanol were similarly evaluated for the IBP, curve profile, and FPB to develop a map of ethanol's azeotropic behaviors with all five hydrocarbon classes. Figure 5 highlights the effect of olefins, aromatics, and naphthenes. Notably, naphthenes were observed to promote significant azeotropic behavior with initial boiling point suppression, a shift in the distillation curve to the right, and an FBP suppression of high ethanol concentrations to that of (or below) ethanol's pure boiling point. However, the effect was not quite as strong as compared to *iso*-paraffins. The olefin and aromatics had minor boiling point suppression, although the distillation curves were shifted.

It can be noted that the olefin (1-octene) utilized in this study was a straight-chain olefin similar to the *n*-paraffin (*n*-octane). Very similar results were obtained for both species, indicating that the double-bond had little impact. Similarly, the naphthene used in this study (methylcyclohexane) was comparable in structure to the aromatic (toluene) with the absence of the aromaticity inside the ring structure. These two molecules behaved significantly differently. This finding suggests the aromatic-ethanol neighboring intermolecular interactions were similar in strength to the pure ethanol intermolecular interactions, whereas naphthene-ethanol blends lowered the intermolecular attractive forces. The following order of deviation from the ideal behavior of the ethanol blends can be summarized as: *iso*-paraffin > naphthene > aromatic > olefin > *n*-paraffin.

To further elucidate ethanol's impact on azeotropic behavior in the hydrocarbon classes, SAFT models were used to obtain bubble point and dew point curves for these binary systems to view the azeotropic composition and expand upon the ASTM D86 limitations. The study of binary systems offers an advantage in that T-X-Y curves can be generated for binary systems, which are instrumental in visualizing the azeotropic phenomena. By employing SAFT models (inherent intermolecular interactions modeled), we were able to extract valuable information from these T-X-Y plots. These plots (Figure 6), which include the bubble point curve (black) and the dew point curve (red) across the compositional fraction of the binary mixture, show a remarkable correspondence with the D86 results, thereby validating the effectiveness of these models.

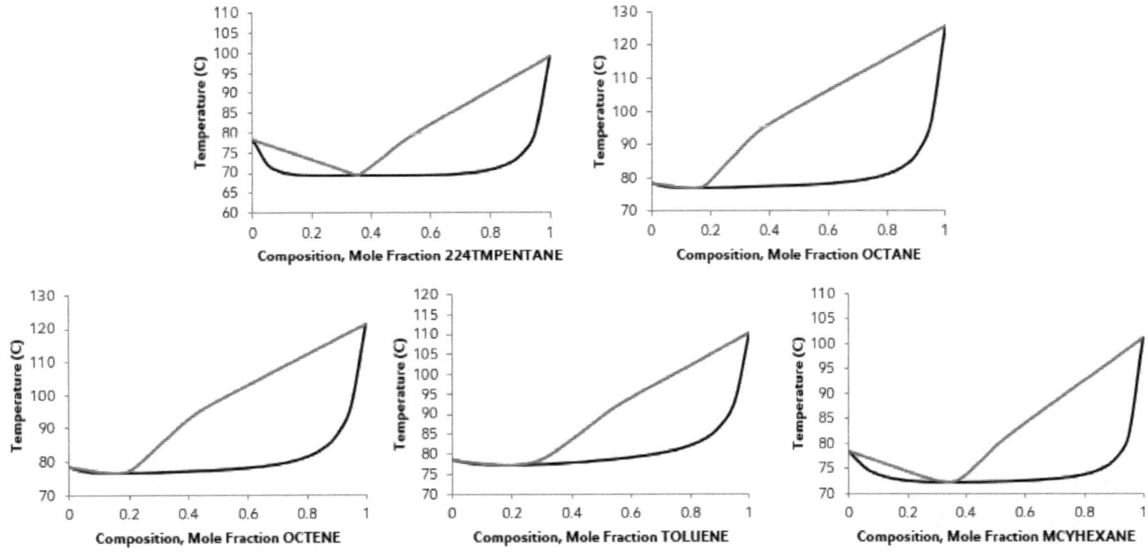

Figure 6. T-X-Y plots of: (a) isooctane:ethanol, (b) octane:ethanol, (c) 1-octene:ethanol, (d) toluene:ethanol, (e) methylcyclohexane:ethanol. Thermodynamic model ESAFT, 14.7 psi. Red = dew point, black = bubble point.

The *iso*-paraffin and naphthene fractions indicate apparent azeotropic behavior, demonstrating both boiling point suppression and a significant flattening of the bubble point curve. Notably, the azeotrope point matches the ASTM D86 curves. As discussed previously, the azeotrope point for the *iso*-paraffin blend was considered to be near the 60% blend concentration (by volume), as indicated by the tightest boiling point range (flattest curve, Figure 3). Here, the T-X-Y plot shows approximately 35% (molar), which is indeed in agreement.

The remaining *n*-paraffin, olefin, and aromatic curves indicate minor boiling point suppression; however, an overlapping of the bubble point and dew point curves was observed at high concentrations of ethanol, indicating that at higher concentrations, the distillation profile would boil off at constant temperature. This is indeed what was observed in the ASTM D86 results described previously. Notably, these T-X-Y plots help explain the right-shift in the distillation profile from the ASTM D86 results. Although no boiling point suppression is easily viewable, an azeotropic behavior is observed, promoting hydrocarbon distillation simultaneous with ethanol, causing a right shift in the distillation profile.

With certain fuel systems development (e.g., spray characteristics), expanding the range of conditions (i.e., temperatures, pressures) is required to see if this azeotropic behavior is present. A phase envelope can be modeled using the thermodynamic SAFT model to understand the bubble point and dew point curves across a range of temperatures and pressures. A true azeotrope will demonstrate its azeotropic behaviors over a significant portion of the phase envelope with overlapping bubble point and dew point curves, indicating direct phase change from liquid to gas with compositional retainment. Indeed, zeotropic mixtures can also indicate azeotropic behavior at specific conditions, often called partial azeotropes or near-azeotropes. By viewing the phase envelope, the azeotropic behavior across the entire pressure-temperature plot can be obtained (at a specific composition), which can be useful for various fuel systems and product development efforts.

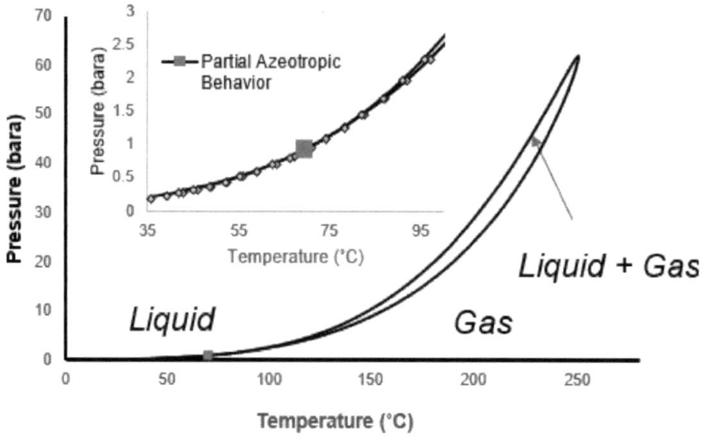

Figure 7. Phase envelope of methylcyclohexane:ethanol. Mixture of 0.35 molar volume methylcyclohexane.

Figure 7 shows the phase envelope for a binary mixture of methylcyclohexane and ethanol (35 mol% methylcyclohexane). It is clear that the azeotropic behavior is not present at higher temperatures and pressures; however, it is observed for a portion of the curve that is of interest for many fuel system aspects. From 40-90°C (at varying pressures), this azeotropic behavior is present. A 3-D map including composition can also be applied to reveal additional information.

The experimental and modeling results show that individual structural classes that comprise gasoline do not afford identical VLE effects. Specific structural classes produce varying electrostatic interactions with ethanol that require a detailed knowledge to formulate fuels with appropriate properties. Vapor pressure and boiling point knowledge of either pure component is insufficient to understand their behavior once mixed. Several notable concepts can be ascertained from these findings.

For example, the IBP lowering effect in ethanol-gasoline blends is the principal reason these fuels are recognized for improving cold start properties. Indeed, this study reveals that the *iso*-paraffin blend component (typically 20-40%) is involved in this IBP lowering effect. This indicates that a high-octane component may be one of the molecular constituents in the vapor phase at initial engine start. Additionally, the FBP lowering effect has implications for meeting final boiling point specifications while maintaining heavier molecular weight hydrocarbon components. These details are significant for formulation, particularly in terms of engine performance and an understanding of the general azeotropic trends each class of hydrocarbons promote is helpful for fuel development.

Iso-butanol and Hydrocarbon Fuel Blends

Isobutanol was found to behave quite differently compared to ethanol, with all hydrocarbon classes displaying azeotropic behavior (Figure 8), thus highlighting that not only do hydrocarbon species have varying effects (as seen with ethanol), but the oxygenate structure also has an impact. This implies that not only is there tunability in the hydrocarbon classes but additionally in the oxygenate structures. The distillation results in all five hydrocarbon classes show significant boiling point suppression and alterations to the curve profile. The extent of the effect still varies across the

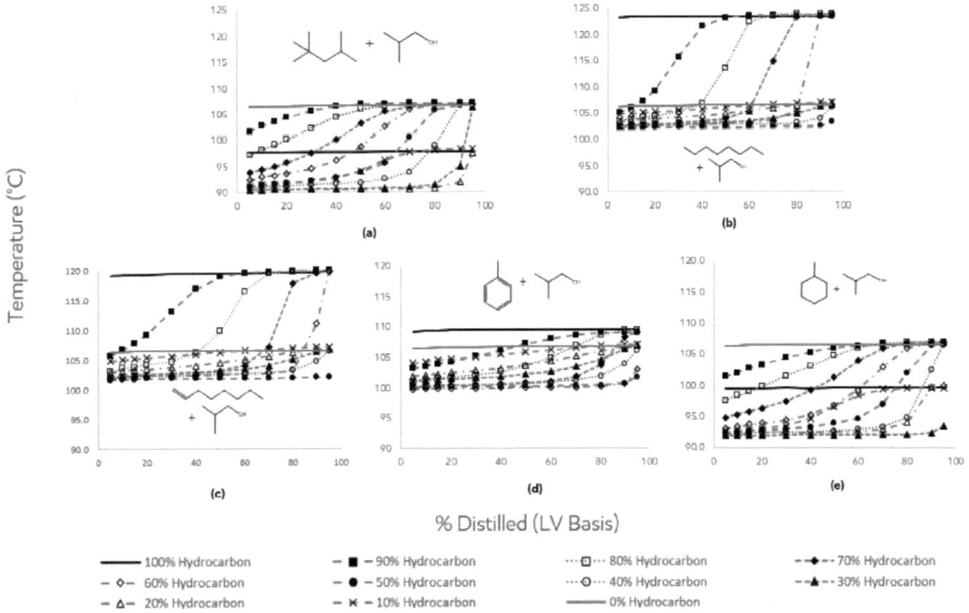

Figure 8. ASTM D86 distillation curves of: (a) isobutanol:*iso*-paraffin blend, (b) isobutanol:*n*-paraffin blend, (c) isobutanol:olefin blend, (d) isobutanol:aromatic blend, (e) isobutanol:naphthene blend. Binary blends were produced by volume percent. Barometric pressure was approximately 1 bar.

hydrocarbon classes. The order of deviation from ideal behavior in this blend sequence can be summarized as: aromatic > olefin > *n*-paraffin > naphthene > *iso*-paraffin.

Interestingly, aromatics, olefins, and *n*-paraffins disrupt the isobutanol intermolecular interactions the most. These were the classes with lowest deviation from ideal in the ethanol case. Similar to ethanol blends, the *iso*-paraffin had notable differences compared to the straight-chain paraffin (*n*-octane). Likewise, *n*-octane and 1-octene appear to have similar extent of effect, which was observed in the ethanol blends.

The significant change in these trends highlights the importance of structural attributes to the oxygenate class. Presumably, various oxygenates moieties will have critical impacts on this behavior. Sterics, surface area, and electronic alterations can all be tuned for targeted performance.

Ether and Hydrocarbon Fuel Blends

As expected, due to the lack of ether's ability to act as a hydrogen bonding donor (HBD), azeotropic behavior was not apparent (Figure 9) in these blends. Overall, the boiling point curves stayed within the bounds of the pure component boiling points with minor non-linear effects. These minor non-linear effects are often observed when similar hydrocarbon species are blended. It can be noted that although ethers do not act as HBDs (unlike alcohols and acids), they can act as hydrogen bonding acceptors (HBAs) that can impact systems that include HBDs. This introduces additional complexities that can be used to tune properties with regard to molecular interactions.

In order to understand ether's distillation profile shifts further, an octane-hexane blend to highlight a similar distillation profile of non-azeotropic behavior with some alterations to the distillation

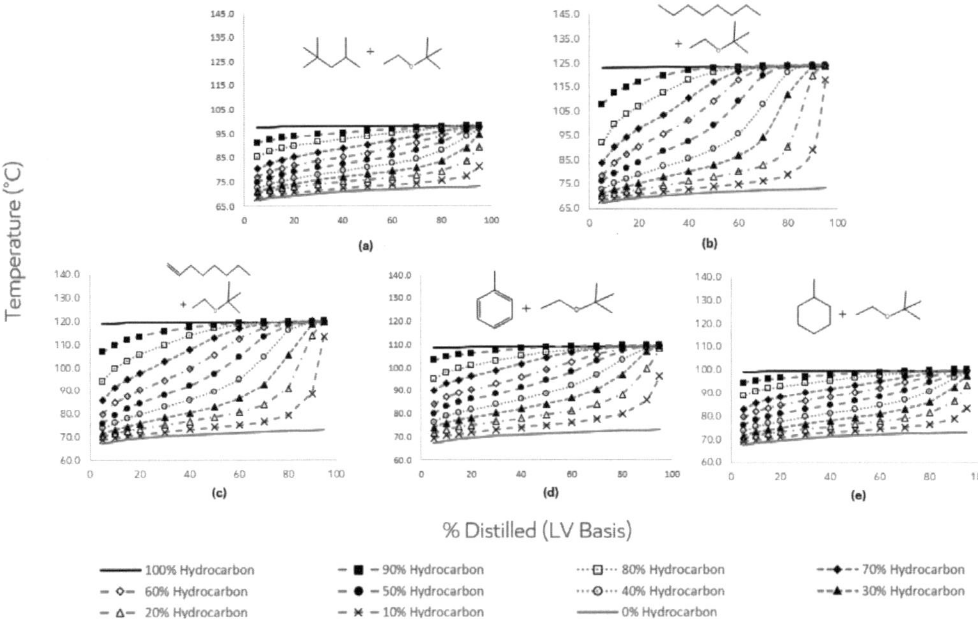

Figure 9. ASTM D86 distillation curves of: (a) ETBE:*iso*-paraffin blend, (b) ETBE:*n*-paraffin blend, (c) ETBE:olefin blend, (d) ETBE:aromatic blend, (e) ETBE:naphthene blend. Binary blends were produced by volume percent. Barometric pressure was approximately 1 bar.

profile was obtained (Figure 10a). This blend demonstrates typical hydrocarbon-hydrocarbon interactions. Figure 10a depicts the distillation profile which resembles the overall observations in the case of ether blends. Figure 10b depicts the T-X-Y curve, showing no azeotropic behavior and no flattening of the bubble point curve. The distillation shifts in these cases are typical non-idealities.

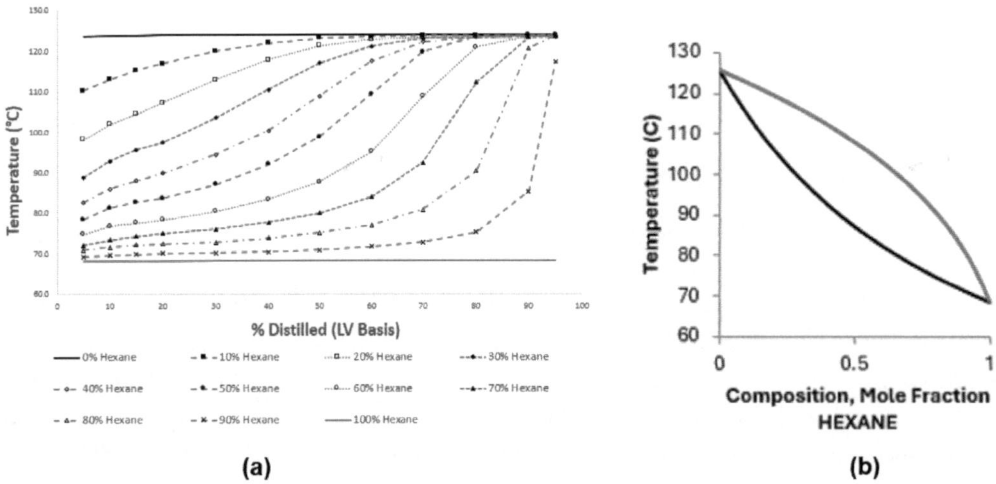

Figure 10. ASTM D86 distillation curves of *n*-hexane:*n*-octane blend. Binary blends were produced by volume percent. Barometric pressure was approximately 1 bar.

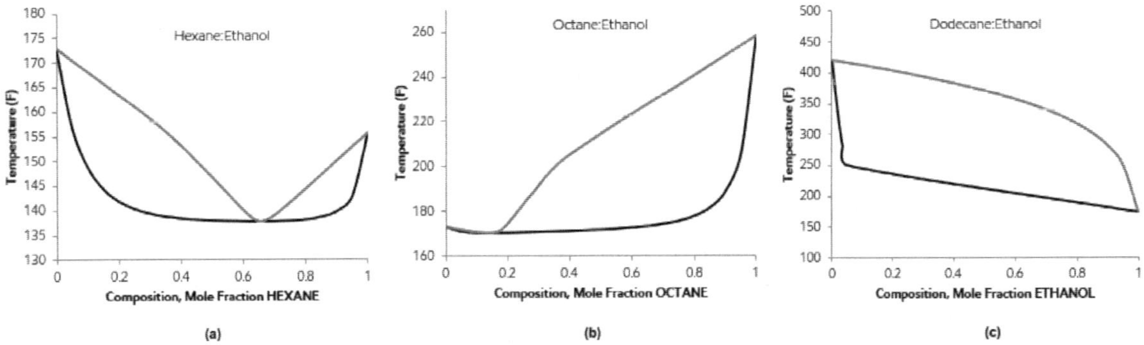

Figure 11. Molecular Weight impact of Interaction (SAFT). Thermodynamic model ESAFT, 14.7 psi. Red = dew point, black = bubble point.

The results of these oxygenate-hydrocarbon blends begin to form a map of the structural effects that can be tuned to increase or decrease azeotropic behavior. The methods used reveal how simple experimental designs can be used to develop such a map. Future work to build a more detailed map of various oxygenate effects on hydrocarbon species is ongoing.

Homologous Series Within Hydrocarbon Classes

Expansion to homologous series within each hydrocarbon class is explored to determine additional important structural effects. In typical gasoline blends, a series of similar molecules within each structural class is inherent. To understand practically, the impact of these classes, trends within these homologous series must be understood. Initial insight into molecular weight and isomeric variations have been investigated.

Figure 11 depicts a series of T-X-Y plots of ethanol and increasing molecular weight *n*-paraffins. The plots indicate a clear trend in azeotropic behavior of increasing molecular weight; as the molecular weight increases, the azeotropic behavior decreases. This agrees with the assumption that the increase in electrostatic forces will decrease the azeotropic effect. The *n*-hexane blend displays a significant azeotropic behavior with a boiling point suppression of almost 15°C and a flattening of the bubble point curve. *n*-Octane has previously been discussed, with only a minor azeotropic behavior at very high ethanol concentrations. Once the molecular weight is increased to dodecane, the T-X-Y plot displays no overlap of bubble point – dew point curves.

Figure 12 depicts a series of T-X-Y plots of ethanol and decreasing alkyl branching. The series shows *n*-octane, 2-methylheptane, 2,2-dimethylhexane, and 2,2,4-trimethylpentane (isooctane) providing a series of 0-3 alkyl branches. As the *n*-paraffin content is increased, a decrease in azeotropic behavior is observed as anticipated from previous results. Additional branching positions (e.g., internal vs terminal) and alkyl chain length is needed to further elucidate the effect.

The exploration of the homologous series of *n*-paraffin chain length and *iso*-paraffin alkyl branching reveals practical implications. Within the hydrocarbon classes, not all molecules will behave the same. This observation is crucial as it allows us to reason on which molecules might show azeotropic behavior and to what extent, thereby making our research directly applicable to real-world scenarios.

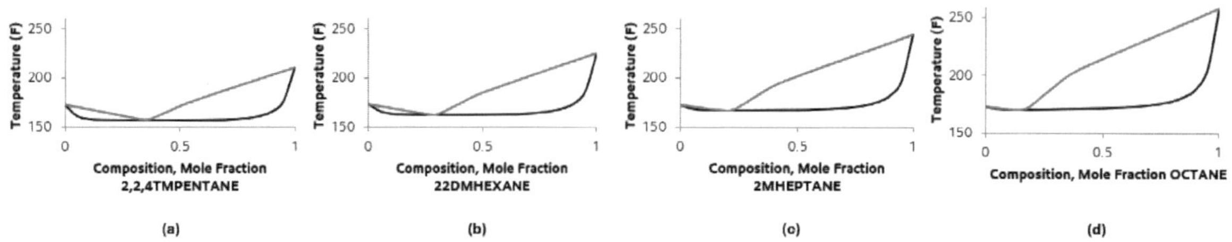

Figure 12. Isomeric Impact of *iso*-Paraffin (SAFT). Thermodynamic model ESAFT, 14.7 psi. Red = dew point, black = bubble point.

Utilizing Thermodynamic Models to Predict Distillation Curves

As the thermodynamic SAFT models have proven helpful in the design of experiments, the ability to produce distillation curves through a series of flash processes in PRO/II was developed. Figure 13 shows the overlaying SAFT distillation curve (red) with the experimental ASTM D86 curves (black) for the ethanol-isooctane series. Indeed, all three aspects of interest were accurately modeled, including IBP suppression, alterations to the curve profile, and FPB suppression, particularly in the higher ethanol concentration blends. The reduction in experimental burden through the use of these models will provide useful in elucidating the numerous VLE trends between various oxygenate-hydrocarbon interactions.

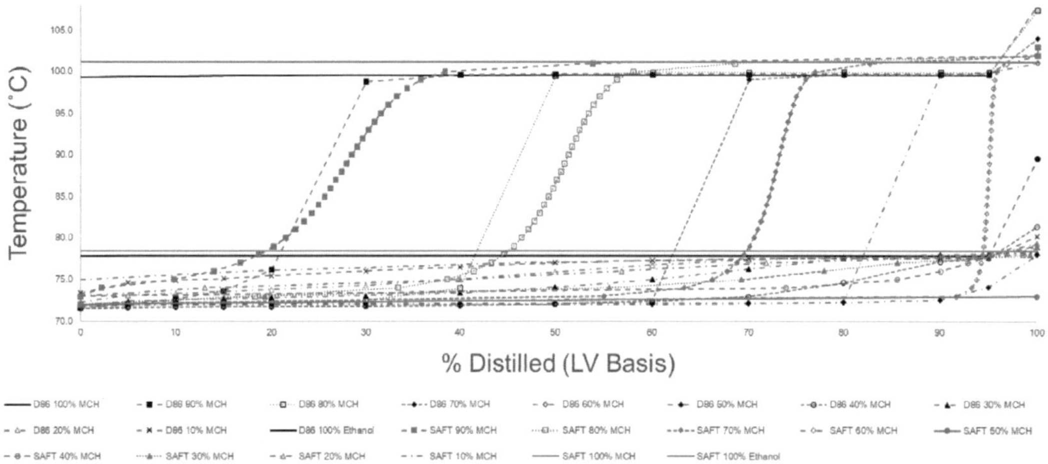

Figure 13. SAFT – D86 distillation comparison.

Conclusions

This study highlights a methodology for evaluating the azeotropic behavior of oxygenated blend components with various hydrocarbon classes utilizing binary blends. ASTM D86 distillation profiles and SAFT modeling tools can provide substantial information regarding fuel blend attributes. The results obtained help to fundamentally understand how structural tuning can be utilized for fuel property benefits in terms of VLE phenomena. Results validate that not all oxygenated species interact with these hydrocarbons similarly, with some revealing significant azeotropic behavior and others minimal. Additionally, unexpected trends in these hydrocarbon classes reveal unique structural attributes that can be targeted. Future work will continue to elucidate homologous series effects within these classes and extend to additional oxygenate species.

References

1. Elvers, B., Schutze, A., Handbook of Fuels. 2022.
2. Bielaczyc, P., et al., Ethanol as an automotive fuel – a review. Combustion Engines, 2016. 166(3): p. 39-45.
3. Awad, O.I., et al., Alcohol and ether as alternative fuels in spark ignition engine: A review. Renewable and Sustainable Energy Reviews, 2018. 82: p. 2586-2605.
4. Melendez, J.R., et al., Perspectives in the production of bioethanol: A review of sustainable methods, technologies, and bioprocesses. Renewable and Sustainable Energy Reviews, 2022. 160: p. 112260.
5. Eaves, J. and S. Eaves, Renewable corn-ethanol and energy security. Energy Policy, 2007. 35(11): p. 5958-5963.
6. Kunwer, R., et al., Economic assessment and production of ethanol-A review. Materials Today: Proceedings, 2022. 69: p. 543-548.
7. Kofler, R., et al., Techno-Economic Analysis of Dimethyl Ether Production from Different Biomass Resources and Off-Grid Renewable Electricity. Energy & Fuels, 2024. 38(10): p. 8777-8803.
8. Falano T, Jeswani HK, Azapagic A. Assessing the environmental sustainability of ethanol from integrated biorefineries. Biotechnol J. 2014 Jun;9(6): p. 753-765
9. Styring, P., G.R.M. Dowson, and I.O. Tozer. Synthetic Fuels Based on Dimethyl Ether as a Future Non-Fossil Fuel for Road Transport From Sustainable Feedstocks. in Frontiers in Energy Research. 2021.
10. Awad, O.I., et al., Overview of the oxygenated fuels in spark ignition engine: Environmental and performance. Renewable and Sustainable Energy Reviews, 2018. 91: p. 394-408.
11. Lim, C.-S., et al., Comparative effects of oxygenates-gasoline blended fuels on the exhaust emissions in gasoline-powered vehicles. Journal of Environmental Management, 2019. 239: p. 103-113.
12. Huang, Y., G. Hong, and R. Huang, Investigation to charge cooling effect and combustion characteristics of ethanol direct injection in a gasoline port injection engine. Applied Energy, 2015. 160: p. 244-254.
13. Yanowitz, Janet, and Robert L. McCormick. "Review: fuel volatility standards and spark-ignition vehicle driveability." SAE International Journal of Fuels and Lubricants, vol. 9, no. 2, June 2016, pp. 408+.
14. Johnson, Caley, Kristi Moriarty, Teresa Alleman, and Daniolo Santini. 2021. History of Ethanol Fuel Adoption in the United States: Policy, Economics, and Logistics. Golden, CO: National Renewable Energy Laboratory. NREL/TP-5400-76260.
15. Patil, V., et al., Applicability of ketone-gasoline blended fuels for spark ignition engine through energy-exergy analyses. Fuel, 2023. 339: p. 127416.
16. Usman M, Khan T, Riaz F, Ijaz Malik MA, Amjad MT, Shah MH, Ashraf WM, Krzywanski J, Nowak W. Acetone-Gasoline Blend as an Alternative Fuel in SI Engines: A Novel Comparison of Performance, Emission, and Lube Oil Degradation. ACS Omega. 2023 Mar 13;8(12): p. 11267-11280.

17. Kocakulak T, Uyumaz A, Arabacı E, Dağoğlu Y, Çamoğlu C. Experimental Investigation of the Effects of Gasoline-Methyl Ethyl Ketone Fuel Blends on Engine Performance and Exhaust Emissions. J. Mater. Mechat. A. 2023;4(2): p. 397-408.
18. Oppong, F., et al., Esters as a potential renewable fuel: A review of the combustion characteristics. Fuel Processing Technology, 2022. 229: p. 107185.
19. Tack, R. D., Fuel Additives Chemistry and Technology. 2022.
20. Abdollahipoor, B., et al., Near-azeotropic volatility behavior of hydrous and anhydrous ethanol gasoline mixtures and impact on droplet evaporation dynamics. Fuel Processing Technology, 2018. 181: p. 166-174.
21. Aghahossein Shirazi, S., et al., Effects of dual-alcohol gasoline blends on physiochemical properties and volatility behavior. Fuel, 2019. 252: p. 542-552.
22. Ahmed, A.A., A.M. El-Masry, and Y. Barakat, Azeotrope formation in gasoline–ethanol blends. Part 1 – Impact of nonionic on E10 distillation curve. Egyptian Journal of Petroleum, 2018. 27(4): p. 1167-1175.
23. Furey, R.L. and K.L. Perry, Volatility Characteristics of Blends of Gasoline with Ethyl Tertiary-Butyl Ether (ETBE). 1990, SAE International.
24. Kar, K., et al., Measurement of Vapor Pressures and Enthalpies of Vaporization of Gasoline and Ethanol Blends and Their Effects on Mixture Preparation in an SI Engine. 2008.
25. Michalopoulou, D.-P., et al., Impact of Bio-Ethanol, Bio-ETBE Addition on the Volatility of Gasoline with Oxygen Content at the Level of E10. Fuels, 2021. 2(4): p. 501-520.
26. Oliveira, A.V.d.S., G.H.M. Alegre, and R.G.d. Santos, A comprehensive experimental study on nucleate boiling in gasoline and gasoline-ethanol blends. Experimental Thermal and Fluid Science, 2017. 88: p. 134-144.
27. Li, D., et al., Molecular Mechanism for Azeotrope Formation in Ethanol/Benzene Binary Mixtures through Gibbs Ensemble Monte Carlo Simulation. The Journal of Physical Chemistry B, 2020. 124(16): p. 3371-3386.
28. Sasikumar, P., et al., Impact of physio-thermo properties on the azeotrope formation in benzene – 1-alkanol blends. Journal of Molecular Liquids, 2021. 337: p. 116400.
29. PRO/II V2022.1, Copyright (C) 2020-2024 AVEVA Group Limited
30. Vella, J.R. and B.D. Marshall, Prediction of the Distillation Curve and Vapor Pressure of Alcohol–Gasoline Blends Using Pseudocomponents and an Equation of State. Industrial & Engineering Chemistry Research, 2020. 59(17): p. 8361-8373.

Towards 100% Sustainable Fuels

Discussion Panel with Aviation Focus

What do we want and why?

What needs to be discovered and managed?

IASH 2024, Tuesday 10th September 2024

Speakers and Areas of Interest

Speaker	Interest for this topic
Gurhan Andac	OEM perspective, plus ASTM Task Force leader
Rick Kamin	USER, with MILITARY including NATO fleet operability and specifications
Amy Carico	USER, with COMMERCIAL fleets
Patrick Bosmans	PIPELINE & STORAGE, Logistics, Military with Commercial customers
Joanna Bauldreay	PRODUCER, Synthetic Blend Components +/or conventional jet fuels

IASH 2024, Tuesday 10th September 2024

GE Aerospace

September 8-12, 2024 – Louisville, KY, USA

Fully Synthetic Aviation Turbine Fuel

Gurhan Andac
Engineering Technical Leader
Aviation Fuels & Additives
GE Aerospace
ASTM D02.J06 Chair

ECCN: 9E991 (NLR)

What is (and is not) SAF?

Most only refer to the synthetically and sustainably derived component as SAF (e.g., US SAF Grand Challenge, ReFuelEU, UK Jet Zero Strategy, etc.). Misunderstanding, however, is that many think this synthetic-only component itself is Jet A/A-1.

Prevailing (incorrect) thought:

"*Today, we allow synthetic/sustainable Jet A/A-1 (SAF) to be used only when blended with conventional Jet A/A-1. One day, we will remove this limitation and will use this synthetic/sustainable Jet A/A-1 (SAF) by itself, unblended, at 100%.*"

What really is the case:

Synthetic Blending Component (SBC) + Conventional Blending Component = SAF Blend
(SAF) (typically petro Jet A/A-1) (Jet A/A-1)

What is determined to be equivalent to Jet A/A-1 is the SAF blend, not the SBC;
SBC (which is what people refer to when they say SAF) is not necessarily Jet A/A-1 equivalent.

2

Drop-in vs non-drop-in fully-SATF

Drop-in: not just compatible with a particular engine and/or aircraft, but fleet-wide and infrastructure-wide compatible

	Drop-in (Jet A/A-1 equivalent; main focus)	**non-Drop-in** (paraffinic, longer-term)
Composition:	Fully formulated Jet A/A-1	**Subset** of Jet A/A-1
Fleet applicability:	Fleet wide drop-in	**Designated** aircraft/engines only
Example pathways:	FT-SKA (D7566 Annex A4), CHJ (D7566 Annex A6), ATJ-SKA (D7566 A8), future: HEFA-SKA, multi-blend, others	FT-SPK (D7566 Annex A1) HEFA-SPK (D7566 Annex A2) ATJ-SPK (D7566 Annex A5) *certain types*
Specification:	ASTM D7566	**New standard** needed
Regulatory cert/substantiation:	No change	**Required** for each intended aircraft/engine model
Infrastructure:	No impact	**Separate** supply chain/handling/storage required

ASTM Task Force est. Apr '21
Chair: G. Andac (GE), Vice-Chair: M. Rumizen (Air Company)
Approval of use of conforming fully-SATF as Jet A/A-1
(No change to the semi-synthetic fuel requirements)

ASTM Task Force est. Apr '22
Chair D. Parmenter (Airbus), Vice-Chair: A. Hobday (Rolls-Royce)
NOT approval of use; to be used for testing and certification

OEMs are active via ASTM, IAEG, and internally

GE Classification: Approved for Public Use

General characteristics of fully-SATF

- More stringent requirements than conventional and semi-synthetic blends
  - Aromatics
  - Acidity
  - Distillation
  - Naphthalene content
  - Sulfur content
  - Viscosity
  - Derived Cetane Number
  - Relative Permittivity
  - Lubricity
  - JFTOT
  - more...
- Allow blending of multiple SBCs
- Only D7566 annexed SBCs could be used to derive a fully-SATF
- Once certified to D7566 *drop-in* fully-SATF will be subject to D1655 requirements (just like semi-synthetic today); non-drop-in fully-SATF is not a D1655 fuel

4

GE Classification: Approved for Public Use

Thank you!

5

GE Classification: Approved for Public Use

Getting to 100% Sustainable Fuels: A Military View

IASH 2024

Rick Kamin
Naval Air Systems Command
10 September 2024

NAVAIR

NAVAIR Public Release 2024-0146
Distribution Statement A - Approved for public release; distribution is unlimited

Fuels and Lubricants

Sustainable Fuels Through A Military Lens

- **Mission Capability**
  - Performance and Maintainability
    - Legacy, Emerging, Future
  - Tactical Distribution
  - Availability
  - Interoperability
  - Cost
- **Risk vs Opportunity**

Getting to 100% Sustainable Fuels: Moving Forward

- **Clear Definition of Requirements and Timeframe**
- **Quantification of Mission Capability Risks and Opportunities**
- **Expand OEM Engagement**
- **Optimize the Power of the Military Collective**
  - **OEMs, Industry Specification Organizations, Data Sharing**
- **Maintain Continuity of Effort/Resources**
- **Acknowledge the Cost Differential**

Airlines for America®

We Connect the World

Airline Perspective

Toward 100% Sustainable Fuels

10 September 2024

AIRLINES FLY GREEN

COMMITTED TO NET-ZERO CARBON EMISSIONS BY 2050

The major U.S. airlines have committed to working across the aviation industry and with government leaders in a positive partnership to achieve **net-zero carbon emissions by 2050**.

Airlines for America®
We Connect the World

airlines.org

U.S. Domestic SBC Volumes

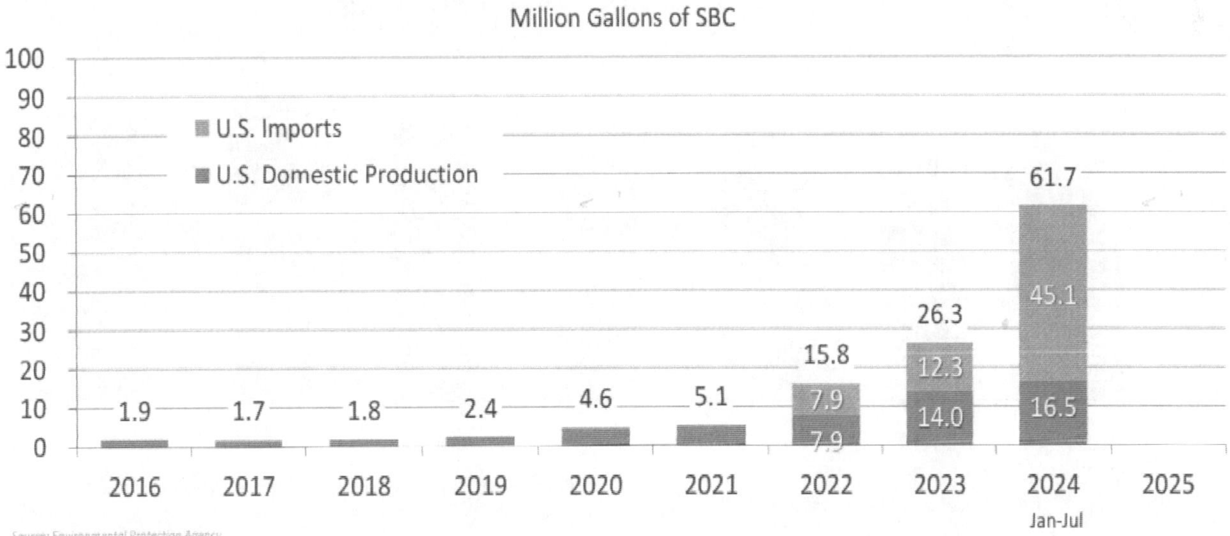

Source: Environmental Protection Agency

Airlines for America®
We Connect the World

airlines.org

3 US Airports Dispensed More Than 1B Gallons in 2023

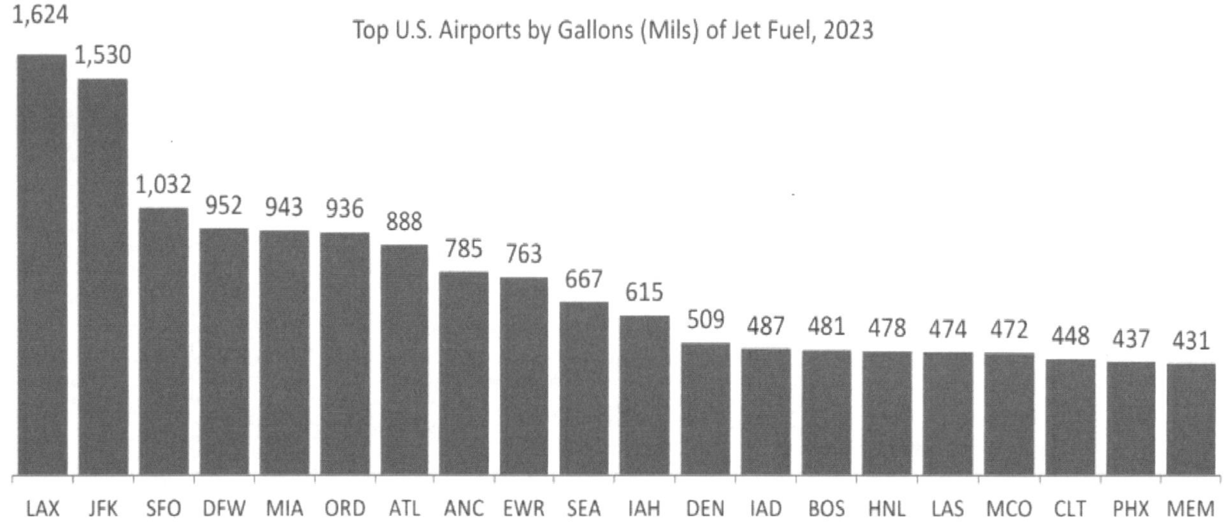

Source: A4A and airports or fuel system operators

Airlines for America®
We Connect the World

airlines.org

EIA Projects 2.6% Annual Jet Fuel Growth until 2050

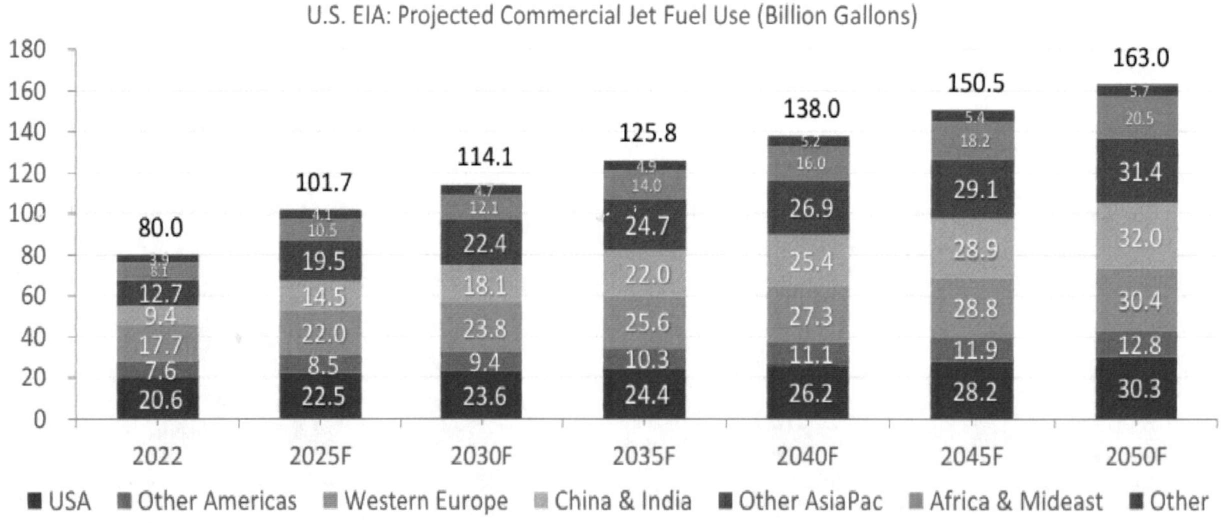

Source: U.S. Energy Information Administration (EIA) International Energy Outlook

Airlines for America®
We Connect the World

airlines.org

More Than 10yrs for RD to Achieve "Commercial" Scale

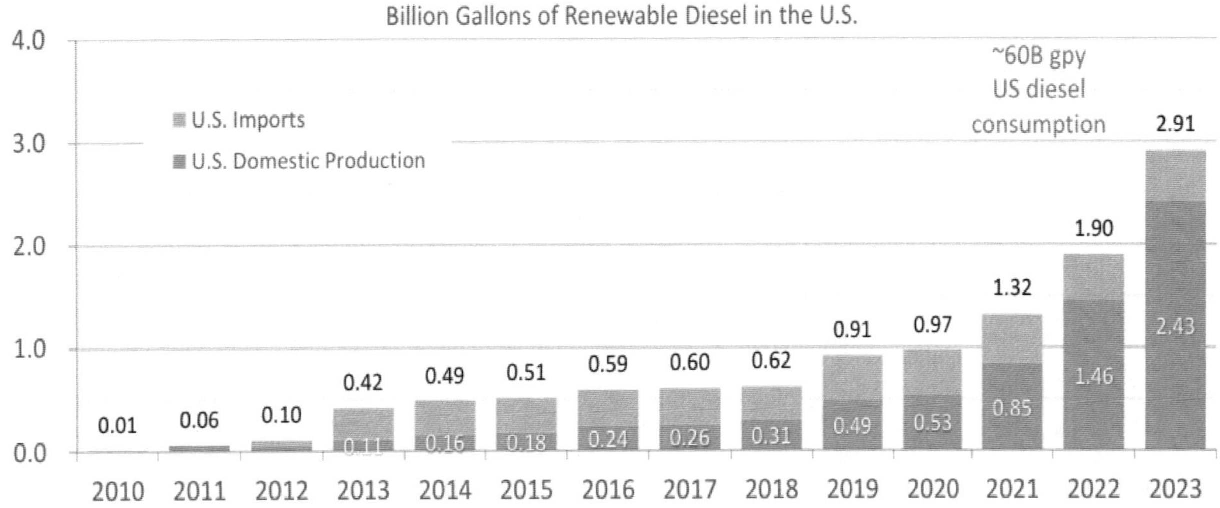

Source: Environmental Protection Agency (Note: EPA does not disclose data broken down by producer)

Airlines for America® We Connect the World

airlines.org

100% SAF

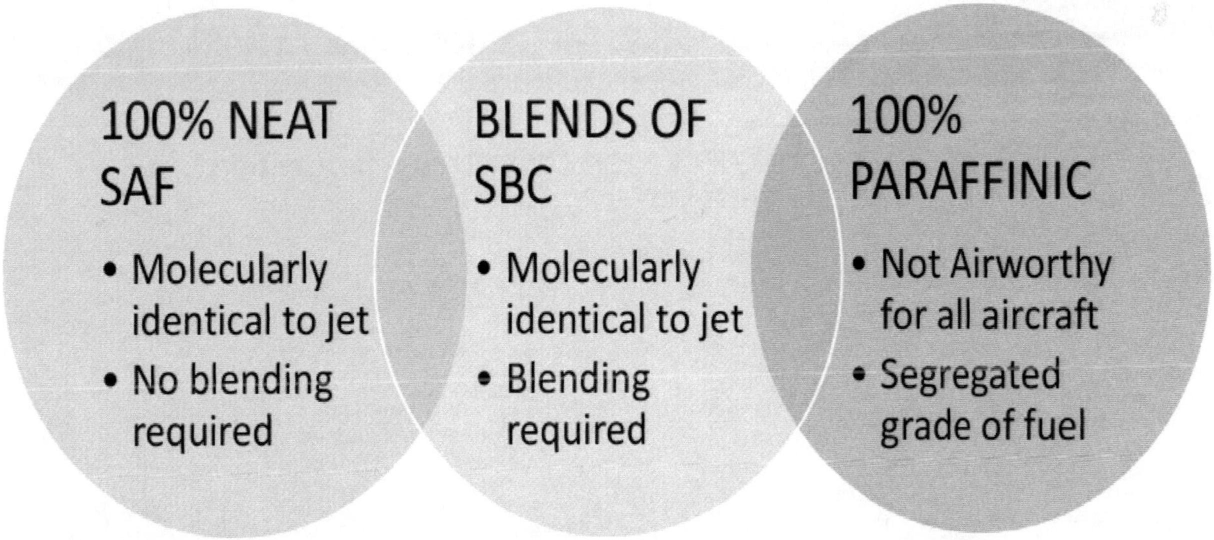

Airlines for America® We Connect the World

airlines.org

Airlines for America®

We Connect the World

www.airlines.org

NATO OTAN

NATO Support and Procurement Agency (NSPA)

IASH
TOWARDS 100% ALTERNATIVE FUELS

Patrick Bosmans - CEPS

09-12 Sept 2024, Louisville, Kentucky, USA

EFFICIENT.EFFECTIVE.RESPONSIVE.

NATO UNCLASSIFIED

SYNTHETIC 100 % ALTERNATIVE FUELS

Fuel no longer contains conventional components!

Compatibility has to be evaluated for the aircraft & the fueling infrastructure

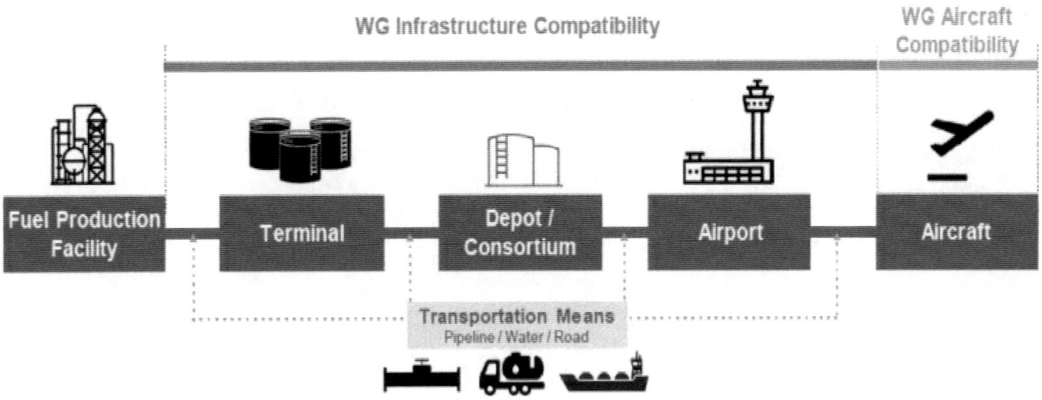

DROP-IN SYNTHETIC JET FUEL (JET A-1)

Aircraft compatibility

- Test flights have been demonstrated, but is it safe for *all* aviation equipment?
  - SBC (mostly HEFA) + 'Synthetic Aromatic Kerosine' (seal swell)
- Are the current test methods suitable?
  - Although combustion, lubrication, heat exchange etc. have been studied for decades, unknowns remain => any uncertainty may impact safety!
  - New requirements to be added to the Jet A-1 specification (e.g. CN; minimum limit for total (8%?) or mono and di- aromatics)?
  - New issue of alignement of the different jet fuel specs (ASTM, DefStan...)

NATO/CEPS

- CEPS (Jet A-1 banking system) : new approval of *all* 6 Member Nations for all their platforms
- NATO:
  - New approval (joint operations, in flight fueling...)
  - Possible issues for jet fighters (e.g. afterburner) and for NATO's 'Single Fuel Policy' (e.g. CN)

DROP-IN SYNTHETIC JET FUEL (JET A-1)

Logistics compatibility

- \+ Less blending restrictions and associated risks (Cfr EI 1533)
- − MBC sensitivity to be further studied
- − Valves and hoses: still compatible due to less aromatics?
- − PL interfaces/transmix: handling issues if new requirements/limits are added to the spec

CEPS:

- \+ Will be commingled with other jet fuels in its Jet A-1 banking system
- − CI/LI additive (lubricity): not allowed at ingress, since it can cause an overdose when transformed into F-34 with remote injection systems at fix rates

NON-DROP-IN SYNTHETIC JET FUEL (JET-X)

Aircraft compatibility

- No aromatics: impact on other characteristics (e.g. MSEP) + replaceable by cycloparaffins?

Logistics compatibility

- It will require *separate* handling, storage and logistics to keep it *safe*
  - Dual fuel system at airports (!!) and separate storage at terminals
  - Positive segregation and an unique identification procedure to eliminate any possible mistakes in fuel identity (+ training of the operators)
- It must be *compatible* with the fuel infrastructure
  - Valves and hoses (lack of aromatics): will it affect the maintenance schedule for fuel handling equipment?
  - PL interfaces/transmix : Jet A-1/Jet X cutting on density (what contamination levels are acceptable)? Handling?

CEPS

- Only point-to-point deliveries ! => more time to deliver & monthly volumes will drop

SYNTHETIC JET FUEL – TAKE AWAY

Still more questions than answers.

Jet-X could be great for environmental reasons, but the opposite for fuel handling

We need to take as much time as needed to make sure that these fuels are safe and fit for purpose for the airframes and … the fuel infrastructure

- This approach needs to be data driven
- Synergy with other existing industry efforts (e.g. IAEG WG13 to facilitate a widespread adoption by 2030 of 100 % paraffinic jet fuel)

NSPA EFFICIENT.EFFECTIVE.RESPONSIVE. NATO UNCLASSIFIED 14

Producer Views on 100% SAF

Joanna Bauldreay

Bauldreay Jet Fuel Consulting Limited (2019+)

Chester, UK

IASH 2024, Tuesday 10th September 2024

Producer Status – 1- Technical

- 8 annexes approved, but so far only 1-2 producing synthetic blending components in commercial volumes.
- Coprocessing is an important route being used too.
- Significant numbers of new SBC candidates are in or approaching the ASTM D4054 process via 3 Clearing Houses and 1 OEM group.
- No shortage of bright ideas but are there preferred chemistries for 100% SAF that need to be prioritised?

IASH 2024, Tuesday 10th September 2024

Producer Status – 2 - Production

- Streamlining approval process still further (to speed up process) will help new (and conventional) players finance their projects.
- Incentives and Mandates can help producers (and their bankers) invest in production at scale, but even 10% SAF by 2030 targets will be a stretch.

IASH 2024, Tuesday 10th September 2024

Producer Questions

- When will new fuels be defined enough to commit £/$ to production facilities?
- Could rules for "Sustainable" change, leaving feedstocks out in the cold?
- Are all drivers for 100% fuels valid? Could we make new problems (technical, availability/supply, economical, etc.) and is anyone checking on this?
- Can we link "new feedstock" SBCs with repurposed conventional refineries? (Sustainable includes jobs, infrastructure, etc.)
- Do we have enough technical experts to audit new fuel supply systems, even for SAFs as defined now?

IASH 2024, Tuesday 10th September 2024

What do you think?

IASH 2024, Tuesday 10th September 2024

IASH 2024, Tuesday 10th September 2024

IASH 2024, the 18TH INTERNATIONAL CONFERENCE ON STABILITY, HANDLING AND USE OF LIQUID FUELS
Louisville, Kentucky USA
8 – 12 September 2024

AN INNOVATIVE MICROFLUIDIC REACTOR FOR TESTING THE OXIDATION STABILITY OF FUELS: APPLICATION TO A SAF SURROGATE AND COMPARISON WITH THE PETROOXY TEST

Ryma Benrabah[1], Emilien Girot[1], Philippe Arnoux[1], Jean-Marc Commenge[1], René Fournet[1], Pierre-Alexandre Glaude[1], Baptiste Sirjean[1]

[1]*Université de Lorraine, CNRS, LRGP, F-54000 Nancy, FR.* ryma.benrabah@univ-lorraine.fr

ABSTRACT

One potential solution to diminish the aviation carbon footprint is to blend Sustainable Aviation Fuel (SAF) into jet fuels. However, the thermal stability and oxidation stability of standard jet-fuels may be affected by the addition of SAF [1]. In this context, we developed a Si-glass microfluidic reactor with a fluid supply made of inert polyether ether ketone (PEEK). The experimental rig PEEK-ACHU (Autoxidation µChip Heated Unit) was used to investigate the oxidation stability of a SAF surrogate and the results were compared to Induction Period (IP) measures, obtained in a standard PetroOxy device. A mixture of *n*-decane and iso-octane was used as the surrogate SAF [2]. The thermal oxidation stability of the mixtures was measured in the microreactor, working at conditions up to 453 K and 30 bar, with a gas-liquid segmented flow that allows the liquid oxidation to occur in homogeneous, self-agitated liquid slugs. PEEK-ACHU enables residence times as long as 3.8 hours to be reached on a 4-inch diameter Si/glass wafer, visual access to the reactor, and a well-defined kinetic study of the liquid fuel autoxidation. The total hydroperoxide (ROOH) concentration was quantified at the exit of the microreactor using a specifically designed HPLC device with a post-column reactor [3]. For comparison, the same analysis was conducted on the SAF oxidized in the PetroOxy apparatus, with the same residence time. The results of this study show that the microfluidic apparatus is able to perform reliable autoxidation measurements and allows a better understanding of the kinetics. The self-stirring ensures a homogenous phase, and the visual access of the reactor is relevant for other applications, such as spectroscopic monitoring of deposit formation.

KEYWORDS: microfluidic, Sustainable Aviation Fuel (SAF), oxidation, High performance Liquid Chromatography (HPLC), hydroperoxide (ROOH).

INTRODUCTION

The environmental issues regarding the impact of the transportation sector on the emissions of CO_2 and their consequences lead to the establishment of rules and regulations to limit carbon footprints. For the aviation sector, the diminution of emissions remains a challenge because of the increase in worldwide air traffic which is predicted to be multiplied by 2 within 15 years, in a business as usual scenario [4]. Although the development of aircraft fueled with decarbonized jet fuel, such as hydrogen, is a promising solution, technical obstacles are delaying their deployment for at least 20 years [5]. For this reason, the most effective way of rapidly limiting the global carbon impact of the aviation sector is to develop Sustainable Aviation Fuel (SAF) as a drop-in fuel for current jet engines [2].

The composition of SAF varies according their feedstock and their production processes, which leads to different distributions of chemical compounds compared to traditional petroleum jet fuels. These changes affect their physical and chemical properties, either used as pure fuels or blended with traditional jet fuels, and can make it difficult to comply with standard norms dedicated to jet fuels, such as those defined by the ASTM. One of the specific features of jet fuels is that they are used both as an energy carrier and as a coolant, which subjects them to thermal and oxidative stress. The formulation of alternative fuels must therefore take this constraint into account, which also apply for storage [6]. Thermal stability refers to the tendency of fuels to degrade and generate soluble or insoluble oxidation by-products, depending on the temperature. There are several ASTM standards that determine the oxidation or storage stability of jet fuels, such as ASTM D3241[7] relying on the Jet Fuel Thermal Oxidative Tester (JFTOT) procedure, ASTM D7739 [8] that use the Quartz Crystal Microbalance [9] (QCM) and ASTM D7545 [10] based on the Rapid Small Scale Oxidation Test (RSSOT). The JFTOT and QCM methods measure the deposits induced by the autooxidation of fuels, which form at significant extent of reaction in the oxidation process. The PetroOxy apparatus, a RSSOT, is widely used for the kinetic study of jet fuels oxidation because it measures quantitative Induction Periods (IP) that characterize the oxidation phenomenon at lower extent of reaction, in principle before deposit formation [6].

The PetroOxy is a gold-plated batch reactor, which uses a small quantity of liquid fuel (5 mL) and 7 bar of gas oxygen, at the beginning of the test, and monitors the total pressure in the cell as a function of time. The decrease of pressure corresponds to the consumption of oxygen in the liquid phase of the fuel. The fastest oxygen is consumed, the lowest the IP and therefore its oxidation stability. Since, this method usually does not yield any deposit formation, this reactor is practical to study the homogeneous liquid phase oxidation of petroleum or synthetic fuels. It requires a small quantity of fuel and, the IP measure is quantitative and repeatable in a short period of time.

However, even if the PetroOxy is increasingly used to study fuel autoxidation kinetics in the literature [11–18], it fundamentally remains a rapid test to compare the relative stability of different traditional fuels. The opaque nature of PetroOxy, without optical access, makes it more difficult to study the oxidation stability of synthetic fuels. These fuels are usually made up of a mixture of a few compounds, and demixing or evaporation of only the lightest compound (changing the composition of the liquid fuel) can occur but is impossible to observe when the device is running at the set temperature under high-pressure conditions. In addition, although several studies have used the PetroOxy to carry out kinetic studies of the autoxidation of surrogate fuels, with or without simulations based on kinetic models, the reactor remains not well defined for kinetics. The heating time is a few minutes and the cooling time is several tens of minutes which allow the oxidation reaction to take place at different rates. The opaque, closed cell of the PetroOxy makes it difficult to make rigorous *in situ* analyses. The test cell is not stirred, which can induce concentration gradients or oxygen mass transfer limitations. Even if the quantity of fuel used is small, there are safety issues linked with high-temperature and oxygen rich conditions that restrain the maximum pressure in the cell (18 bar). The characteristic time of the IP can be several hours. Given the intrinsic leakage rate of the device, uncertainty increases for fuels that are more oxidatively stable, which is generally the case with SAF compared with fossil fuels [5].

The goal of this study is to develop a new experimental rig to study the oxidation stability of fuels which is kinetically well-defined, with no concentration gradient or mass-transfer limitations, chemically inert and transparent for *in-situ* analyses, versatile for a large range of temperature, pressure, residence time, and safe, even beyond explosive limits. Microfluidics tools and theories are a solution of choice to achieve this goal. Thanks to this well-defined reactor, it will be possible to perform rigorous homogeneous kinetic simulations of fuel autoxidation with kinetic models to better understand this phenomenon. It will also be possible to rapidly and efficiently compare the reactivity of different fuels within tens of minutes.

BACKGROUND

Numerous studies have shown that it is possible to follow the kinetics of oxidation of liquid fuels with PetroOxy [9,11,14,17,19]. These studies are based on experiments with varying degrees of characterization and kinetic analyses, ranging from global rate constants derived from experiments to simulations based on detailed kinetic models containing hundreds of reactions. One of the most comprehensive characterizations of the kinetics of formation of liquid and gaseous products in the literature, measured during the oxidation of *n*-dodecane in a PetroOxy, is the recent study by Aminane et al. [12]. The wide range of quantitative chemical analyses enabled the authors to establish the different stages in the autoxidation mechanism and to propose rate constants of several reactions. This study allowed a better understanding of the oxidation mechanism before the formation of solid gums. Chatelain et al. [15] or recently Le et al. [19] used detailed chemical kinetic models to simulate IPs measured in PetroOxy and proposed different approaches for simulating this value that is fundamentally biphasic (the liquid phase oxidation rate is measured as a gas pressure drop) with a homogenous liquid reactor model. Chatelain et al. [15] defined the IP as the simulation time needed to reach 5% of fuel conversion. Le et al. [19] defined it as the time required in the simulation to reach the total ROOH concentration measured experimentally at the IP. Auzani et al.[20] proposed a simplified python code to predict the pressure drop of the PetroOxy, trying to account for the liquid phase oxidation in the liquid with detailed kinetic simulations and the mass transfer of O_2 from and into the liquid.

The complexity of the simulation of the IP with detailed or semi-detailed kinetic models led to the search of more well-defined reactor for the validation of kinetic models. Alborzi et al. [21] proposed to use a near isothermal tubular reactor to study the oxidation of n-paraffinic solvent used as a model aviation fuel and simulate their results with a reduced kinetic model. They used a 1 m long stainless-steel tube with a 0.65 cm inner diameter, for temperature ranging between 135 and 150°C, to follow the variation in dissolved O_2 concentration in the paraffinic fuel for residence times up to 20 minutes. However, this system cannot achieve characteristic residence times similar to those achieved with the PetroOxy and the fuel oxidation kinetics is difficult to follow because the reactor material is not chemically inert and the availability of dissolved oxygen can be the rate limiting step at early reaction extent.

One of the best solutions in the literature for studying hydrocarbon oxidation in a chemically inert material, at high temperature and high pressure, with transparent access throughout the reactor, is the use of Si-glass microreactors. Indeed, microfluidics makes it easy to use with less reactants and allow a modularity (versatility) of use. The micro scale enables a fine control of well-defined experimental conditions that remain safe beyond explosive limits, especially for the use of pure oxygen with organic compounds. The possibility of establishing a Taylor flow for a gas-liquid system is also a considerable advantage of these systems, which ensures oxygen saturation in the liquid slugs throughout the reactor. The market for ultra-small scale reactors, which employs microfluidic devices for the oxidation of organic compounds, is developing in parallel with the emergence of the microfluidics market, which is witnessing a growing interest since 2010 [22–25].

For example, Leclerc et al. [22] performed a study of the gas–liquid oxidation of cyclohexane at high temperature (>473K) and pressure (up to 25 bar) in a microfluidic Silicon / glass reactor with a residence time in the reactor of 1 min. The consumption and the extent of the fuel oxidation is visible thanks to the transparent glass and the gas bubble consumption was monitored. The goal of the authors was to test such device for the intensification of the production of cyclohexanone and cyclohexanol in the cyclohexane autoxidation process. In 2014, Fräulin et al. [24] also used a microchip to study the liquid phase oxidation of cyclohexane, with dissolved oxygen, and were able to perform *in situ* analyses using Raman spectroscopy, recording simultaneously the concentrations of dissolved molecular oxygen, cyclohexane and of oxidation products. They achieved a residence time of 10 seconds at 503 K.

If the Si-glass microreactors of the literature used to study the autoxidation of hydrocarbons can withstand high temperatures and high pressures, achieving long residence times, typically of the order of magnitude of the IP in PetroOxy, remains a challenge never before met in the literature. In this work, we present a Si-glass microfluidic device able to operate at high pressure and temperature, with a gas-liquid Taylor flow, and able to reach long residence time. The experimental rig PEEK-ACHU (Autoxidation μChip Heated Unit) also allows the oxidation of hydrocarbons with proportions of pure oxygen beyond the explosive limits. To develop this microfluidic pilot, several preliminary studies were carried out to predict as much as possible the hydrodynamics of the microreactor, given the physico-chemical properties of the hydrocarbon reactants. In this work, we do not focus on the work performed to design the microreactor but we present oxidation results and compare them with those obtained in a PetroOxy. Figure 1 gives an overview of the PEEK-ACHU rig.

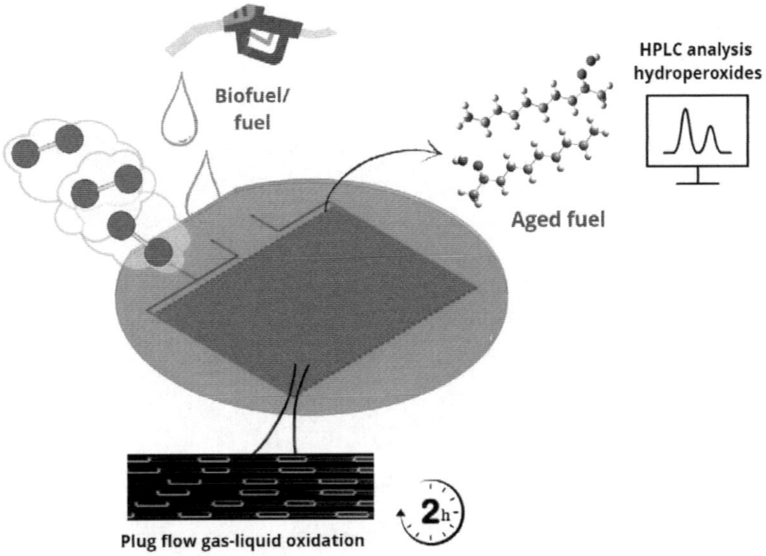

Figure 1: General overview of the Si-glass microreactor of PEEK-ACHU and the analyses of the oxidized fuel.

We first present the results of the oxidation of a single component surrogate fuel, *n*-decane, with the PEEK-ACHU rig. Oxidation is characterized by measuring the total hydroperoxide content in the oxidized liquid at the exit of the reactor. These results are compared to the one measured in the standard PetroOxy test. In the second part of the paper, we explore the oxidation kinetics of a surrogate SAF, which is defined as a blend of *n*-decane and isooctane, using the PEEK-ACHU and PetroOxy devices. Comparisons of thermal oxidation stability of *n*-decane and the SAF surrogate are discussed with both experiments.

EXPERIMENTAL

Methods:

1. Oxidation of fuel: experimental devices

The PetroOxy method

The PetroOxy from Anton Paar is a Rapid Small-Scale Oxidation Test (RSSOT) and is designed to determine the thermal oxidation stability of liquid fuels (gasoline, diesel, biodiesel/FAME and blends). The oxidation of the liquid is made in a 25 mL gold-plated cell, where 5 mL of fuel is introduced and pressurized with 7 bar of O_2. The gas pressure is monitored in the cell and the IP is defined as the time required to reach a 10% decrease of the maximum pressure in the cell. This test meets requirements for standardized norms ASTM D7525, D7545, EN 16091, IP 595 for liquid fuels.

The IP measurements in our PetroOxy have a reported uncertainty of around 10% [19]. At the end of the PetroOxy test, the total amount of hydroperoxide was systematically determined. It can be noted that the cell can take up to 20 min to cool down and it can impact the rates of formation or decomposition of oxidation products, especially for short IPs. In this work, we also measured the total hydroperoxide content of oxidized fuels as a function of time. For these experiments, the test was pre-set to a defined residence time and the apparatus stopped after this test duration and the liquid was withdrawn after the cool down of the cell.

PEEK-ACHU: a new gas-liquid oxidation microreactor

The microfluidic reactor and the whole experimental setup were designed to be able to reach long residence times. The different constraints regarding the oxidation process were taken into consideration in the design. The use of PEEK, an inert polymer, was chosen to prevent the catalysis of fragile compounds, such as ROOH, by metal or inox material. The process was dimensioned for extremely low flows, at high temperatures and pressures, and modular, i.e., easy to connect and disconnect from fluid supply system. This process also allows to reach conditions beyond the explosivity limits compared to macroscale processes.

The Si/glass is chemically inert and withstand high-pressure/high-temperature conditions. In the literature, this system was reported to reach 80 bar and 503 K [24]. The serpentine shaped microchannel is 3.5 m long and was designed as a square cross-section channel and a simple T-junction was used to generate a Taylor flow [26] as described in Figure 2.

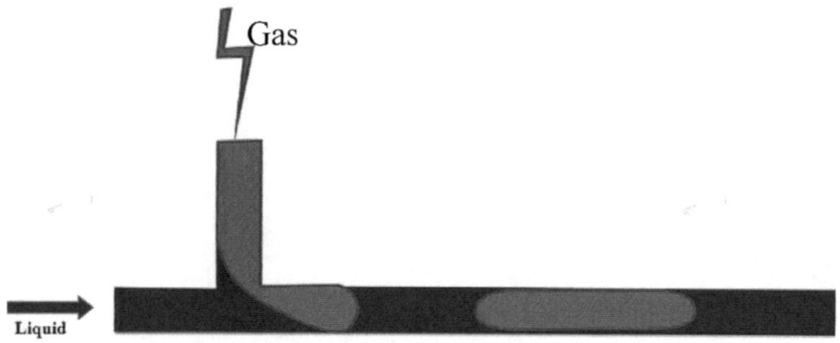

Figure 2: Gas-liquid Taylor flow formation (adapted from Zhang et al. 2021 [27]).

Taylor flow is a type of gas–liquid flow pattern. It consists of elongated gas bubbles with a length many times that of the channel width. The bubbles adopt a capsular shape and can fill the channel. This flow was chosen due to its ability to produce small perfectly agitated reactors of liquid surrounded by gas bubble ensuring a constant concentration of dissolved gas in the liquid phase.[28]

The formation of Taylor flow and the serpentine shaped reactor induce a homogeneous recirculation in the liquid slug as shown in Figure 3 [29–31].

Figure 3: Illustration of the recirculation inside the liquid slug in microfluidic reactor at Taylor flow (adaptation of Abadie et al. 2013 [29]).

The channel length was optimized to have the longest residence time in the microfluidic reactor. The microreactor was manufactured by the MIMENTO technology center of the FEMTO-ST laboratory (Besancon, France), and is shown in Figure 4, along with the experimental set-up.

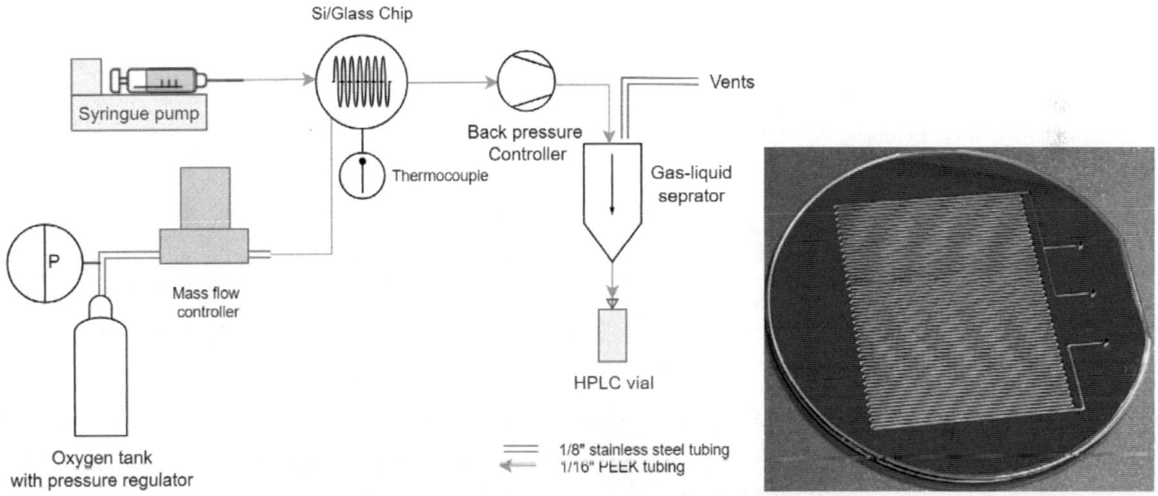

Figure 4: Experimental setup overview and photography of the microreactor 4-inches in Si/glass.

The liquid fuel is cooled down in the PEEK tubing at the exit of the reactor with an ice cooling system, so that the oxidation reactions are frozen, up to the gas-liquid separator. The temperatures, pressures and residence time ranges reached by the experimental set up are given in Table 2. The residence time is determined as the time of the fuel spent in the Taylor flow.

Table 1: Range of use of PEEK-ACHU rig

Residence time (min)	16-230
Temperature range (K)	298-458
Pressure range (bar)	1-40

2. Hydroperoxide content

Following the concentration of hydroperoxides is a good indicator for the extent of the fuel autoxidation process since hydroperoxides are the primary oxygenated products formed following the simplified mechanism shown in Figure 5 [32].

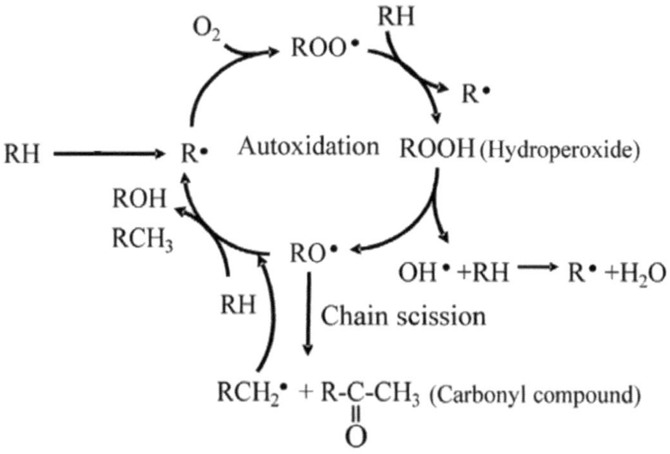

Figure 5: Autoxidation mechanism (adapted from Nakatani et al. [32]).

The total hydroperoxide content was measured using a high-performance liquid chromatography (HPLC) device based on a separation column in the normal phase (silica) combined with a post-column reactor (P-cR). This system designed by our group in 2024 [3] was shown to be able to quantify hydroperoxides for concentrations ranging between 5 10^{-7} and 1 10^{-3} mol. L^{-1}. The HPLC-P-cR, schematically presented in Figure 6, was used at a flow of 1 mL per minute of solvent (isooctane:absolute ethanol 95:5). 10 µL of the sample of aged surrogate fuels and fuel blends was injected using an autosampler through the system. A selective iodometric reaction takes place in the reactive column that permit a precise quantification and a separation of the different hydroperoxides contained in the samples.

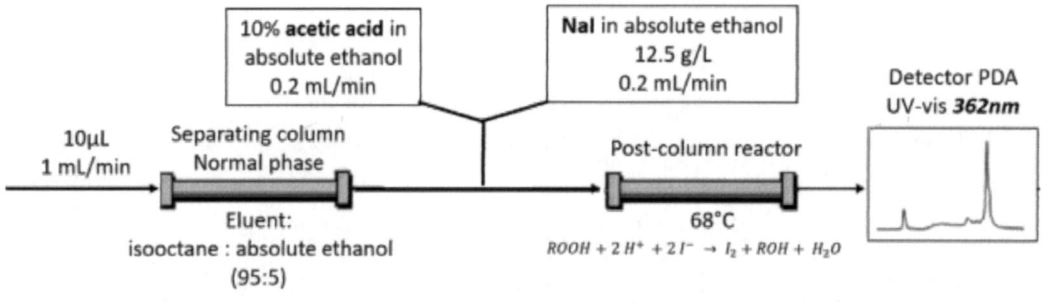

Figure 6: Representation of the HPLC-P-cR to determine the ROOH content of each samples

The final chromatogram combined with a calibration curve is used for the quantification of the hydroperoxide content of each sample. [3] This method was found to be more efficient, and more reproducible than traditional hydroperoxide quantification methods.[3]

Materials

n-Decane was purchased from Acros Chemicals. Among reactants necessary for the iodometric titration of peroxides, absolute ethanol was purchased from VWR Chemicals, whereas sodium iodide (NaI) (> 99 %), acetic acid (≥ 99.8 %) and 2,2,4-trimethylpentane (isooctane; HPLC quality ≥ 99.5 %) were purchased from ThermoFisher Chemicals. A bottle of oxygen (purity 6.0) was obtained from Air Liquide.

The reference induction period (IP) measured with the PetroOxy device of *n*-decane used in this work at 413K and 7 bar of oxygen was 189 min.

The SAF surrogate was chosen as a blend of isooctane and *n*-decane at 50/50 %mol and 55/45 %weigth as proposed by Mawid[33]. The blend for SAF surrogate was made with a precision scale of Mettler Tolendo with an accuracy of +/- 0.01mg. Table 1 summarizes surrogate data.

Table 2: Characteristics of the SAF used in this study.

	298 K | 1.01325 bar
Isooctane (iC_8)	50 %mol
n-decane (nC_{10})	50 %mol
Density (kg/m^3)	710.65
IP measured in PetroOxy (413K/7 bar O_2)	286 min

The IP of the SAF is more than 100 minutes longer than the IP of *n*-decane. The addition of a branched alkane, isooctane, to *n*-decane is known to increase its oxidation stability. The observed increase under the conditions of Table 1 is significant.

RESULTS AND DISCUSSION

Validation of the PEEK-ACHU rig for gas-liquid fuel oxidation

As mentioned previously, it is important to establish a Taylor flow regime in the microreactor to ensure homogeneity and dissolved oxygen saturation of the liquid slugs. Problems of mass transfer limitation from gas to liquid and oxygen depletion in the liquid are thus avoided. Multiple experiments were conducted to map of the different flow regimes as a function of the gas and liquid volume flow rates. The map of flow regimes obtained is presented in Figure 7.

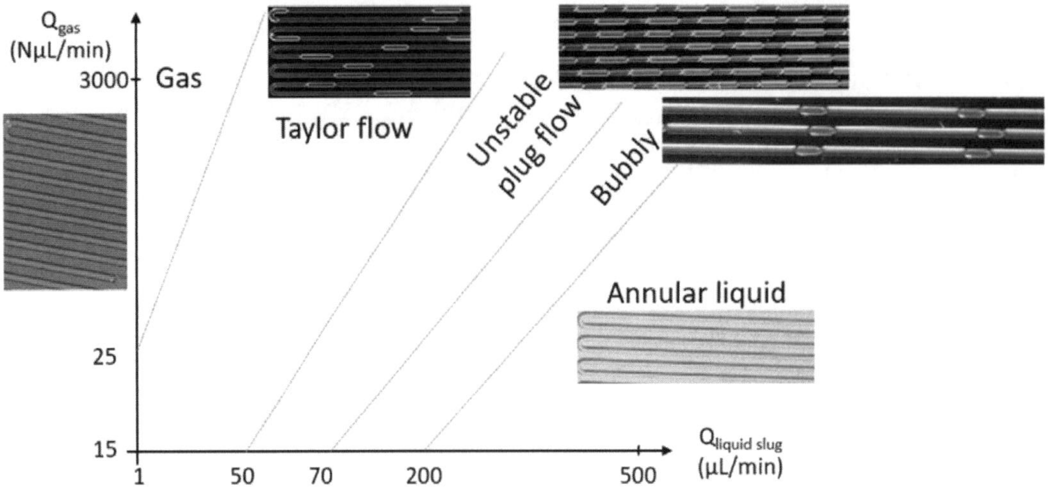

Figure 7: Gas-liquid flow regimes map in our microreactor at 413K and 20 bar.

In Figure 7, it can be seen that a stable Taylor flow regime can be established for a wide range of flow rates. The lowest gas and liquid flow rates, enabling a Taylor-type flow to be maintained, impose limits on the minimum residence time that can be achieved in the reactor. Using the rig configuration presented in the paper, the minimum time is 16 minutes and the maximum 230 minutes. The rigorous characterization of the microreactor was conducted as a function of temperature and pressure. Figure 8 gives an example of a Taylor flow during a fuel oxidation experiment conducted at 413 K and 20 bar.

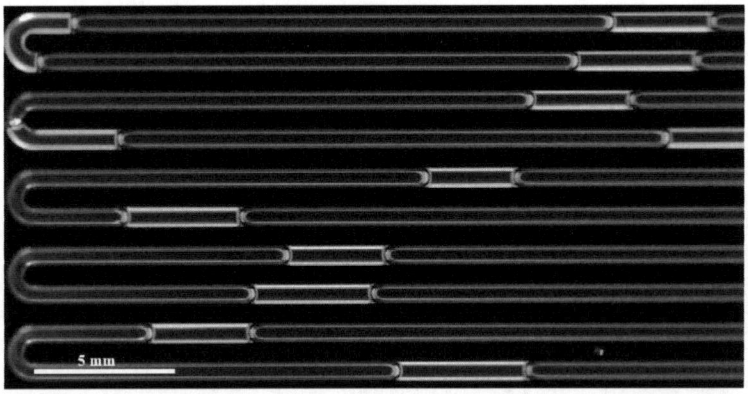

Figure 8: Photography of the gas-liquid *n*-decane autoxidation (residence time: 45 min | 20bar | 413 K | O_2 / C_{10} volumic ratio: 3.02) in the microreactor. White slugs: liquid fuel; grey bubbles: oxygen.

All the autoxidation experiments were performed on microfluidic gas-liquid flows looking like Figure 8, with a Taylor flow based on a significant molar and volume excess of oxygen over fuel (a ratio of 3). The relative size of liquid slugs and gas bubbles ensure a permanent saturation of dissolved oxygen in the liquid slugs.

At the exit of the reactor, the oxidized liquid was collected in vials cooled by a cryostat operating at 263 K. The vials were then diluted in the HPLC eluent (isooctane:absolute ethanol 95:5) and analyzed in the HPLC-P-cR to probe the types and total quantity of hydroperoxides. Figure 9 presents a comparison between chromatograms of aged *n*-decane obtained in experiments performed in the PetroOxy and in PEEK-ACHU.

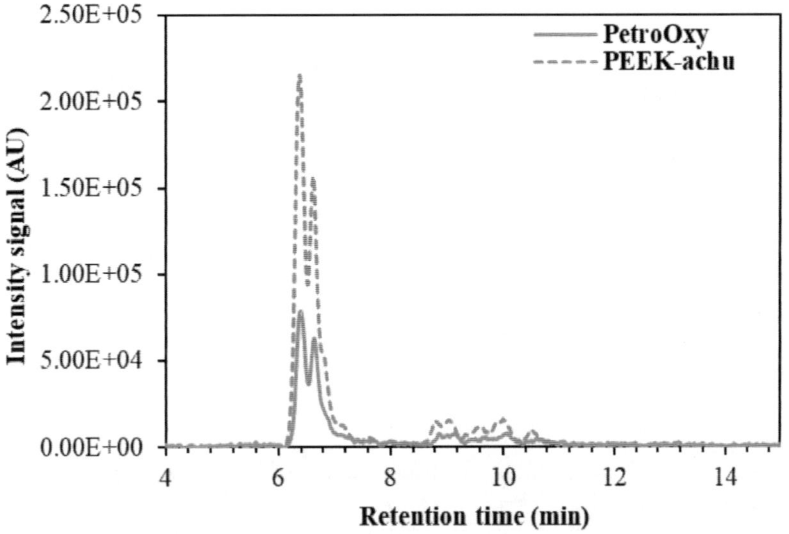

	[ROOH] (mol/L)	Deviation
Peekachu	7.07E-03	30%
PetroOxy	5.07E-03	

Figure 9: Chromatograms of hydroperoxides formed in *n*-decane ageing in the PetroOxy and in PEEK-ACHU (PetroOxy: 413 K, 7 bar initial O_2 pressure yielding $P_{max} \approx 10$ bar, 45 min experiment | PEEK-ACHU: 413 K, 10 bar, 45 min residence time). The Table included in the Figure is the total concentration of hydroperoxides (ROOH) calculated from the chromatograms.

In Figure 9, the pressure in PEEK-ACHU was set to 10 bar to mimic the pressure inside the PetroOxy (P_{max}). Figure 9 shows that different types of ROOH are formed, those with a retention time of 7 minutes and those with a retention time around 10 minutes. The ROOH eluting at 7 min are decane hydroperoxides and the chromatogram show that two groups of molecules can be separated according to their polarity. Even more polar hydroperoxide elute around 10 minutes and we make the hypothesis that they are dihydroperoxides, hydroperoxide dimers or smaller molecular weight hydroperoxides. A clear difference (above the uncertainty limits) in the total [ROOH] concentration of 30% is observed between the two reactors, with experiments performed in PEEK-ACHU yielding more hydroperoxide than in the PetroOxy. This can be explained by the better oxygen mass transfer achieved in the microfluidic device and the self-agitation inside the liquid slugs. Indeed, the auto-agitation in this type of microfluidic systems is well established and allows a better gas-liquid surface contact and ensures a homogeneity in the liquid fuel slug during the oxidation process.

The apparatus is more functional at high pressure since the regulation of microfluidic parameter are more straightforward. The influence of pressure on the measured [ROOH] in oxidized *n*-decane was tested to check for possible oxygen mass-transfer limitations. To do so, experiments were conducted at different pressures (10 bar and 20 bar) for the same residence time in the microreactor with *n*-decane. Figure 10 gives the peroxide concentrations as a function of pressure, with the corresponding chromatograms.

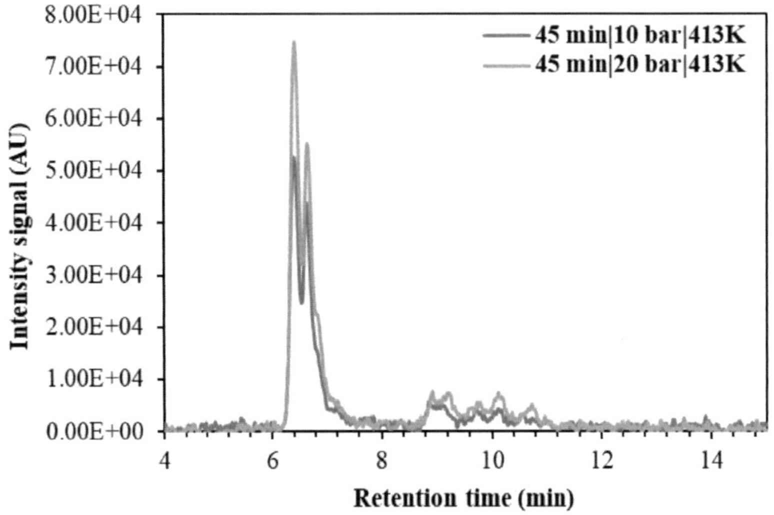

	[ROOH] (mol/L)	Deviation
45 min 20 bar	8.08E-03	*13%*
45 min 10 bar	7.07E-03	

Figure 10: Hydroperoxides chromatograms obtained for *n*-decane autoxidation at two different pressures (10 and 20 bar) in PEEK-ACHU; at 413 K temperature and 45 min residence time.

The two chromatograms presented in Figure 10 are very similar and the deviation between the hydroperoxides concentrations obtained at two pressures is 13%. Since this difference lies within the error of the method of analysis, it can be concluded that we are not in a regime limited by the mass transfer of oxygen in the liquid in the microreactor.

Comparison of the oxidation of the SAF surrogate in the PetroOxy and in PEEK-ACHU

The comparison of the total ROOH content after oxidation of the surrogate SAF in the two experimental devices was performed as a function of residence time. The chromatograms are presented in Figure 11.

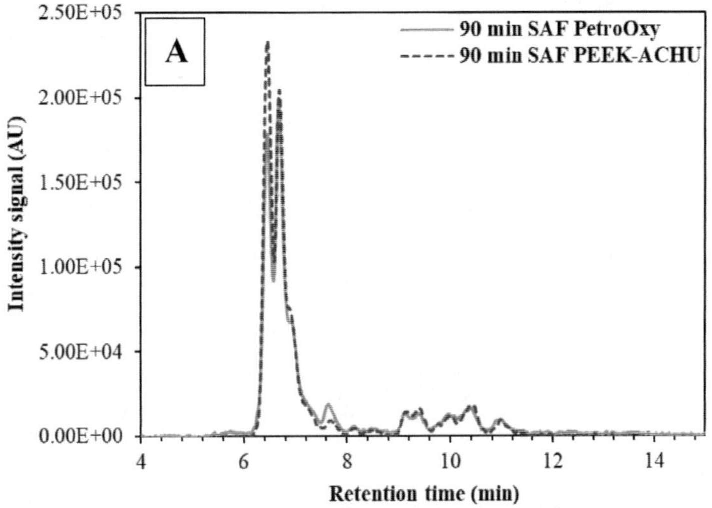

	[ROOH] (mol/L)	Deviation
90min SAF Peekachu	7.37E-03	*12%*
90min SAF PetroOxy	6.51E-03	

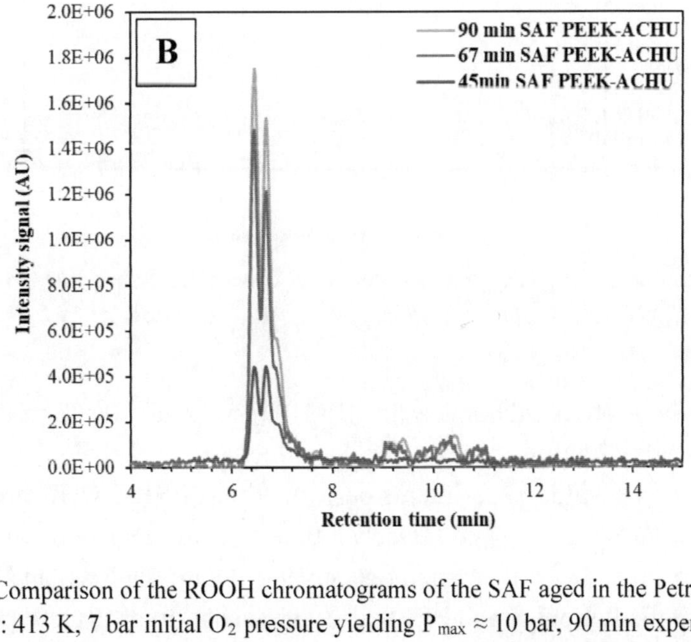

Figure 11: **A** panel: Comparison of the ROOH chromatograms of the SAF aged in the PetroOxy and PEEK-ACHU devices. (PetroOxy: 413 K, 7 bar initial O_2 pressure yielding $P_{max} \approx 10$ bar, 90 min experiment | PEEK-ACHU: 413 K, 20 bar, 90 min residence time). The Table gives the total concentration of hydroperoxides (ROOH) calculated from the chromatograms. **B** panel: evolution of the hydroperoxides chromatograms as a function of residence time in PEKK-ACHU (same temperature and pressure as in the **A** panel).

Chromatograms in Figure 11 show the various types of hydroperoxides that are formed during the oxidation of the SAF surrogate. This can be observed by the number of peaks that are present. Each peak represents a different hydroperoxide, depending on its polarity. It can be observed in Figure 11 that the hydroperoxides that are formed with PEEK-ACHU and the PetroOxy have the same peaks and feature close area (12% deviation in the total amount of ROOH). The chromatograms in the right panel of Figure 11 demonstrate that peaks around 10 minutes of retention time appear only when the residence time is long enough (67 and 90 minutes) and are characteristic of secondary hydroperoxides.

The total ROOH content as a function of time is presented in Figure 12, for both experiments performed in the PetroOxy and PEEK-ACHU.

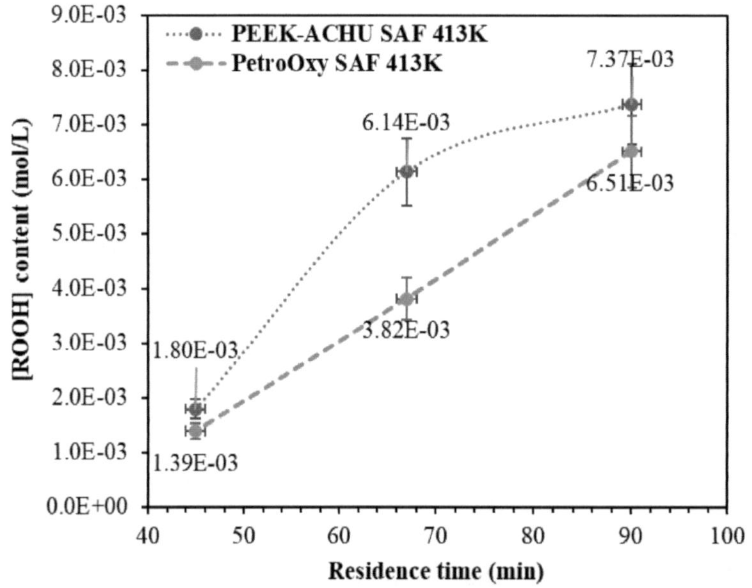

Figure 12: Comparison of total ROOH content at different residence times in PEEK-ACHU and PetroOxy (PetroOxy: 413 K, 7 bar initial O_2 pressure yielding $P_{max} \approx 10$ bar, | PEEK-ACHU: 413 K, 20 bar).

Average deviation of 33% is observed between the [ROOH] content of the sample in PEEK-ACHU and the sample of PetroOxy, with systematic higher values in the fuel aged in the microreactor. From a practical point of view, the experiments performed in PEEK-ACHU are highly versatile, with changes of temperature, pressure and residence time without interruption of the experiments compared to PetroOxy measures that requires several manipulations and interruptions. A comparison between all the PetroOxy and PEEK-ACHU results (total [ROOH] content at the exit of the reactor) is presented in Figure 13.

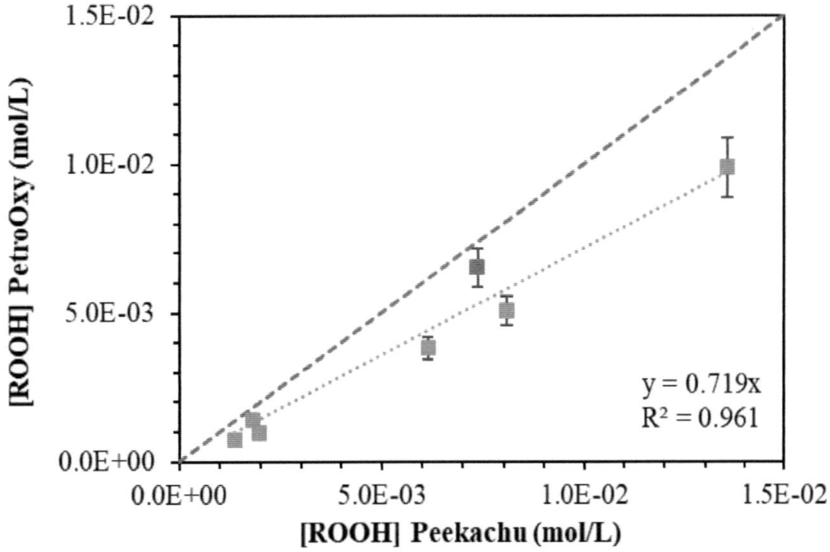

Figure 13: Parity curve of total [ROOH] from SAF and *n*-decane oxidized in PEEK-ACHU and the PetroOxy (413 K and 7 bar for PetroOxy and 20 bar for PEEK-ACHU). Blue symbols: residence times from 16 to 65 min; red symbol: 90 min residence time.

We can observe in Figure 13 that for low residence times (16-65 min), the correlation between the concentration of hydroperoxides from samples aged with PEEK-ACHU and the PetroOxy is almost linear. The data are all below the grey curve y=x, which reflects the fact that more hydroperoxide are formed in the microreactor, by about 30%. The red symbol of Figure 13 appears to be closer x=y curve. It corresponds to the [ROOH] measured for 90min of residence time in both reactors which is the longer time we measured in this study. For longer residence time, the uncertainties associated with the analysis of the samples oxidized in the PetroOxy may diminish or partially cancel.

Our reactor has been shown to be capable of oxidizing liquid fuels at least as effectively as the PetroOxy, but with more efficient O_2 mass transfer into the liquid, with no visible depletion of dissolved oxygen along the reactor, and in homogeneous controlled conditions for the liquid slugs. The PetroOxy is widely used as a certification test that provides a binary response (pass / fail) and its quantitative measurement enables it to compare the oxidation stabilities of different fuels and thus give reactivity rankings. In this work, we propose a way to quantitatively compare the reactivity of two different fuels with the new PEEK-ACHU device, by comparing the ROOH data of oxidation measured for *n*-decane and the SAF surrogate. To do this, we first established the rate laws in our microreactor for low extent of reaction.

Liquid fuel oxidation kinetics in PEEK-ACHU

From a detailed kinetic mechanism approach, we can trace the rate laws and validate these laws on the basis of experimental results. In particular, we can show the influence of oxygen and fuel concentration on the rate of formation of ROOH at low reaction extent, where the fuel oxidation mechanism may be approximated by the following primary mechanism (at the beginning of the autoxidation process):

Initiation	$RH + O_2 \rightarrow R\bullet + HO_2\bullet$	(k_A)
Transfer	$HO_2\bullet + RH \rightarrow H_2O_2 + R\bullet$	(k_x)
Propagation	$RH + RO_2\bullet \rightarrow R\bullet + RO_2H$	(k_1)
Propagation	$R\bullet + O_2 \rightarrow RO_2\bullet$	(k_2)
Termination	$2\,RO_2\bullet \rightarrow P$	(k_T)

where each elementary reaction of the above mechanism has a given rate constant k_i.

Applying the Quasi-Steady-State Approximation (QSSA) on the radicals of the reactions above, the well-known relation between the rate of initiation (r_A) and the rate of termination (r_T) is obtained:

$$r_A = r_T \tag{1}$$

with $r_A = k_A[RH][O_2]$ and $r_T = k_T[RO_2\bullet]^2$. Noting that the rate of formation of ROOH is $r_1 = k_1[RH][RO_2\bullet]$ in the first reaction of propagation, the rate of formation of ROOH can be rearranged as:

$$r_{ROOH} = k_1\sqrt{\frac{k_A}{k_T}}\;[RH]^{3/2}[O_2]^{1/2} \tag{2}$$

This result shows that for low extent of reaction, when the decomposition of hydroperoxides is negligible, their rate of formation depends on the concentration of O_2 and a global rate constant involving the H-abstraction by $RO_2\bullet$ as well as initiation and termination. Within this approach, we conducted a kinetic study at low residence time for the SAF surrogate and *n*-decane fuels to validate these laws. Figure 14 presents the ROOH concentration as a function of residence time.

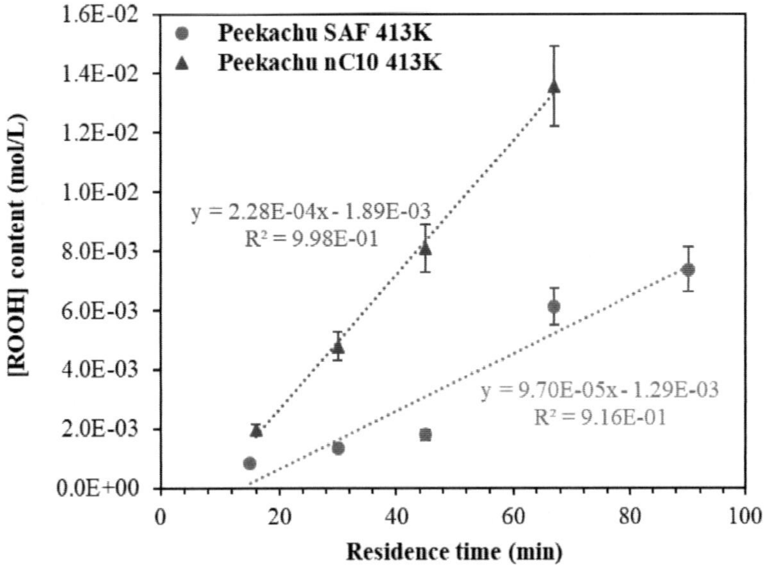

Figure 14: ROOH content of *n*-decane and SAF surrogate as a function of residence time (SAF blend (in blue)/ *n*-decane (in purple) | 413 K | 20 bar).

Figure 14 shows that a linear correlation is obtained for the building of hydroperoxide concentration as a function of time, as expected from the rate law given in equation 2 corresponding to a slow oxidation regime, with no self-acceleration phenomena induced by the decomposition of hydroperoxides. The slopes of the curves are very different, with a slope twice as steep for *n*-decane compared with the SAF surrogate, meaning the rate of formation of ROOH during the SAF oxidation is slower than for *n*-decane. This observation allows us to compare the oxidation stability of the two fuels and to conclude that the SAF is more oxidatively stable than *n*-decane. This conclusion is observed in the of PetroOxy tests, where *n*-decane has an IP of 189 minutes at 413K, while the SAF surrogate has an IP of 286 minutes. This first comparison of fuel stability is promising and need to be extended to other fuels. One of the very strong advantage of PEEK-ACHU for this comparison is that the residence time required to perform this comparison could be reduced to a handful of minutes, compared to hours in a PetroOxy and could be enforce by including variations in temperatures (instantaneous in the microreactor).

CONCLUSIONS

In this work, the validation of a gas-liquid microreactor for studying the kinetics of oxidation of fuels in the liquid phase is presented. The experimental set-up consists of a Si-glass reactor and PEEK fluid supply and recovery systems, both of which are chemically inert. The system developed makes it possible to reach temperatures of up to 458 K and pressures of 1 to 40 bar. Within the reactor, a Taylor-type flow alternating liquid fuel slugs and oxygen gas bubbles is ensured over all operating ranges. This type of flow maximizes oxygen mass transfer in homogeneous liquid slugs, which are self-stirred, thanks to the serpentine design of the microchannel, and isothermal. For the first time in the literature, residence times up to 230 minutes can be reached on a 4 inches diameter silicon wafer. The size ratio between the oxygen bubbles and the size of the liquid plugs is maximized to ensure a constant oxygen concentration throughout the reactor. In order to characterize the capabilities of this new experimental device for oxidizing liquid fuels, comparisons were made with model liquid fuels oxidized with PetroOxy. The basis for comparison is the total amount of hydroperoxide measured after each fuel oxidation in the two experimental set-ups. ROOH quantification was carried out using a method developed in our group using HPLC combined with a post-column reactor. The results showed that the new PEEK-ACHU rig is capable of oxidizing liquid fuels at least as effectively as the PetroOxy, but with more efficient O_2 mass transfer into the liquid, with no visible depletion of dissolved oxygen along the reactor, and in homogeneous controlled conditions for the liquid slugs. These advantages make it a well-defined reactor to study the chemical kinetics of liquid fuel oxidation. At low extent of reaction, we demonstrated that a fast comparison of the thermal oxidation stability of the SAF and *n*-decane fuels is possible and agree with the standardized results of IP in the PetroOxy.

The use of alternative fuel such as sustainable aviation fuel can challenge current standard norms and our new approach could be interesting to explore, even if more tests need to be done.

PEEK-ACHU was found to be a reproducible, versatile rig that allow a high range of use, this rig can be used as to study oxidation kinetics before the formation of solid, but can also be used to study the formation of gums and deposits thanks to its modularity and its transparent access to the reactor. This optical access is also an advantage to perform *in situ* analyses with, for example Raman spectroscopy.

ACKNOWLEDGMENTS

The project leading to this application has received funding from the European Research Council (ERC) under the European Union's Horizon 2020 research and innovation program (grant agreement No 101003318).

REFERENCES

(1) Dewitt, M. J.; West, Z.; Zabarnick, S.; Shafer, L.; Striebich, R.; Higgins, A.; Edwards, T. Effect of Aromatics on the Thermal-Oxidative Stability of Synthetic Paraffinic Kerosene. *Energy and Fuels* **2014**, *28* (6), 3696–3703. https://doi.org/10.1021/ef500456e.

(2) Yang, Z.; Boehm, R. C.; Bell, D. C.; Heyne, J. S. Maximizing Sustainable Aviation Fuel Usage through Optimization of Distillation Cut Points and Blending. *Fuel* **2023**, *353*, 129136. https://doi.org/10.1016/j.fuel.2023.129136.

(3) Benrabah, R.; El Sayah, Z.; Glaude, P.-A.; Arnoux, P.; Fournet, R.; Sirjean, B. Quantification of Hydroperoxides in Liquid Fuels: A Systematic Comparison of Titration and Absorption Methods and Their Coupling to HPLC. *Fuel* **2024**, *365*, 131218. https://doi.org/10.1016/j.fuel.2024.131218.

(4) Sicard, M.; Quignard, A. Carburants aéronautiques. *Techniques de l'Ingénieur* **2021**, *TRP4054 V1*. https://doi.org/10.51257/a-v1-trp4054.

(5) Sicard, M.; Quignard, A. Carburéacteurs alternatifs aéronautiques et leur déploiement. *Techniques de l'Ingénieur* **2023**, *TRP4058 v1*. https://doi.org/10.51257/a-v1-trp4058.

(6) Jia, T.; Zhang, X.; Liu, Y.; Gong, S.; Deng, C.; Pan, L.; Zou, J. J. A Comprehensive Review of the Thermal Oxidation Stability of Jet Fuels. *Chemical Engineering Science* **2021**, *229*, 116157. https://doi.org/10.1016/j.ces.2020.116157.

(7) Standard Test Method for Thermal Oxidation Stability of Aviation Turbine Fuels., 2019.

(8) ASTM. D7739 Standard Practice for Thermal Oxidative Stability Measurement via Quartz Crystal Microbalance, 2022.

(9) Zabarnick, S.; Grinstead, R. R. Studies of Jet Fuel Additives Using the Quartz Crystal Microbalance and Pressure Monitoring at 140.Degree.C. *Ind. Eng. Chem. Res.* **1994**, *33* (11), 2771–2777. https://doi.org/10.1021/ie00035a029.

(10) ASTM. D7545 Standard Test Method for Oxidation Stability of Middle Distillate Fuels—Rapid Small Scale Oxidation Test (RSSOT), 2019. https://www.astm.org/d7545-14r19e01.html (accessed 2024-04-19).

(11) Sicard, M.; Boulicault, J.; Coulon, K.; Thomasset, C.; Ancelle, J.; Raepsaet, B.; Ser, F. Oxidation Stability of Jet Fuel Model Molecules Evaluated by Rapid Small Scale Oxidation Tests. *13th International Conference on Stability, Handling and Use of Liquid Fuels 2013* **2013**, 346–358.

(12) Aminane, S.; Sicard, M.; Melliti, Y.; Ser, F.; Sicard, L. Experimental Study of the Kinetics of Degradation of N-Dodecane under Thermo-Oxidative Stress at Low Temperature and Mechanism Inferred. *Fuel* **2022**, *307* (February 2021), 121669. https://doi.org/10.1016/j.fuel.2021.121669.

(13) El-Sayah, Z.; Glaude, P.-A. A.; Fournet, R.; Sirjean, B. Comparison of the Effects of Different Biofuels on the Oxidation Stability of a Hydrocarbon Fuel. *SAE Technical Papers* **2020**, No. 2020. https://doi.org/10.4271/2020-01-2101.

(14) Benrabah, R.; El Sayah, Z.; Duy Le, M.; Derrick Warren, Y. A.; Glaude, P.-A.; Fournet, R.; Sirjean, B. Experimental Study of the Impact of Alcohols on the Oxidation Stability of a Surrogate Jet-Fuel. *Fuel* **2024**, *361*, 130750. https://doi.org/10.1016/j.fuel.2023.130750.

(15) Chatelain, K.; Nicolle, A.; Ben Amara, A.; Starck, L.; Catoire, L. Structure–Reactivity Relationships in Fuel Stability: Experimental and Kinetic Modeling Study of Isoparaffin Autoxidation. *Energy Fuels* **2018**, *32* (9), 9415–9426. https://doi.org/10.1021/acs.energyfuels.8b01379.

(16) Duong, S.; Lamharess-Chlaft, N.; Sicard, M.; Raepsaet, B.; Galvez, M. E.; Da Costa, P. New Approach for Understanding the Oxidation Stability of Neopolyol Ester Lubricants Using a Small-Scale Oxidation Test Method. *ACS Omega* **2018**, *3* (9), 10449–10459. https://doi.org/10.1021/acsomega.8b00370.
(17) Li, D.; Fang, W.; Xing, Y.; Guo, Y.; Lin, R. Spectroscopic Studies on Thermal-Oxidation Stability of Hydrocarbon Fuels. *Fuel* **2008**, *87* (15–16), 3286–3291. https://doi.org/10.1016/j.fuel.2008.05.006.
(18) Bacha, K.; Ben-Amara, A.; Vannier, A.; Alves-Fortunato, M.; Nardin, M. Oxidation Stability of Diesel/Biodiesel Fuels Measured by a PetroOxy Device and Characterization of Oxidation Products. *Energy Fuels* **2015**, *29* (7), 4345–4355. https://doi.org/10.1021/acs.energyfuels.5b00450.
(19) Le, M. D.; El Sayah, Z.; Benrabah, R.; Warth, V.; Glaude, P.-A.; Privat, R.; Fournet, R.; Sirjean, B. An Experimental and Detailed Kinetic Modeling of the Thermal Oxidation Stability of N-Decane as a Jet Fuel Surrogate Component. *Fuel* **2023**, *342*, 127754. https://doi.org/10.1016/j.fuel.2023.127754.
(20) Auzani, A. S.; Clements, A. G.; Hughes, K. J.; Ingham, D. B.; Pourkashanian, M. Assessment of Ethanol Autoxidation as a Drop-in Kerosene and Surrogates Blend with a New Modelling Approach. *Heliyon* **2021**, *7* (6), e07295. https://doi.org/10.1016/j.heliyon.2021.e07295.
(21) Alborzi, E.; Dwyer, M. R.; Parks, C. M.; Sheikhansari, A.; Mielczarek, D. C.; Zanganeh, M.; Meijer, A. J. H. M. H. M.; Blakey, S. G.; Pourkashanian, M. Construction of a Reduced Chemical Kinetic Mechanism for Autoxidation of N-Paraffinic Solvent – A Model for Aviation Fuel. *Fuel* **2021**, *294* (January), 120170. https://doi.org/10.1016/J.FUEL.2021.120170.
(22) Leclerc, A.; Philippe, R.; Houzelot, V.; Schweich, D.; De Bellefon, C. Gas–Liquid Taylor Flow in Square Micro-Channels: New Inlet Geometries and Interfacial Area Tuning. *Chemical Engineering Journal* **2010**, *165* (1), 290–300. https://doi.org/10.1016/j.cej.2010.08.021.
(23) Marre, S.; Baek, J.; Park, J.; Bawendi, M. G.; Jensen, K. F. High-Pressure/High-Temperature Microreactors for Nanostructure Synthesis. *SLAS Technology* **2009**, *14* (6), 367–373. https://doi.org/10.1016/j.jala.2009.06.005.
(24) Fräulin, C.; Rinke, G.; Dittmeyer, R. Characterization of a New System for Space-Resolved Simultaneous in Situ Measurements of Hydrocarbons and Dissolved Oxygen in Microchannels. *Microfluid Nanofluid* **2014**, *16* (1), 149–157. https://doi.org/10.1007/s10404-013-1223-8.
(25) Gemoets, H. P. L. L.; Su, Y.; Shang, M.; Hessel, V.; Luque, R.; Noël, T. Liquid Phase Oxidation Chemistry in Continuous-Flow Microreactors. *Chemical Society Reviews* **2016**, *45* (1), 83–117. https://doi.org/10.1039/c5cs00447k.
(26) Peng, D.; Xin, F.; Zhang, L.; Yu, H.; Zhang, W. Experiments and Modeling on Bubble Uniformity of Taylor Flow in T-Junction Microchannel. *Chemical Engineering Science* **2015**, *128*, 11–20. https://doi.org/10.1016/j.ces.2015.01.064.
(27) Zhang, Y.; Zhang, J.; Tang, Z.; Wu, Q. Regulation of Gas-Liquid Taylor Flow by Pulsating Gas Intake in Micro-Channel. *Chemical Engineering Journal* **2021**, *417*, 129055. https://doi.org/10.1016/j.cej.2021.129055.
(28) Angeli, P.; Gavriilidis, A. Taylor Flow in Microchannels. In *Encyclopedia of Microfluidics and Nanofluidics*; Li, D., Ed.; Springer US: Boston, MA, 2008; pp 1971–1976. https://doi.org/10.1007/978-0-387-48998-8_1526.

(29) Abadie, T.; Xuereb, C.; Legendre, D.; Aubin, J. Mixing and Recirculation Characteristics of Gas–Liquid Taylor Flow in Microreactors. *Chemical Engineering Research and Design* **2013**, *91* (11), 2225–2234. https://doi.org/10.1016/j.cherd.2013.03.003.
(30) Angeli, P.; Gavriilidis, A. Hydrodynamics of Taylor Flow in Small Channels: A Review. *Proceedings of the Institution of Mechanical Engineers, Part C: Journal of Mechanical Engineering Science* **2008**, *222*, 737–751. https://doi.org/10.1243/09544062JMES776.
(31) Gupta, R.; Leung, S. S. Y.; Manica, R.; Fletcher, D. F.; Haynes, B. S. Hydrodynamics of Liquid–Liquid Taylor Flow in Microchannels. *Chemical Engineering Science* **2013**, *92*, 180–189. https://doi.org/10.1016/j.ces.2013.01.013.
(32) Nakatani, H.; Ohshima, Y.; Uchiyama, T.; Motokucho, S. Degradation and Fragmentation Behavior of Polypropylene and Polystyrene in Water. *Sci Rep* **2022**, *12* (1), 18501. https://doi.org/10.1038/s41598-022-23435-y.
(33) Mawid, M. Development of a Detailed Chemical Kinetic Mechanism for Mixtures of JP-8 Fuel and Fischer-Tropsch-Based Synthetic Jet Fuel. In *Joint Propulsion Conferences*; 2007. https://doi.org/10.2514/6.2007-5668.

IASH 2024, the 18TH INTERNATIONAL CONFERENCE ON STABILITY, HANDLING AND USE OF LIQUID FUELS
Louisville, Kentucky, USA
September 8 – 12, 2024

UNRAVELLING SUSTAINABLE AVIATION FUEL INTERACTIONS: A STABILITY PERSPECTIVE

Dr. David Evans*, Rachelle Hewitt, Nathan Matheson

Defence Science and Technology Group, 506 Lorimer Street Fishermans Bend, 3207 Australia

ABSTRACT

The burgeoning demand for Sustainable Aviation Fuels (SAF) has prompted exploration into various production pathways, yielding eight distinct synthetic blending components (SBCs) vital to SAF supply. While prior demonstrations primarily involved single SBC utilization, our research, presented at the 2022 IASH conference, delves into the production of a neat SAF comprising SBCs from multiple pathways. Leveraging fuel derived through catalytic hydrothermal conversion as the base, chosen for its favourable aromatic levels, we crafted blends incorporating FT-SPK, HEFA-SPK, SIP-SPK, and ATJ-SPK at maximum blend ratios. Thorough testing against D7566 requirements for finished fuel, alongside evaluations of thermal conductivity, enhanced thermal stability, and chemical and physical ignition delay properties, revealed an unforeseen interaction in one blend adversely affecting thermal stability, undetectable in isolated SBC testing. Building upon these initial insights, this paper focuses on isolating polar components to elucidate their impact on fuel stability, offering crucial insights for the sustainable aviation industry.

*Corresponding author: Dr. David Evans, david.evans29@defence.gov.au

IASH 2024, the 18TH INTERNATIONAL CONFERENCE ON STABILITY, HANDLING AND USE OF LIQUID FUELS
Louisville, Kentucky, USA
September 8 – 12, 2024

STUDY OF MOLECULAR INTERACTIONS IN THE LATER STAGE OF JET FUEL SURFACE DEPOSITION: COMBINING AB-INITIO DENSITY FUNCTIONAL THEORY WITH EXPERIMENTAL ANALYSIS

Rahimakhanim Babayeva*, Ehsan Alborzi, Kevin J. Hughes, Anthony J. H. M. Meijer, M. Pourkashanian

The University of Sheffield, 219 Myrtle Road, Sheffield S2 3HH United Kingdom

ABSTRACT

The rapid expansion of the aviation industry requires improvements in operational efficiency and environmental sustainability. For example, modern aircraft engines employ a higher pressure ratio to enhance efficiency, leading to an elevated thermal load on the engine lubrication system. This, in turn, increases heat sink requirements. Recognising that aviation fuel serves a secondary role as a heat sink prior to combustion, subjected to thermal stress, undergoes thermal degradation. This situation results in the generation of various soluble chemical species and insoluble macromolecules, contributing collectively to the formation of carbon deposits on surfaces in contact with aviation. The formation of surface deposits is linked to the presence of heteroatomic species such as nitrogen, sulphur, and oxygen-containing species, as well as dissolved metals. This research focuses on the combining quantum chemistry simulations using the Vienna Ab Initio Simulation Package (VASP) and small scale testing, namely PetroOxy/RapidOxy to study the molecular interactions between some of the heteroatomic species and surface in the course of deposition. We utilised PetroOxy and RapidOxy devices to examine the deposition propensity of specific compounds and surrogate fuels, with a particular emphasis on the later stages of deposition when the C-C interactions become more noticeable. This approach allows for a thorough exploration of the stability and reactivity of adsorbates on various metal surfaces, providing insights into the complex physico-chemical interactions that govern deposit formation.

*Corresponding author: Rahimakhanim Babayeva, rbabayeva1@sheffield.ac.uk

IASH 2024, the 18TH INTERNATIONAL CONFERENCE ON STABILITY, HANDLING AND USE OF LIQUID FUELS
Louisville, Kentucky, USA
September 8 – 12, 2024

OXIDATION MECHANISM OF SULFUR-CONTAINING COMPOUNDS AND ANTIOXIDANT DEPLETION DYNAMICS: INSIGHTS INTO INTERACTIONS

Jinxia Fu*, Scott Turn

University of Hawaii at Manoa, 1680 Eastwest Road, Post 109, Honolulu, Hawaii 96826 USA

ABSTRACT

Sulfur-containing compounds (SCCs) are the most abundant heteroatom compounds in petroleum fuel. This study investigates the thermal stability of SCCs in an air atmosphere and their influence on the depletion of commercial antioxidants (AOs). A total of 11 SCCs, encompassing both reactive (thiols, sulfides, and disulfides) and nonreactive (thiophenes and benzothiazole) species, were subjected to analysis. The reaction and degradation processes of SCCs were tracked by monitoring changes in their concentrations over an extended period, while the resulting products after thermal treatment were analyzed to elucidate oxidation and degradation mechanisms. Additionally, the study scrutinized the depletion kinetics of commercial fuel AOs and quantitatively evaluated the impact of SCCs on AO depletion, uncovering antagonistic interactions. Furthermore, the reactions and interactions among the 11 selected SCCs were investigated to explore potential synergetic or antagonistic effects during the thermal treatment. These findings offer comprehensive insights into the behavior of SCCs and AOs under oxidative conditions, providing valuable guidance for the development of more effective AO formulations and sulfur mitigation strategies for the fuel industry.

*Corresponding author: Jinxia Fu, jinxiafu@hawaii.edu

IASH 2024, the 18TH INTERNATIONAL CONFERENCE ON STABILITY, HANDLING AND USE OF LIQUID FUELS
Louisville, Kentucky, USA
September 8 – 12, 2024

FURTHER STUDIES OF JET FUEL OXIDATIVE DEPOSIT FORMATION IN QUARTZ CRYSTAL MICROBALANCE EXPERIMENTS

Krege M Christison*, Gregory B Boursalian

Chevron, 100 Chevron Way Richmond,California 94801 USA

ABSTRACT

When analyzed by Quartz Crystal Microbalance (QCM) under oxidative conditions there is no deposit formation in hydrocracked jet fuel components due to the severe hydrotreating used in their production, which generally eliminates reactive, polar, and heteroatom-containing molecules. Previous research has shown that the addition of quinolines, aliphatic amines, and cyclohexylpyridines to a hydrocracked jet fuel component leads to increased deposit formation during QCM analysis. Furthermore, the addition of reactive sulfur compounds, specifically mercaptans and sulfides, along with the nitrogen compounds, increases the deposit formation observed. It was also observed that the delayed oxidation from the addition of alkyl phenols led to lower overall deposit formation when combined with reactive sulfur and nitrogen compounds.

These previous studies have inspired additional avenues of investigation. A larger quantity of the previously identified cyclohexylpyridine compound has been synthesized for further experimentation. The addition of reactive sulfur compounds on their own are of interest, especially since they are prevalent in unhydrotreated jet fuel components. Certain refinery processes lead to oxidation of mercaptan molecules, forming disulfides, and possibly other sulfur oxides. Therefore, these compounds have been explored, both on their own and in the presence of nitrogen compounds that are known to form deposits. Nitrogen oxides have not been explored previously but could be present due to oxidative processes in refineries and are therefore added to this test matrix.

*Corresponding author: Krege M Christison, kregechristison@chevron.com

IASH 2024, the 18TH INTERNATIONAL CONFERENCE ON STABILITY, HANDLING AND USE OF LIQUID FUELS
Louisville, USA
8-12 September 2024

IMPACT OF SBCS ON THE BEHAVIOUR OF POLYMER MATERIALS AND EFFECT OF FUEL CYCLING

Mickaël Sicard*, Bruno Raepsaet, Duc Minh Le, Yanis Melliti, Frédéric Ser

ONERA, Chemin de la Huniere Palaiseau, 91123 France

ABSTRACT

As it travels through the fuel system, kerosene comes into contact with a wide range of materials. To avoid any leaks or malfunctions, it is essential to ensure fuel/material compatibility. A large number of tests are carried out as part of the ASTM D4054 certification process for alternative fuels (SBC). As a result, SBCs can be used safely in blends of up to 50 % in fossil fuels. The next step is to study their behavior in a pure state. The compatibility of three SBCs and three types of O-ring was assessed using the ASTM D4054 procedure. The three SBCs tested were HEFA, ATJ and an alkane/cycloalkane mixture. One of the interests of this study was also to observe the behavior of cycloalkanes, which had not yet been characterized. The results show that these molecules behave more like alkanes.

The other objective was to assess the impact of fuel cycling on materials. It is clear that on a daily basis the chemical composition of kerosene varies, especially if SBCs are used. So we also need to assess the behavior of materials subjected to successive fuels. As a first step, simple Jet/SBC or SBC/Jet cycling was evaluated. Cycling had no significant effect on the mechanical properties of the nitrile and fluorosilicone O-rings. Cycling did, however, degrade the properties of fluorocarbon seals. Mass variation was also affected. The last fuel used for cycling had a very marked impact on the variation in mass, which returned to values close to those of the pure fuel.

*Corresponding author: Mickaël Sicard, mickael.sicard@onera.fr

IASH 2024, the 18TH INTERNATIONAL CONFERENCE ON STABILITY, HANDLING AND USE OF LIQUID FUELS
Louisville, USA
8-12 September 2024

DIGITAL TRANSFORMATION OF FUEL PROPERTY AND COMPOSITION DATA FOR THE RAPID ASSESSMENT OF CONVENTIONAL AND SYNTHETIC FUELS

Terrence W. Dickerson*, Fernando Morales, Fausto Reyher, Richard A. Kamin

US Navy, 22229 Elmer Rd. Bldg. 2360 Patuxent River, Maryland 20670 USA

ABSTRACT

As it travels through the fuel system, kerosene comes into contact with a wide range of materials. To avoid any leaks or malfunctions, it is essential to ensure fuel/material compatibility. A large number of tests are carried out as part of the ASTM D4054 certification process for alternative fuels (SBC). As a result, SBCs can be used safely in blends of up to 50 % in fossil fuels. The next step is to study their behavior in a pure state. The compatibility of three SBCs and three types of O-ring was assessed using the ASTM D4054 procedure. The three SBCs tested were HEFA, ATJ and an alkane/cycloalkane mixture. One of the interests of this study was also to observe the behavior of cycloalkanes, which had not yet been characterized. The results show that these molecules behave more like alkanes.

The other objective was to assess the impact of fuel cycling on materials. It is clear that on a daily basis the chemical composition of kerosene varies, especially if SBCs are used. So we also need to assess the behavior of materials subjected to successive fuels. As a first step, simple Jet/SBC or SBC/Jet cycling was evaluated. Cycling had no significant effect on the mechanical properties of the nitrile and fluorosilicone O-rings. Cycling did, however, degrade the properties of fluorocarbon seals. Mass variation was also affected. The last fuel used for cycling had a very marked impact on the variation in mass, which returned to values close to those of the pure fuel.

*Corresponding author: Terrence W. Dickerson, terrence.dickerson@navy.mil

IASH 2024, the 18TH INTERNATIONAL CONFERENCE ON STABILITY, HANDLING AND USE OF LIQUID FUELS
Louisville, Kentucky, USA
September 8 – 12, 2024

GENOMICS AND TRANSCRIPTOMICS ANALYSIS UNVEIL CONSERVED FUNGAL ADAPTIVE MECHANISMS ESSENTIAL FOR THRIVING AND PROLIFERATING IN TOXIC HYDROCARBON ENVIRONMENTS

Osman Radwan*[1], Oscar N. Ruiz[2]

[1]*University of Dayton Research Institute, 1529 Brown Street, Dayton, Ohio 45469 USA*
[2]*Biomaterials Branch, Materials and Manufacturing Directorate, Air Force Research Laboratory*

ABSTRACT

Fungal contamination in fuel has substantial effects on production, transport, and storage. This project aimed to comprehend fungal adaptation to toxic fuel environments through advanced genomics, transcriptomics, and bioinformatics tools. Genomes of filamentous fungi and yeast species, including Byssochlamys sp., Lecanicillium sp., Fusarium fujikuroi, Aspergillus versicolor, Graphium sp., Yarrowia lipolytica ATCC20496, and Meyerozyma guilliermondii, were sequenced using next-generation sequencing, followed by bioinformatic analysis for genome assembly and annotation. Comparative genomics and local BlastP searches against databases such as Pfam, COG, and KOG revealed that fungal genomes harbor gene functional categories linked to energy production, carbohydrate, amino acid, and lipid transport, and metabolism. Additionally, RNA-Seq technology was employed to profile gene expression in fungal species growing in either Sodium Acetate or Jet A fuel. The study included three fungal species: Hormoconis resinae ATCC22711, Byssochlamyss sp., and Y. lipolytica ATCC20496. RNA-Seq data unveiled differentially expressed genes and pathways crucial for microbial adaptation to hydrocarbon fuels, demonstrating similar regulatory patterns across diverse fungal species. These genes involved in the uptake and extrusion of hydrocarbon via membrane protein channels and efflux pumps, iron acquisition, hydrocarbon degradation using monooxygenases and hydroxylases, and biofilm formation. Results suggest that fuel fungal adaptation take place via conserved global mechanisms that provide the capacity to the fungal cell for survival and proliferation to the toxic environments. The global mechanisms identified through this study may become important targets for potential applications in developing biomanufacturing and fungal detection-mitigation technologies for fuel system biodeterioration prevention.

*Corresponding author: Osman Radwan, *oradwan1@udayton.edu*

IASH 2024, the 18TH INTERNATIONAL CONFERENCE ON STABILITY, HANDLING AND USE OF LIQUID FUELS
Louisville, Kentucky, USA
September 8 – 12, 2024

CORRELATION BETWEEN MICROBIAL BIOMASS IN AIRCRAFT FUEL TANKS AND AVIATION FUEL SAMPLES WITH CFU COUNTS OBTAINED BY CULTURE METHODS

Lucy Owen*, Gareth Williams, Graham Hill,

ECHA Microbiology Ltd., ECHA Microbiology Ltd., Units 22 & 23 Willowbrook Technology Park, Llandogo Road, Cardiff CF3 0EF, UK.

ABSTRACT

IATA recommend microbiological testing of drain samples from aircraft fuel tanks to detect microbial growth. IATA guidance classes microbial contamination results from culture-based (colony forming units; CFU) tests as Negligible, Moderate or Heavy. Moderate and Heavy results instigate preventative and/or remedial action. Other types of tests should show broad agreement with contamination categories defined for CFU tests. However, the relationship between CFU results and the amount and type of microbial biomass/biofilm in fuel tanks is not fully understood. This study aimed to investigate the correlation between quantity and type of microbial biomass in aircraft tanks and aviation fuel samples with CFU results.

The findings of visual inspections of aircraft fuel tanks (n=11) were compared to IP 613 CFU results for drain tank samples taken over several occasions. A correlation between CFU count and tank inspection observations was found; Heavy results were obtained by IP 613 for 3 of 4 fuel tanks with visible microbial contamination.

Moreover, 346 aviation fuel samples were assessed by CFU tests (IP 385 and IP 613) and results compared with a graded severity of biomass contamination determined by phase contrast microscopy. Correlations between CFU counts/IATA contamination level and biomass contamination grade were confirmed. Fuel samples with Moderate and Heavy contamination are likely to contain significant microbial biomass, and, in the case of Heavy contamination, clumps of fungal hyphae and/or microbial extracellular polymeric substances.

Overall, the results of this study support the justification for remedial actions triggered by Moderate and Heavy CFU test results (IATA limits).

*Corresponding author: Lucy Owen

IASH 2024, the 18TH INTERNATIONAL CONFERENCE ON STABILITY, HANDLING AND USE OF LIQUID FUELS
Louisville, Kentucky USA
8 – 12 September 2024

MICROBIAL CONTAMINATION COMPARISON BETWEEN SUSTAINABLE AVIATION FUEL (SAF) AND JET A-1

Danika Nicoletti[1], Timothy Carter[1], Paul Bjorndahl[1, 2, 3], and Jordan Schmidt[1].

[1]*LuminUltra Technologies Ltd., 520 King Street,[1] Fredericton, NB E3B 6G3, Canada.* *danika.nicoletti@luminultra.com*
[2] *Centre for Water Resources Studies, Department of Civil and Resource Engineering, Dalhousie University, 1360 Barrington Street Halifax, Nova Scotia, Canada, B3H 4R2. bjorndahl@dal.ca*
[3] *Department of Mathematics & Statistics, Dalhousie University, 6316 Coburg Road PO BOX 15000 Halifax, Nova Scotia, Canada. bjorndahl@dal.ca*

ABSTRACT

The use of Sustainable Aviation Fuel (SAF) in commercial aviation can reduce CO_2 emissions by up to 80%, playing an important role in the International Air Transport Association's (IATA) goal of achieving net zero carbon emissions by 2050. As demand grows, production volumes will continue to increase for SAF which will require clean storage to maintain product quality. Previous studies have investigated microbial contamination in Jet A-1 fuel storage, however, there is a gap in knowledge for the susceptibility of SAF to microbial contamination. In this laboratory study, the microbial community dynamics of contaminated fuels were monitored and characterized. Microcosms were set up with either SAF or Jet A-1 and Bushnell Haas medium to simulate tank bottoms water. The same hydrocarbon-degrading inoculum was applied to each microcosm along with a no-substrate control. The microcosms were incubated over 12 weeks and were sacrificed and analyzed at regular intervals for testing adenosine triphosphate (ATP) concentration, qPCR for total prokaryotes, and 16S rRNA gene sequencing for total microbial community analysis. ATP concentration results revealed higher microbial activity in SAF microcosms over the course of the experiment by an order of magnitude compared to Jet A-1 microcosms, while qPCR and 16S rRNA gene sequencing analyses revealed a comparable microbial community structure over time between the two substrates. This study aims to address the gap in knowledge of SAF susceptibility to microbial contamination providing insight into the risks involved in its storage compared to Jet A-1 fuel.

KEY WORDS: microbial contamination, SAF, biodegradation, qPCR, ATP, aviation fuel

INTRODUCTION

As demand for Sustainable Aviation Fuel (SAF) increases to meet CO_2 reduction goals set forth by the International Air Transport Association (IATA), safe storage for SAF will become increasingly important. Fuel has long been a topic of research for microbial contamination due to the settling of bottoms water in storage tanks or fuel tanks that then harbor microbial populations capable of deteriorating tank walls by microbiologically influenced corrosion (MIC) and biodegrade the hydrocarbons of fuel[1]. Microbial growth within aircraft fuel tanks can cause a plethora of issues if left unaddressed, including filter blocking, FQIS malfunctioning due to biofilm growth on fuel probes, and wear on fuel injectors[2]. The impact of microbial contamination in Jet A-1 fuel, including biodegradation rates, and microbial community analysis has been well studied and characterized[3-7]. On the other hand, microbial contamination research of SAF is still in its infancy due to its more recent introduction into the market. Though the types of microbial communities present in SAF are hypothesized to be similar to Jet A-1 and confer similar impacts, it is still important to understand whether new challenges may arise with microbial contamination subjected to a different composition of hydrocarbons as found in alternative fuels[8].

BACKGROUND

In this study, contaminated Jet A-1 and SAF were examined in a microcosm lab-scale experiment to investigate the differences in microbial community dynamics and activity between the two fuels over time, given the same microbial inoculum and incubation conditions. There are two central considerations as to the threat of microbial contamination in fuel that were focused on in this study. Firstly, the types of microorganisms present – different microbial taxa can confer different deleterious impacts on a system based on their metabolic capabilities or cell physiology. For example, there are certain microorganisms that are implicated in MIC, including sulfate-reducing microorganisms or acid producing microorganisms. There are also microbial taxa that excrete extra polymeric substances (EPS) and facilitate biofilm formation, which can then cause filter plugging or fuel probe malfunctioning. The second consideration is the activity of those microorganisms that are present. If a MIC implicated taxon is present in low abundance as part of an inactive or lowly active microbial community, the risk is not as high as in a highly active microbial community. These two considerations were used to compare and contrast Jet A-1 and SAF amended microcosms.

This experiment was designed to assess the differences between Jet A-1 and SAF in the microbial community that the hydrocarbon matrices support in a controlled environment and under equivalent conditions. The results from this study will inform future laboratory and field-based studies on SAF or SAF blends as it begins to be commonly used and produced in larger volumes to support demand.

MATERIALS AND METHODS

A hydrocarbon degrading culture sourced from an O-ring of an aircraft fuel tank was used as inoculum in the study. This microbial community had been maintained in Bushnell-Haas medium (artificial tank bottoms water)[9]. Ahead of the experiment, the culture was transferred to fresh Bushnell-Haas medium with 10 mM of sodium acetate to ramp up cell density 5 days before inoculation of the experiment.

Microcosms were prepared by adding 25 mL of sterile Bushnell Haas medium, 70 mL of filter-sterilized fuel substrate (Jet A-1 or SAF), and 5 mL inoculum (5% volume) to sterile 125 mL serum bottles. Uninoculated controls were also prepared, along with a no substrate control which contained no added fuels. After the microcosms were prepared, they were capped and incubated at 23 °C in the dark. Fifteen microcosms were set up per condition for sacrifice in triplicate at 5 time points for analysis. Microcosms were analyzed at 2-week intervals for the first 3 times points, then 4-week intervals for the final 2 timepoints for a total of 12 weeks. This method of sacrificial microcosms was chosen to avoid variability due to lack of homogeny in a bulk incubation when sampling at different time points.

At each time point, triplicate microcosms per condition including controls were removed from the incubator, mixed thoroughly, then aliquoted for various analyses. Adenosine triphosphate (ATP) concentration was measured immediately upon mixing the sample to record microbial activity (BugCount Fuel, LuminUltra Technologies, Fredericton, NB). Portions of the sample were then preserved for genomic analysis by quantitative polymerase chain reaction (qPCR) and next generation sequencing (NGS) using a commercially available preservation kit (GeneCount qKit Preserve, LuminUltra Technologies, Fredericton, NB). Total prokaryote concentration was determined by qPCR and microbial community analysis was performed by next generation sequencing of the V4 variable region of the 16S rRNA gene (Primers 515F and 806R)[10] using sample DNA purified by a magnetic bead based purification method (GeneCount qKit Purify Auto, LuminUltra Technologies, Fredericton, NB). Bioinformatics processing of the sequencing data was performed using Mothur (v. 1.36.1) and taxonomic identification was completed using SILVA SSU ARB database (Version 132). Additional beta-diversity analyses were conducted using Poisson measurement error and sequencing bias corrected principal component analysis (PCA) as well as non-metric multidimensional scaling (NMDS) with permutational MANOVA[11].

Hydrocarbon analysis is to be performed via gas chromatography – mass spectrometry (GCMS) on Jet A-1 and SAF samples before and after incubation to characterize the biodegradation that had occurred, if any, during the incubation period.

RESULTS AND DISCUSSION

Before inoculation, the hydrocarbon-degrading culture was grown up to an ATP concentration of 2.02×10^6 pg cellular ATP mL^{-1} (Log_{10} 9.3 microbial equivalents (ME) mL^{-1}). This ATP concentration is within high contamination range for fuel and fuel-associated water, selected to model an advanced contamination scenario for Jet A-1 and SAF. This culture was also analyzed by 16S rRNA gene sequencing alongside the raw SAF and Jet A-1 samples to determine the microbial community structures before the time series (Figure 1). *Thermomonas* and *Thalassospira* were the only taxa above 1% relative abundance present in the inoculum but not in the raw Jet A-1 or SAF microbial communities. The hydrocarbon-degrading culture used for inoculum had 84.6% shared taxa to the Jet A-1 and SAF samples.

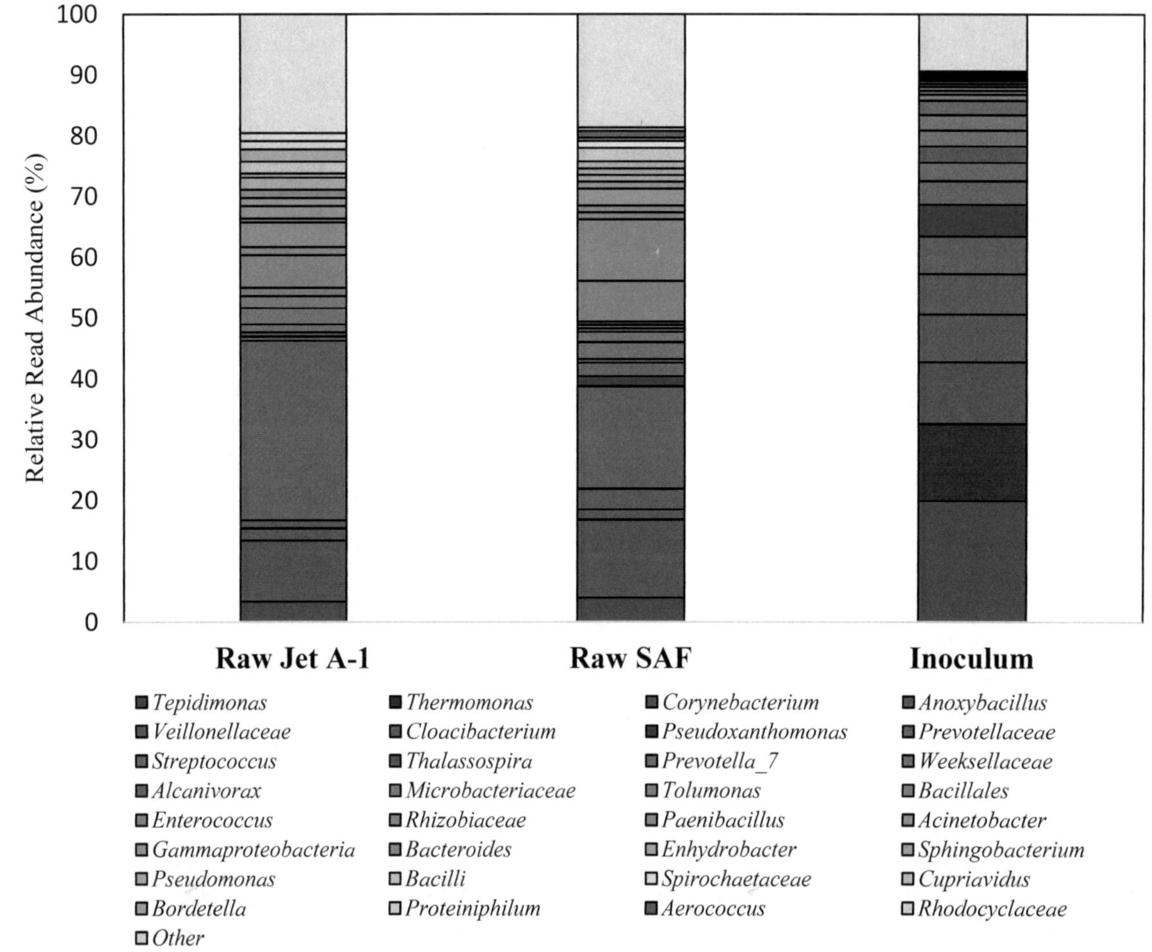

Figure 1: 16S rRNA gene sequencing data from the raw fuel samples used for the timecourse experiment and the hydrocarbon-degrading culture used as inoculum. Shades of green represent taxa that are present in all three samples, shades of blue represent taxa present in only the inoculum. Purple and red indicate taxa that are present in Jet A-1 but not SAF and SAF but not Jet A-1, respectively.

The microcosms were monitored for ATP concentration for each condition at 5 time points. Inoculated SAF (I SAF), inoculated Jet A-1 (I Jet A-1) and the no substrate control all measured at the same order of magnitude at T1, immediately post inoculation (Figure 2). Inoculated SAF maintained the highest ATP concentration at time points 2 through 5. The no-substrate control microcosms had an overall lower ATP concentration compared to SAF, however this was higher than in Jet A-1 throughout. This result suggests that SAF sustained the high microbial activity of the initial inoculum over the course of the experiment compared to Jet A-1 and control conditions. Similar to the no hydrocarbon substrate microcosms, the exposure of the inoculum to Jet A-1 decreased microbial activity upon inoculation (time zero). ATP concentration decreased over time in all experimental conditions over the course of incubation respective to their initial ATP concentration, indicating the total biomass and activity of those microorganisms present decreased from T1 through to T5 in the presence of SAF or Jet A-1.

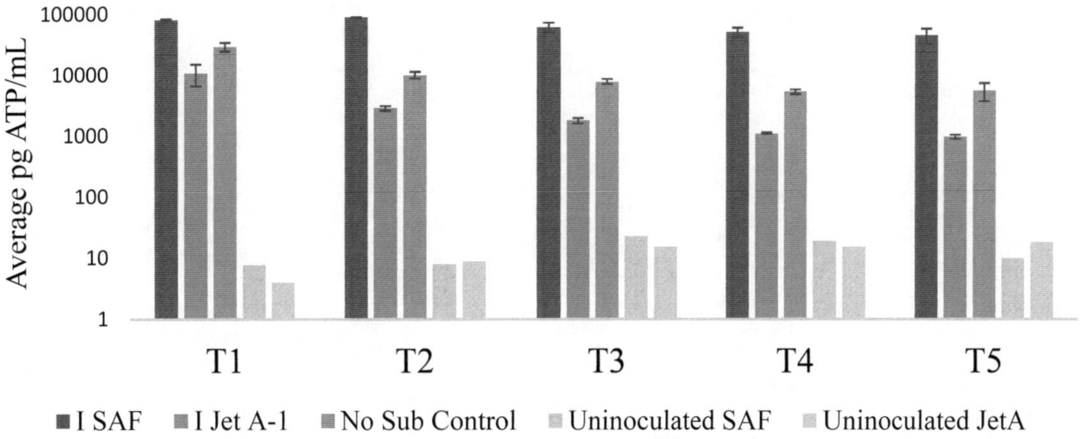

Figure 2: Average ATP concentration at 5 time points for each condition. Error bars indicate the standard deviation among triplicate microcosms.

The 16S rRNA gene sequencing microbial community analysis revealed a comparable microbial community structure over time between the two fuel-amended conditions (Figure 2). The most dominant taxa in both conditions were hydrocarbon-degrading marine microorganisms *Thalassospira* and *Alcanivorax* which were present above 50% at most time points. *Thalassospira* species have been commonly identified in polycyclic aromatic hydrocarbon (PAH) exposed microbial communities and have been found to degrade intermediates produced by other PAH degrading taxa[12, 13]. Similarly, *Alkanivorax* species have been identified as abundant hydrocarbon-degraders in hydrocarbon exposed environments and are important in early stages of biodegradation[14-16]. The microbial community structure became more diverse as time increased in both SAF and Jet A-1 amended microcosms. Both fuel-amended conditions had approximately 10^6 to 10^7 cells/mL determined by total prokaryote qPCR over the course of the experiment. Total fungi qPCR was also completed on the samples however no fungi was detected.

Table 1: Relative read abundance (%) and Total Prokaryote qPCR of microbial taxa in SAF and Jet A-1 amended microcosms over five time points. Relative abundance and qPCR values are the average of three replicate microcosms.

Taxa	SAF					Jet A-1				
	T1	T2	T3	T4	T5	T1	T2	T3	T4	T5
Total Prokaryote qPCR (cells/mL)	1.84E+07	9.60E+06	3.12E+07	1.68E+07	1.26E+07	1.60E+07	1.82E+07	3.37E+07	1.88E+07	3.18E+07
Alcanivorax	46.9	63.0	13.1	54.7	49.5	35.0	45.8	7.2	45.5	46.5
Thalassospira	31.0	30.2	4.2	12.0	9.3	42.3	49.4	4.9	21.3	15.5
Tepidimonas	0.5	0.2	12.5	3.4	2.4	0.8	0.1	14.6	3.3	2.6
Burkholderiaceae	0.8	0.2	7.5	2.5	2.3	0.4	0.1	7.6	1.6	2.3
Bacteroides	0.1	0.0	8.3	2.3	1.7	0.1	0.0	8.8	2.1	1.3
Paenibacillus	0.1	0.1	6.5	1.7	1.2	0.3	0.1	7.1	1.8	1.5
Weeksellaceae	0.1	0.0	0.0	0.2	3.3	0.2	0.0	0.0	0.0	2.5
Cloacibacterium	0.3	0.1	2.1	0.8	0.8	0.5	0.1	2.4	0.5	0.8
Enterococcus	0.4	0.2	0.6	0.4	1.3	0.3	0.1	0.6	0.6	0.5
Schlegelella	0.3	0.0	2.8	1.1	0.8	0.2	0.0	2.7	1.0	0.6
Bacillales	1.2	0.3	0.1	0.4	0.7	2.3	0.3	0.1	0.3	0.3
Campylobacterales	0.0	0.0	1.7	0.6	0.5	0.1	0.0	2.0	0.4	0.6
Pseudomonas	0.1	0.0	0.5	0.2	0.4	0.0	0.0	0.0	0.0	0.0
Corynebacterium	0.3	0.1	0.0	0.0	1.1	0.2	0.0	0.0	0.0	1.4
Tepidiphilus	0.2	0.0	2.0	0.6	0.5	0.5	0.0	2.2	0.7	0.4
Deinococcus	0.0	0.1	1.1	0.6	0.4	0.0	0.0	1.3	0.3	0.4
Acinetobacter	0.2	0.1	0.9	0.7	0.4	0.1	0.1	1.0	0.5	0.3
Tolumonas	0.1	0.2	1.7	0.4	0.4	0.2	0.1	1.8	0.6	0.7
Ruminococcus_1	0.0	0.0	2.7	0.6	0.5	0.0	0.0	3.1	0.7	0.6
Rhodocyclaceae	0.1	0.1	1.2	0.4	0.3	0.1	0.0	1.5	1.2	0.2
Treponema	0.0	0.0	1.7	0.4	0.3	0.0	0.0	1.6	0.2	0.3
Desulfomicrobium	0.0	0.0	1.9	0.5	0.2	0.0	0.0	1.9	1.0	0.3
Martelella	0.8	0.5	0.3	0.7	0.3	0.8	0.8	0.3	1.9	2.5
IheB3-7	0.0	0.1	1.5	0.3	0.3	0.0	0.0	1.5	0.5	0.1
Wolinella	0.0	0.1	1.4	0.3	0.3	0.0	0.1	1.5	0.2	0.3
Burkholderia-Caballeronia-Paraburkholderia	0.0	0.0	0.7	0.3	0.1	0.0	0.0	0.8	0.1	0.2
Pseudoxanthomonas	1.0	0.3	0.4	0.1	0.2	1.5	0.1	0.4	0.0	0.2
Desulfovibrio	0.0	0.0	0.0	0.0	0.0	0.0	0.0	1.0	0.4	0.5
Methylophilaceae	0.0	0.0	0.0	0.0	0.0	0.5	0.0	0.0	0.1	0.0

Measurement error and bias corrected PCA results (Figure 3) showed variation in the data confirms similar community transitions over time for both SAF and Jet A-1 microcosms. Distinct clusters were observed at each point of the time course as well as between inoculated and uninoculated controls. Community wide variation was similar among substrates as no discernable clusters were evident within any groupings of SAF vs Jet A-1 samples Figure 3 bottom). This was supported by alpha diversity analysis (result not shown). Genera with largest PCA loadings (i.e., influential taxa) and/or their putative metabolic functions were labelled as vectors in biplots (Figure 3 top) so that relevant putative metabolic groupings could be identified in relation to sample types. Earlier time points (1 and 2) had higher abundance of sulfate/sulfur reducing genera *Desulfomicrobium* and acid producer *Bacteroides*. Genera capable of hydrocarbon degrading functions were also more abundant in inoculum samples earlier on. Moreover, biofilm-producing *Anoxybacillus* was more predominant in the uninoculated microcosms of both fuel substrates.

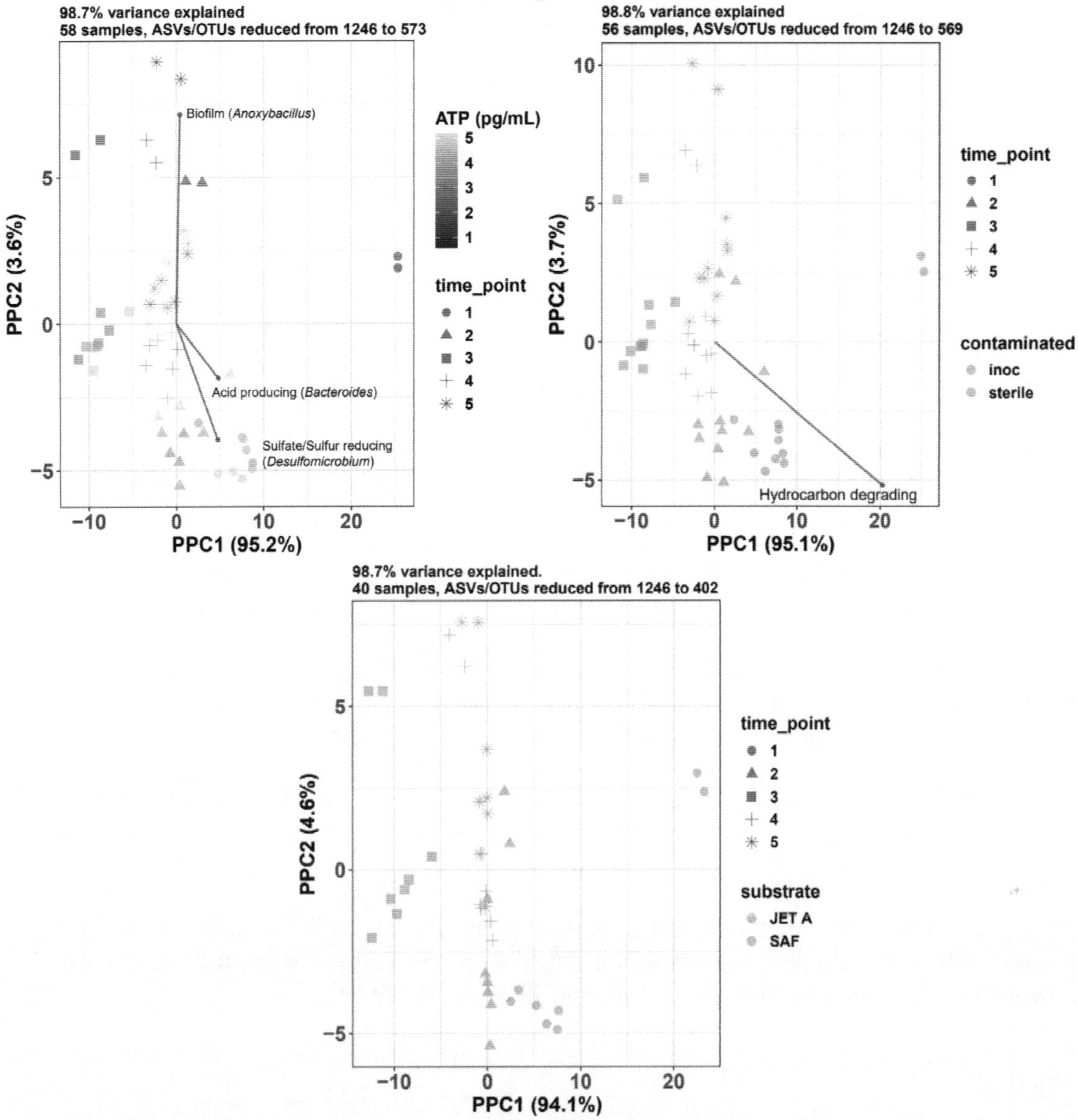

Figure 3: (Top) PCA biplots showing time course of microcosm sample community clustering based on ATP (left) and inoculated (right). Gray vector lines represent largest loading taxa and/or their most relevant metabolic functions. (Bottom) PCA scores plot of substrate sample clustering over time.

Hydrocarbon analysis by gas chromatography/mass spectrometry (GC/MS) will identify the biodegradation that occurred in SAF and Jet A-1 and will provide important insight into the impact that the microbial communities had on the two fuel types (analysis under way).

CONCLUSIONS

This laboratory comparison study of SAF and Jet A-1 under high microbial contamination conditions found that SAF maintained higher total microbial activity over time, however both conditions supported a comparable microbial community taxonomic composition. These findings indicate that SAF and Jet A-1 storage may confer similar microbial threats that are tied to presence of particular microorganisms (i.e., microbiologically-influenced corrosion (MIC)), however the microbial community may be more metabolically active in SAF storage scenarios compared to Jet A-1. Suggested future research includes characterizing the microbial community structure and activity of field-derived microbially contaminated SAF and Jet A-1, and characterization of biodegradation rates of SAF.

REFERENCES

1 Passman, F. J., Microbial contamination and its control in fuels and fuel systems since 1980 - a review, *International Biodeterioration & Biodegradation,* vol. 81, pp. 88-104, 2013.

2 International Air Transport Association, Microbiological Contamination in Aircraft Fuel Tanks Edition 6 Guidance Material, International Association for Air Transport (IATA), Montreal, 2023.

3 Chiciudean, I., Mereuţă, I., Ionescu, R., Vassu, T., Tănase, A., Stoica, I., Jet A-1 Bacterial Contamination: a Case Study of Cultivable Bacteria Diversity, Alkane Degradation, and Biofilm Formation, *Polish Journal of Environmental Studies,* vol. 28, no. 6, pp. 4139-4146, 2019.

4 Itah, A., Brooks, A., Ogar, B., and Okure, A., Biodegradation of International Jet A-1 Aviation Fuel by Microorganisms Isolated from Aircraft Tank and Joint Hydrant Storage Systems, *Bulletin of Environmental Contamination and Toxicology,* vol. 83, pp. 318-327, 2009.

5 Hu, D., Lin, W., Zeng, J., Wu, P., Zhang, M., Guo, L., Ye, C., Wan, K., and Yu, X., Profiling the microbial contamination in aviation fuel from an airport, *The Journal of Bioadhesion and Biofilm Research,* vol. 35, no. 8, pp. 856-869, 2019.

6 Rauch, M., Graef, H., Rozenzhak, S., Jones, S., Bleckmann, C., Kruger, R., Naik, R., and Stone, M., Characterization of microbial contamination in United States Air Force aviation fuel tanks, *Journal of Industrial Microbiology and Biotechnology,* vol. 33, no. 1, pp. 29-36, 2006.

7 McComb, J. P., A Metagenomic Analysis of Microbial Contaminatino in Aviation Fuels, *PhD. Thesis,* 2009.

8 Pires, A., Han, Y., Kramlich, J. and Garcia-Perez, M., Chemical composition and fuel properties of alternative fuels, *BioResources,* vol. 13, no. 2, pp. 2632-2657, 2018.

9 Bushnell, L. D. and Haas, H. F., The utilization of certain hydrocarbons by microorganisms, *Journal of Bacteriology,* vol. 41, no. 5, pp. 653-673, 1941.

10 Apprill, A., McNally, S., Parsons, R. and Weber, L., Minor revision to V4 region SSU rRNA 806R gene primer greatly increases detection of SAR11 bacterioplankton, *Aquatic Microbial Ecology,* vol. 75, pp. 129-137, 2015.

11 Kenney, T., Gu, H. and Huang T, Poisson PCA: Poisson measurement error corrected PCA with application to microbiome data, *Biometrics,* vol. 77, no. 4, pp. 1369-1384, 2021.

12 Kayama, G., Kanaly, R. A. and Mori, J. F., Comprehensive genomic characterization of marine bacteria Thalassospira spp. provides insights into their ecological roles in aromatic hydrocarbon-exposed environments, *Microbiology Spectrum,* vol. 10, no. 5, 2022.

13 Santisi, S., Zoccali, M., Catania, V., Quatrini, P., Mondello, L., Genovese, M. and Cappello, S., Biodegradation Potential of Oil-degrading Bacteria Related to the Genus Thalassospira Isolated from Polluted Coastal Area in Mediterranean Sea, *Soil and Sediment Contamination: An International Journal,* vol. 31, no. 3, pp. 316-332, 2022.

14 Vila, J., Nieto, J. M., Mertens, J., Springael, D. and Grifoll, M., Microbial community structure of a heavy fuel oil-degrading marine consortium: linking microbial dynamics with polycyclic aromatic hydrocarbon utilization, *FEMS Microbiology Ecology,* vol. 73, pp. 349-362, 2010.

15 Yakimov, M., Golyshin, P., Lang, S., Moore, E. R., Abraham, W. R., Lunsdorf, H. and Timmis, K. N., Alcanivorax borkumensis gen. nov., sp. nov., a new, hydrocarbon-degrading and surfactant-producing marine bacterium, *International Journal of Systematic and Evolutionary Microbiology,* vol. 48, no. 2, 1998.

16 Shapiro, T., Chekanov, K., Alexandrova, A., Dolnikova, G., Ivanova, E. and Lobakova, E., Revealing of Non-Cultivable Bacteria Associated with the Mycelium of Fungi in the Kerosene-Degrading Community Isolated from the Contaminated Jet Fuel, *Jounral of Fungi (Basel),* vol. 7, no. 1, p. 43, 2021.

IASH 2024, the 18TH INTERNATIONAL CONFERENCE ON STABILITY, HANDLING AND USE OF LIQUID FUELS
Louisville, Kentucky, USA
September 8 – 12, 2024

FORMULA 1 RACING FUELS EVOLUTION INFERRED FROM REGULATION CHANGES

David J. Abdallah[1] and Eric M. Bunnelle[2]

[1]*ExxonMobil Technology and Engineering Company, 1545 Route 22 East, Annandale, NJ 08801, USA, david.j.abdallah@exxonmobil.com*
[2]*ExxonMobil Product Solutions Company, 22777 Springwoods Village Parkway Spring, TX 77389, USA, eric.m.bunnelle@exxonmobil.com*

ABSTRACT

Formula 1 fuels are tailored to a highly specific use and deliver more power from an F1 engine than is achievable from an all-purpose commercial fuel. The different applications, priorities and use of the F1 fuels compared to commercial fuels has led to their different compositions. These differences have narrowed over the years by specifications that force harmonization. A part of a competitive constructors' championship, the actual F1 fuels are a tightly guarded secret. This paper aims to review the evolution of F1 fuel regulations over the years that seek to limit the use of certain components and to better harmonize F1 fuels with commercial fuels. From these changes we can gain insights and recognize how F1 fuel compositions were optimized over the history of the sport.

INTRODUCTION

Although it has been said many times that they are similar, F1 fuels have long differed from commercial pump fuels because F1 fuel suppliers are tasked with optimizing their fuel for a specific engine, chassis, and operation range, rather than an entire commercial fleet with a wide range of vehicles, engine types, and operational ranges. Table 1 highlights the application, priorities, and operational expectations that drive the divergence between these two fuels.

The details of F1 fuel formulations are kept under tight wraps as fuel suppliers, along with the teams they support are exploring emerging technologies that they believe can give them competitive advantage both in the sport and in their commercial endeavors. The highly competitive and secretive nature means that understanding F1 fuels and their differences has been a challenge. However, some details can be gleaned from evolving fuel regulations over the years that seek to limit the use of certain components and rein-in compositional departures from commercial fuels shedding light on the Formula 1 race fuels used throughout the history of the sport without speaking about the specifics of any individual fuel suppliers' formulations.

Table 1. Comparison of F1 and commercial gasoline expectations

	Application	Top priorities	Engine operational range
Commercial Gasoline	✓ General formulation to cover all types of engines and vehicle types available in the market	✓ Affordability ✓ Availability ✓ Drivability ✓ Compliance with emissions regulations	✓ Wide range of engine operating conditions ✓ Rarely operate at wide open throttle apart from the occasional merge onto a freeway
F1 Gasoline	✓ Specific formulation to cover a single vehicle and short-stroke, high-revving, high-compression ratio engine	✓ Power ✓ Fuel efficiency	✓ Narrow range, tuned to specifically operate at or near wide open throttle

F1 FUEL SPECIFICATION BACKGROUND

The name 'Formula One' is often shortened to 'F1' and refers to the premier motorsport series. The term "formula" in the name refers to a set of rules for all aspects of the competition, including the fuels. Over the years the specification governing body has changed names from Commission Sportive Internationale (CSI) until October 1978 to the Fédération Internationale du Sport Automobile (FISA) until October 1993 to the Fédération Internationale de l'Automobile (FIA) up to the present time.[1] The FIA website maintains the digital versions of the F1 technical and sporting regulations dating back to 2018,[2] the full history is maintained in printed documents in the FIA archive in Valleiry France. Online databases not affiliated with the FIA capture the digital versions of the fuel regulation dating further back to 1994.[3] The technical regulations cover fuel properties and composition, and the sporting regulations cover fuel use during a race.

The fuel regulations consist of definitions, expectations and rules for physical properties and chemical compositions. Often it is more practical to confine a property rather than the compositional space as there are potentially hundreds of hydrocarbons that can be present in fuel. However, in a few specific cases, defining exact hydrocarbons and types can be the more expedient way in Formula 1 to prevent the use of certain hazardous compounds, exotic components, or disproportional levels of certain hydrocarbons from what is observed in commercial fuel.

Changes to the fuel specification come from the FIA but are the result of a process which includes input from F1 teams, fuel suppliers, and engine makers. Changes can have different objectives, including to relax regulations and expand performance or alternatively to tighten regulations and harmonize with commercial fuels. The latter has historically been more common than the former. In the early days of the sport the fuel requirement started off simple and stated that the fuel needs to be "like commercial fuel". However, this definition was open to more than one interpretation, and the fuel formulations evolved to include a variety of exotic chemicals that, while technically a fuel, were uncommon or completely absent in commercial motor gasoline. Because the simple statement failed to achieve its intent, the F1 fuel spec added new requirements to bridge the gap which significantly expanded overtime. In 1958 the F1 fuel specification consisted of only 75 words, in 1994 it grew to 345, in 2023 to 1236, and 2026 is shaping up to be ~1700 words[4]. This expansion of verbiage in the racing fuel regulations reflects an accrual of attempts to mainly narrow the gap between commercial and race fuel compositions which bans the enrichment of high performance components and exotic components not found in commercial gasoline.

Even with constraints and excluded classes of hydrocarbons, there is still a very large space with which one can blend a technology advantaged compliant fuel. Each hydrocarbon component will impart specific properties that are important for how that fuel behaves under any given set of conditions. Although the hydrocarbons may be found in both F1 and commercial fuels it is the proportion of the different components that make all the difference and fuel suppliers have found clever ways to meet the standards while simultaneously increasing the power of vehicles. Furthermore, each F1 team brings a unique combination of engine, chassis, and race strategy. Racing fuel formulations are co-optimized to the specific engine and teams methodology for peak performance and the fuel properties that drive peak performance will differ in relative priority from one team to another, leading to an optimized fuel for one race team that will likely not be optimal for another team.

EVOLUTION OF THE F1 FUEL SPECIFICATION

Up to the 1960s

Specialty racing fuels containing high levels of alcohols (relative to commercial fuels) were being used at Grand Prix (GP) races prior to 1940. The inaugural Formula 1 World Championship was the British GP held on May 13, 1950 at the Silverstone Circuit in Silverstone, England. The first F1 regulations allowed free choice of fuel between 1950 and the end of 1957 and was generally alcohol based. Table 2 highlights some of the exotic race fuels during this time.

Table 2. Examples of earlier Grand Prix racing fuels (v%)

	1939[5]	1939	1950[6]	1951[7]	1951[7]	1957[8]
	Auto-Union Mercedes-Benz	Alfa Romeo Maserati	Silverstone GP	Ferrari-Shell	Maserati-Shell	Fuel used by Mercedes
Methanol	60	34.5	98.5	85	85	25
Ethanol	10	49.5	-	-	-	-
Benzene	22	-	-	-	5	45
Nitrobenzene	3*	-	-	-	-	2
Acetone	-	-	-	-	10	3
Petroleum ether	5	-	-	-	-	-
Aviation gasoline	-	-	-	15	-	23
Castor Oil	-	-	1	-	-	-
Water	-	0.5-3	0.5	-	-	-
Not provided	-	12.5-15.5	-	<1	-	2

* Mixture with castor oil and toluene

Post WW2, many areas of the world were well-equipped to produce high-octane fuels developed for the high compression ratio (CR), piston-driven engines used in military aircrafts, which had to perform under extreme conditions. Availability of these fuels allowed these engines to be pushed to their performance maximum, taking full advantage of optimal spark-advance, higher (effective) CR, and higher operating temperatures without suffering from catastrophic effects of engine-knock in the air. In contrast, ground-based consumer and commercial road vehicles operated under more modest conditions and had a much greater tolerance for engine knock. As such, motor (road) gasoline produced at the time were mainly derived from the fractional distillation of petroleum and often was inadequate for high performance engines.

As in all high-performance engines, Formula One cars had high CR engines akin to aircraft engines that performed best with high octane fuels.[9] However, during the early years of Formula 1, from 1958 through the 1960s, the specification required the fuels to comply with commercial fuels meant for use on road and sold at service stations, unless a special exemption was granted for the organizers. Exempt fuel was common especially in countries where commercial fuel was of insufficient quality. These fuels had much better performance as they typically consisted of high-octane aviation gasoline or commercial fuels blended with alcohols or fuels specifically designed for F1 containing high levels of methanol as shown in Table 2. The use of aviation gasoline also provided uniformity at GPs held around the world as the composition and properties are very similar country to country as they comply with international standards. On the other hand, commercial motor gasolines were made to local standards that vary country to country and were not high octane.

High octane fuels have always been a top priority throughout F1's history. There are two fuel measurements commonly used for motor gasoline to characterize the fuels resistance to engine knock: Research Octane Number (RON), and Motor Octane Number (MON). While some exceptions exist, RON has tended to be the more meaningful metric for F1 racing. Alcohols in general exhibit beneficial impact to RON and interestingly they have what is called a "synergistic nonlinear blending effect", which means that they can have an outsized positive impact on the RON measurement relative to the amount that was included in the fuel formulation. Several other oxygenated components like furans, ethers and ketones also have this effect.

The first restriction on fuel additives came in 1958 where the fuel specification stated, "Unless otherwise stipulated in the rules of a competition, the upper cylinder lubricants or lubricants for two-stroke engines are permitted, provided that their addition to the fuel does not increase its octane". This requirement highlights the constant desire by the racing teams to find ways to improve the octane of the fuel from the sports beginnings.

1970s

In the 1970s[10] the FIA reasserted its intention to promote the use of commercial fuels by placing more restricted controls on the use of waivered fuels. During this period, commercial fuels became more complex as new molecular conversion technologies emerged to meet the growing demand for gasoline and the imminent removal of TEL. Higher octane "premium" grades, composed of synthetic components from a refinery were marketed to meet the popular and high performing "muscle cars" of the generation. But even these fuels were suboptimal for the demands of F1 racing. Also, during this time, much like aviation fuels, tetraethyl lead (TEL) additive was the focus of increasing octane providing anti-knock properties for high CR F1 engines.

1979 was a significant year for F1 fuel specification changes as it marked the beginning of numeric-based limits on fuel properties and compositions. In particular,

- 102 RON maximum
- 2 m% oxygen maximum
- 1 m% nitrogen maximum

The remainder of the fuel was required to consist exclusively of hydrocarbons and not contain any alcohols, nitro-compounds or other power boosting additives.[11] Some necessary fuel additives like detergents and antioxidants contain oxygen and nitrogen. Although 1 m% nitrogen requirement is well above the levels that would be seen through normal levels of detergents and antioxidants fuel additization, it is less than what would be helpful from the standpoint of power boosting nitro-compounds like nitromethane and nitrobenzene.

A maximum regulation on RON of 102 highlights an example of what causes F1 fuel formulations to diverge from commercial fuels. Commercial fuels specify a minimum as the cost is directly proportional to the octane value of the fuel. Unlike commercial gasoline, which is a high volume commodity product, the cost difference is trivial when designing a small volume race fuel used in competitions where there is a high value on winning. Thus, F1 fuel producers maximize the level of octane so the F1 regulations need to apply a maximum regulation to maintain relevance to commercial fuels. However, there is a significant RON gap between a SuperPlus commercial gasoline and a F1 fuel of the order ~3.5 octane. This gap is similar to the steps in RON between normal (91 min), Super (95 min), and SuperPlus (98 min) demonstrating F1 fuels are still a step above the highest octane commercial fuels available.[12]

The FIAs disapproval of alcohols was related to compatibility issues with fuel systems and the ban remained in place until the early 1990s.

Regulations were also added that restricted the use of any other oxidant other than oxygen in the air. The more common F1 racing engines of this period are naturally aspirated and are limited by the amount of air that can be packed into the combustion chamber. Nitrous oxide delivered directly into the air induction system offers an enhancement of the oxidants and was gaining popularity in short duration racing, like drag racing, during this time (and is still used in some drag race series today).

1980s

Although it made its entry in the late 70's, the 1980s saw the rise of turbocharging in F1. 1983 marked the first time a car with a turbocharged engine won the F1 championship.[13] The turbocharger increases the density of the air resulting in more air in the combustion chamber than can be achieved in a naturally aspirated (NA) engine. Thus, turbocharged engines experience greater pressures than NA engines at top-dead-center (TDC) for engines of otherwise equivalent CR. This increased pressure (sometimes expressed as an "effective compression ratio") results in higher tendency to knock, furthering the need for high octane components.

Refueling was unregulated before 1984. Some teams found they could achieve quicker race times when starting with just half a tank of fuel, however, refueling came with the additional costs and more importantly the risk of fires in the pits. FISA banned refueling in 1984 on safety grounds, and this ban would remain in place until 1994. Before the refueling ban, volumetric considerations on the fuel were not particularly important, but mass considerations were as fuel added weight to the vehicle that decreased fuel economy and thus the energy per weight of the fuel should be as high as possible. During this first refueling ban, teams were also limited to a specific fuel volume during a race that decreased from 220 L in 1984 to 150 L in 1988 (Table 3).

Table 3. Fuel volume capacity limitation and permitted refueling periods

	Fuel capacity max	Refueling permitted
1973 to 1981	250 L	-
1982 to 1983	250 L	Yes
1984 to 1985	220 L	No
1986 to 1987	195 L	No
1988	150 L	No
1989 to 1993	-	No
1994 to 2009	-	Yes
2010 to 2013	-	No
2014 to 2017	100 kg	No
2018	105 kg	No
2019 to 2021	110 kg	No

With the refueling ban and tank volume limitations lowering over the years, the amount of energy that could be "carried" onboard became increasingly important and these regulations created an advantage for fuels with higher energy density (MJ/L) in the F1 competition. By example, Table 4 compares pure C7 components. Toluene has a higher density than its heptane counterparts but lower specific energy. Density varies over a wider range than the heating values. Thus, the density difference between heptane and toluene, expressed as a percent of average density is more than twice the difference in energy, expressed as a percent of average energy. This makes toluene better in applications where one wants to maximize the energy when volume is constrained. By extension, fuels high in aromatic content are better for racing when onboard volume of the fuel is limited. Energy density is not particularly relevant to commercial road fuel specification, and this is one key difference that causes the formulations of F1 fuels to diverge from commercial gasolines.

Table 4. C7 hydrocarbon heating value comparison[14]

		Lower Heating Value (LHV)	
C7's	Density @ 60°F (kg/L)	Specific energy (MJ/kg)	Energy density @ 60°F (MJ/L)
Heptane	0.688	44.6	30.7
2,3-Dimethylpentane	0.699	44.5	31.1
1-Heptene	0.702	44.4	31.2
2,4,dimethyl-1,3-Pentadiene	0.741	43.5	32.2
Methylcyclohexane	0.774	43.4	33.6
Toluene	0.872	40.5	35.3

With the rise in turbocharged engines and reduction in fuel volume allowed per race, toluene-based fuel, which in motorsports is often informally termed "rocket fuel" emerged as the most popular in F1. Toluene combines excellent knock resistance and high energy density. Toluene has a RON value that far exceeds the fuel specification maximum limit and therefor for the final blended fuel to meet specifications it needed to also incorporate lower octane rating n-alkanes.[15]

The term "rocket fuel" for these fuels is erroneous as the hydrocarbon fuels used in rockets and missiles are higher boiling range petroleum fractions that mainly consist of saturated hydrocarbons.[16] On the contrary, F1 fuels during this time were constrained volumetrically, and thus disadvantaged if they consisted of highly saturated hydrocarbons which have a relatively low energy density (energy released during combustion per unit volume of fuel)[17].

In the late 80's another innovative approach was used to increase the amount of fuel that can be carried on the vehicle. This practice was called "fuel freezing" which is a misleading colloquialism as the fuel was not actually frozen but rather significantly chilled. Chilled fuel reduces the fuels volume thereby allowing more to be added to the cars. It also increases the amount of energy as chilled fuel is denser and thus more calorific. It also had the add on benefit of more efficient fuel pumping to the engine and a relatively colder air-fuel charge to the combustion chamber which brings in efficiency benefits.

In 1989, turbocharged engines were banned again, and the engines were restricted to 8 to 12 cylinders, naturally aspirated, 3.5 L. With this came the removal of the fuel capacity limits shifting the emphasis away from high aromatic fuels.

1990s

In 1990, the fuel temperature in the tank was restricted to be no more than 10 °C below ambient temperature, removing the opportunity for "fuel freezing" that was widely used in the 80s. These benefits are not relatable to the commercial world, so limits were introduced to keep the practice in check.

In 1991 the first quantitative limit confining a hydrocarbon came with the max limit of 5 v% on benzene (Table 5). Benzene is a high performance component as it has good combustion properties with respect to flame speed, octane and high volatility[18] but was limited as it had been deemed a carcinogen and there is significant risk of contact with fuel under normal operations in the pit. The benzene limit was also a harmonization of the racing fuel spec with commercial fuel.

Table 5. F1 specification maximum limits for octane improvement additives and componentry

	1979 ↓ 1990	1991	1992	1993	1994	1995	1996 ↓ 1998	1999	2000	2001
TEL (mg/L)	-	-	13	-	5	5	5	5	5	5
MMT	-	-	-	-	-	-	-	NP	NP	NP
Nitrogen (m%)	1	0.2	0.2	1	0.2	0.2	0.2	0.2	0.2	0.2
Oxygen (m%)	2	2	2.5	2	3.7	3.7	3.7	2.7	2.7	2.7
Alcohols	NP	NP	-	NP	-	-	-	P	p	p
Benzene (v%)	-	5	5	-	5	5	5	1	1	1
Aromatics (v%)	-	-	-	-	-	50	60	42	35	35

P, permitted; NP, not permitted

TEL was restricted by a Pb max limit of 13 mg/L in 1992 (which is less than 5% of lead concentrations in typical leaded fuels in the 1970s and 1980s). A consequence of TEL being phased-out was the increased reliance on aromatics blending in fuel which would be later limited as well as depicted in Table 5.

The limitation on maximum density was tightened and helped to drive the fuel to lower aromatics content, which had been increasingly relied on as an octane booster in response to the lead phase out. These changes mirror what was happening with premium commercial gasolines. Components consisting of benzene, toluene, ethylbenzene, and xylenes (BTEX) were being used in premium gasoline. BTEX is a high-octane petroleum blend stream product which is produced from a refining technology called "reforming." As a result, BTEX volumes rose in the commercial gasoline pool. In premium gasoline grades, the BTEX volume content could be as high as 50 percent in some gasoline samples.

The changes to the fuel specification in 1993 were a relapse of the fuel specification to 1990 rules for a single year. This was the last year the specification was governed by FISA.

Efforts to harmonize race fuels toward pump fuel standards ramped up in 1994. Language in fuel regulation was changed and stated that commercially produced meant "Petrol of a kind recognised by the FIA as being on general and genuine sale to the public in a country with at least twelve events entered on the International Calendar."

The Pb limit was further reduced to the value consistent with unleaded commercial fuels (5 mg/L). Nitrogen based octane improvers were gaining use such as aniline and N-methyl aniline but required rather large amounts, on the order of 1 or 2 v%, to gain 1 or 2 octane numbers. These components were also further limited in 1994 with a reduction on N maximum weight to 0.2 m%.[19]

The regulation also placed a restriction on the use of components capable of exothermic reaction in the absence of external oxygen (oxygen from the air). A fundamental safety consideration for hydrocarbon fuels is that they only combust when in the presence of both a sufficient ratio of oxidant and a source of ignition. The addition of fuel components that act as oxidants in a hydrocarbon fuel can lead to an unstable mixture as the combination of both in the same blend is now only an unintentional discharge of static away from meeting the requirements for combustion.

Another notable change to the fuel specification was the addition of a final boiling point requirement of 215 °C. Distillation limits would expand in the following years. However, as often is the case, when a new specification test is first introduced the enacted limits are lenient at first.

Specifications are of little value if not adhered to and enforcement was introduced in 1995. In this year a rule was added to ensure compliance of the race fuel with those proposed to be used at the GP's and fuels must comply to the composition of a sample (chemical fingerprint) submitted in advance to the racing authorities for approval to stop the trackside adulteration of certified fuels with banned components. The FIA enforce the new rules with the introduction of the trackside laboratory in this same year, enabling fuel samples taken from the cars to be tested immediately, rather than after the event.

Three new quantitative requirements were introduced aimed at moving F1 fuels from being a narrow set of pure chemicals to a higher compositional distribution.

The first was the introduction of quantitative limits on the hydrocarbon types summarized in Table 6 to include ranges for saturates, aromatics, olefins and di-olefins. This was put in place to better envelope the ranges found in typical commercial fuels. We can also read the unique limit of di-olefins in racing fuels as evidence they were being used to a great extent during this period, presumably to address the need for fuels with high flame speed required by high revving, short stroke engines (*vide infra*).

Table 6. New hydrocarbon limits added to the F1 fuel specification compared to commercial

Hydrocarbon type	F1 regulations change	Commercial fuel[20]
Saturates	60 v% max	
Aromatics	20-60 v% range	Average 29 v%, 12-65 v% range
Olefins	0-35 v% range	Average 8 v%, 0-33 v% range
Di-olefins	1 m% max	Not typically measured

The second new rule stated that the total concentration of individual hydrocarbon components present at less than 5 m% must be at least 50 m% of the fuel, which was referred to as the "5/50 rule." This rule inherently imposed a 12 component minimum in the fuel.[21] It was clearly aimed at fuels that could be blended from a handful of pure chemical components, and pushed fuel suppliers toward using components that were "complex mixtures of hydrocarbons" which is more similar to how commercial fuels are manufactured.

The third rule introduced was constraints on three points in the distillation profile of the fuel highlighted in Table 7. A constrained distillation profile is important in commercial fuels as a way to harmonize the properties of gasolines produced from hundreds of different refineries which must work properly across engines manufactured by dozens of original equipment manufacturers. Additionally, various ranges of a distillation profile have been correlated with specific aspects of commercial gasoline performance in fleet-wide drivability studies: The front end determines ease of starting, the mid range influences smooth running and the tail end governs fuel efficiency.[22] Since fuels blended from a small number of chemical components will tend to have distillation curves with large step changes, this rule further favored "complex mixtures" of hydrocarbons in F1 fuels.

Table 7. Distillation requirements added in 1995 to the F1 fuel specification

Distillation	Minimum	Maximum
E70 (v%)	15	50
E100 (v%)	40	70
E140 (v%)	-	90
E180 (v%)	85	-
Final boiling point (°C)	-	215

Other less impactful changes were also introduced to harmonize commercial and F1 race fuel specifications. For example, a sulfur limit of 0.1 m% was introduced aimed to harmonize the F1 specification with commercial fuel standards in Europe. This didn't have much of a practical impact on the formulation and manufacturing of racing fuels because F1 fuels were mainly blends of chemical streams. Chemical streams involve significantly more processing, which means they are inherently low in Sulfur as the expensive catalysts involved in their manufacturing processes require it be removed prior to processing.

1995 power unit regulations reduced engine displacement to 3.0 L maximum with a maximum of 10 cylinders starting the trend for future engine downsizing.

In 1999, compositional distribution requirements were relaxed to a "5/30" rule which translates in a drop from requiring at least 12 components to at least 8 components in the final fuel.[23] We can speculate that this was a concession to further constraints on specific hydrocarbon types by the addition of a table of maximum allowable on each carbon number in each group of Paraffin, Olefin, Naphthene and Aromatic (PONA) summarized in Table 8. Additionally, naphthenes with chemical "strain energy", that is rings below 5 carbon atoms incorporating ring strain, were no longer permitted; for example, cyclobutane has a 3.5% higher heat of combustion per CH_2 group than cyclopentane.[24] Under the new rules, cyclobutane and derivatives of it were not permitted. The PONA table also requires that components representing at least 90 m% of the total fuel are identified. Thus, the sum of hydrocarbons present at concentrations below 0.5% by mass which cannot be allocated to a particular cell in the table (unallocated) must not exceed 10.0 m% of the total fuel sample.

Table 8. Total concentration of each hydrocarbon group by carbon number in the fuel (m%)

	C4	C5	C6	C7	C8	C9+	Unallocated
Paraffins	10	30	25	25	55	20	-
Naphthenes	-	5	10	10	10	10	-
Olefins	5	20	20	15	10	10	-
Aromatics	-	-	1.2	35	35	30	-
Maximum	15	40	45	50	60	45	10

Additional fuel regulation changes included lowering the following limits:

- Oxygen content lowered to 2.7 m% maximum
- Benzene content lowered to 1 m% maximum
- Total aromatics content lowered to 42 v%
- Total olefin content lowered to 18 v%
- Manganese was not permitted

The significant lowering of olefins from 35 to 18 v% likely hampered performance in high revving engines of the time. Engine power is a function of both how fast it is moving and the force it is exerting (torque, which is most directly proportional to displacement). The higher the RPM, the higher the horsepower, given by the equation Horsepower $\propto$ Torque x Engine RPM. To avoid undesired effects of moving heavy pistons at high linear velocities, the practical way to increase the RPM is by shortening the stroke of the engine, but the downside of doing this is lower fuel efficiency.

With respect to fuels, moving to higher RPM is known to introduce a strong performance correlation to both rapid evaporation and flame speed.[25] A need for rapid evaporation can be explained as follows: the ability to evaporate quickly and fully is beneficial to allow complete atomization within the restricted available time. The lower the carbon number the quicker the evaporation. A positive correlation between power and high flame speed can be explained as follows: To ensure maximum thermal efficiency, the flame must have consumed all the available combustible products as soon after the top-dead-center of the piston's stroke as possible. At high RPM, the time available for complete combustion is shorter, so the fuel flame speed needs to complement the shorter timescales. The drivers of laminar flame speed are complex and include both transport phenomena and chemical kinetics. It is empirically found that for a given carbon number, the further from saturation the higher the flame speed such that alkynes > alkenes > alkanes. Along these lines di-olefins (especially conjugated) were found to have higher flame speeds than similar mono-olefins.[26] Both olefins and di-olefins maximum regulations were introduced in 1995 and were reduced over the years to suppress their use and make the fuel compositions more in line with commercial fuels. Olefins, especially di-olefins, are practically limited in commercial fuels because they introduce undesirable impact on oxidative stability and can form "gums" or peroxides. Oxidative stability is particularly important in commercial fuels because they are exposed to air at many points in the supply chain and may be stored for long durations of time. F1 fuels are stored in sealed containers and do not require a long shelf-life so higher levels of olefins and di-olefins can be used without concern of autooxidation.

Indubitably the restriction of Manganese was to curb the use of Methylcyclopentadienyl manganese tricarbonyl (MMT), an organomanganese compound, that had emerged as an alternative additive to TEL for the purpose of increasing octane. Based on this restriction, we can conclude that at least some F1 fuels may have incorporated MMT prior to 1999.

2000s

Tightening of componentry expanded in the 2000s which is summarized in Table 9. Aromatics were further lowered to 35 v% and sulfur to 50 ppm in 2000 to harmonize with commercial fuel specification changes.

Table 9. Maximum limits for performance improvement additives and componentry

	RON	Styrene	Di-olefin	Non PONA	Una*	Olefin	Aromatics	S	N
		m%	m%	m%	m%			ppm	ppm
1995	102	-	1	-	-	35 v%	50 v%	1000	2000
1996 to 1998	102	-	1	-	-	35 v%	60 v%	1000	2000
1999	102	-	1	-	10	18 v%	42 v%	150	2000
2000 to 2002	102	-	1	-	10	18 v%	35 v%	50	2000
2003	102	1	1	1	5	18 v%	35 v%	50	2000
2004	102	1	1	1	5	18 v%	35 v%	10	2000
2005 to 2007	102	1	1	1	5	18 v%	35 v%	10	500
2008 to 2009	102	1	1	1	5	18 v%	35 v%	10	500
2010 to 2017	-	1	1	-	-	17 m%	40 m%	10	500
2018 to 2021	-	0.1	0.1	-	-	17 m%	40 m%	10	500
2022 to today	-	0.1	0.1	-	-	17 m%	40 m%	10	500

*Unallocated 1999 to 2002. Unassigned 2003 to 2009.

A styrene limit was introduced in 2003. The introduction of a styrene limit points to its use in F1 race fuels in this period. Styrene is a reactive olefin that is not found in crude oil. Its presence is generally avoided in commercial fuel blending components because it has a strong tendency to polymerize, and this can lead to issues with gums or short oxidative stability in finished fuels. The reason styrene would be a desirable blending component for F1 fuel is that styrene combines both high octane and flame speed exceeding its nearest saturated analogue, ethylbenzene.[27] Similar to other reactive olefins, the undesirable properties of styrene for commercial fuels do not necessarily preclude the use in race fuels.

The FIA tightened the rules on unidentified components in the fuel to further reduce the possibility of exotic components that could fall under the 10 m% unallocated in the PONA table (Table 8) by now requiring at least 95 m% of the total fuel to be identified. The concentration maximum limit for an unassigned component was lowered to 0.1 m% and the maximum for the sum of these components was lowered to 5.0 m% of the total fuel. A new Non-PONA maximum limit of 1 m% was also added to the table for components that were identified but outside of the specification definitions or permitted oxygenates.

In 2005 some notable compositional requirement changes were made. Nitrogen m% was further lowered to 500 ppm where it remains to this day. This further limited the use of aniline based octane boosting additives. Sulfur was also lowered to 10 ppm which is where it remains to this day.

In 2006, engine speeds were extremely high and reported to be exceeding 20,000 RPM.[28] The mandated downsized 2.4 L V8 engine drove engine manufactures to focus on high RPMs to gain horsepower. The fortuitous benefits of the high RPMs on fuel development requirements were a decrease in time for knock from pre-flame reactions to develop, lowering the engine octane appetite.

A maximum crankshaft rotational speed limit of 19,000 RPM was introduced in 2008 and would begin the process of driving RPM toward a more commercially relevant value. 2008 also started a move toward the use of non-fossil components with a requirement for a minimum of 5.75 m% of the fuel to be comprised of oxygenates derived from biological sources.[29] This was complemented with the maximum oxygen content being relaxed back to 3.7 m% (like commercial E10 fuel, although this is in v%).

Like fuels, engine regulations were being introduced to drive manufactures toward more commercially relevant designs. In 2009, the crankshaft maximum rotational speed was further dropped to 18,000 RPM. Hybrid technology Kinetic Energy Recovery System (KERS) made its F1 debut but was not mandatory. At the push of a button, drivers can access a boost of horsepower from the electrical motor that is strategically applied around the lap.

2010s

In 2010 refueling was banned again. Notable fuel specification changes included the removal of maximum octane limits which were replaced with an octane minimum (anti-knock index of 87). This relaxation was in preparation of the future transition to turbocharged engines. This allowed fuels with very high octane values to re-enter F1. High octane complements high CR engines but there was no limit to monitor changes to the engines CR.

There was also an expansion of the biofuel requirement to 5.75 m% bio-components, rather than specifically requiring a bio-oxygenate. This presented FIA with a challenge to verify the bio-content used in fuel formulations. Previously, oxygenates could be verified as coming from bio-sources by looking at ratios of C^{12} and C^{13} and Hydrogen and Deuterium on the oxygenate itself.[29] But with hydrocarbons, and potentially complex hydrocarbons, fuels with non-oxygenate bio components now needed to pass a C^{14} test (a radiocarbon dating method that tests for carbon which was recently being exchanged with the atmosphere).

Aromatics max was changed to mass based instead of volume based to harmonize all compositional specs in m%.

In 2014 major specification changes emerged to both the engine and fuel. It is perhaps best to briefly explain how the engine changes impacted the combustion requirements, and then explain fuel specifications in light of the new demands brought by these very different engines.

Engine regulations took a step-change in further downsizing from V8 to V6 engines with 1.6 L engine capacity. This step brought the size of the engine in F1 car to something similar to what one finds in a fuel-efficient compact car. At the same time, the FIA also wanted to encourage competition between the engine makers in *fuel efficiency*. This prompted the addition of an engine specification limiting the fuel mass flow to 100 kg/h max at 10,500 RPM that remained constant above this RPM and was reduced below this RPM according to the equation 0.009 N(RPM)+ 5. Since additional fuel could not be introduced into the engine by increasing RPM beyond 10,500, the practical RPM limits of the engine were indirectly limited to around 12,000. These changes prioritized fuel efficiency as an engine performance objective in F1 racing over maximum power and the engine designs moved away from rich to lean air-to-fuel ratios (AFRs) as shown in Figure 1.[30]

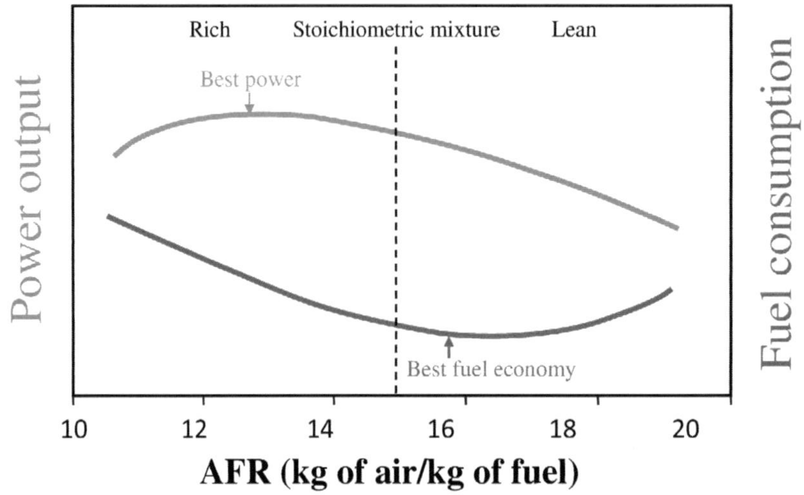

Figure 1. AFR on power and thermal efficiency

In addition to fuel flow requirements, FIA also introduced a stringent regulation on the maximum fuel use for each race. Originally the fuel weight was limited to 100 kg which was a significant reduction than had been used with their V8 predecessors (approximately 150 kg). Unlike prior regulations that restricted the volume of fuel, the new regulations were on fuel weight and were optimized to get the most energy out of a set weight of fuel (MJ/kg). The originally stated goal was to lower this weight limit as efficiency improved. In actuality, the limit went up, first to 105 kg to take account of regulations that made the vehicle heavier and later to 110 kg following a recognition that competition was being stifled by the drivers needing to employ a ‘lift-and-coast’ technique to conserve fuel or drivers were not able to make it through an entire race. Even so, to this day, fuel used by each car in an F1 race remains substantially lower than in the V8 era, even while modern power and performance has now exceeded that of a decade ago.

Deriving power from such a small engine while at the same time under a fuel flow constraint required pushing the CR higher. This was also regulated for the first time in F1 history to 18 maximum. Engine manufactures were keen to push to this limit as a maximum attainable power from an engine is connected to higher CR.[31] Typical gasoline commercial engine CR’s are around 10 to 14. The fact that a CR limit was thought necessary illustrates how F1 engines truly push boundaries of conventional designs and technology.

Along with reducing engine displacement, turbocharging was once again allowed which closely aligns with a strategy that many automakers in the commercial sector had taken to meet aggressive fuel efficiency standards for light duty vehicles. With this transition also came the mandatory use of advanced integrated electrical hybridization. The combination of engine and hybridization is now referred to as a 'Power Unit'. The number of PU replacements was also limited which forced the engine makers to move toward increasing engine reliability a necessary expectation with commercial engines. The amount of PU replacements during a race season was limited to 5 per driver.

Downsized engines with shorter strokes of the piston generally results in reduced combustion quality because of increased radial flame path lengths and reduced chamber height in order to reach a complete fuel burn. This combustion challenge is further exasperated because AFR are required to be very lean given the limitations on fuel use. Lean mixtures cause unstable combustion and partial burns. Both are addressed with ignition and combustion chamber designs not found in commercial engines which ensure a particularly robust ignition and flame front.

The combination of engine changes made the engine more prone to knocking. From a fuels design perspective, engine reliability translates into mitigation of abnormal combustion events, like pre-ignition, knock, superknock, etc being much more important because aside from inefficiency and a loud sound, they also cause damage to engines. When knocking occurs, one can usually retard the ignition timing to avoid the effect. However, this comes at a cost of combustion efficiency and power because the heat release moves away from ideal. Therefore, the fuels formulated needed to be of a much higher octane rating to both avoid knock and keep ignition timing close to ideal. The removal of the maximum octane rules a few years prior set the stage for blending very high octane fuels.

Because fuels in 2014 were subject to a fuel flow constraint, and a constraint on total fuel, the fuel energy per unit mass and fuel efficiency became the two factors in determining total energy available to the power unit. Factors that characterized the combustion kinetics, like octane rating, flame speed, etc were determiners of fuel efficiency. The thermodynamics of combustion determined the fuel energy release per unit mass. Therefore, fuels needed to be highly paraffinic in this regard to maximize the heating value of the fuel. Aromatics, oxygenates, and other unsaturated fuel components all have lower heating values than paraffinic components. However, paraffins tend to have poor octane performance, and the high CR meant the paraffins needed to be highly isomerized (isoparaffins).

Lastly, the RPM was still significantly higher than commercial engines so fuel components that were able to provide flame speed were still required in F1 fuels. The fuel formulator needs to also take into consideration that flame speeds are generally slower in lean mixtures and may need to compensate with fast burning components in the fuel blend. In the end, the challenge for the fuel formulator is to find the best balance between high heating value, high octane, and high flame speed components as often each component will have tradeoffs between these properties.

In summary, there were two key shifts in fuel requirements that occurred because of these engine changes:

- Fuel specific energy (energy per fuel mass) became very important
- Fuel octane rating (and more generally knock resistance) was important again after years in an era of high-speed naturally aspirated V8 engines that did not require as much octane

The changes implemented to move F1 to more efficient lean burn engines clearly took hold as two years in, engine manufactures claimed thermal efficiency had reached 45%[32] to 50%[33] which is a remarkable step from the prior V8 2.4 L engines which were around 29%.

The styrenes and di-olefins maximum limit were lowered to 0.1 m% in 2018.

2020s

In 2022, the first mandated use of a specific component was added to the specification with the inclusion of a bio-ethanol 10 m% minimum requirement. Unlike road fuels, the ethanol must be 'Advanced Sustainable' (second generation[34]). There was a change in viewpoint with respect to limiting oxygenates from imposing an upper limit to imposing a minimum limit. The oxygen requirement changed from a max of 3.7 m% to a min of 3.45 m% to reflect consistency with an

E10 minimum. Previous biofuel regulations permitted any biocomponent, which was preferably hydrocarbon-based given the importance of specific energy content with a fuel flow limit in kg/h. Adoption of E10 aligned with evolution in commercial fuels around the world. The regulations allowed much higher levels of ethanol to be used, but the penalty with respect to specific energy would mean teams would have a significant disadvantage in doing so. Another change to enable the introduction of E10 fuels was that the vapor pressure maximum was increased to account for increased volatility caused by the positive azeotrope of ethanol blends.[35]

After 2022 both fuel and engine specifications remain unchanged until the anticipated major overhaul to both in 2026.

The 2026 engine spec changes are subtle from the internal combustion engine side with a reduction of the CR to 16 but remains a 1.6 L V6 with single turbo. The fuel flow requirement is now specified as energy flow, rather than mass flow and is reduced to a maximum of 3000 MJ/h.

The 2026 fuel spec changes will result in by far the most significant and expansive fuel changes in the history of the sport. The anticipated changes prioritize meeting the FIA definition of a 100% 'Advanced Sustainable' fuel.[36] Underpinning these changes are an extensive description about the feedstocks that can be used to make the sustainable components used in the fuels.

The F1 fuels are required to have a calculated GHG savings of >65% and >75% for Renewable Fuels from Non-biological Origins (RFNBOs) relative to a conventional fossil fuel.[36] These GHG savings requirements are meant to assure there is a benefit to the carbon footprint over fossil-based hydrocarbons. With a wealth of synthetic chemistry tools and reactions available, practically any bio hydrocarbon component can be synthesized from biocarbon feeds if the requirement is just that a molecule be "bio-derived." A good analogy is that even the most complex molecules can be synthesized with precision by the chemicals industry. However, the complex, multistep reaction pathways used to reach the target molecules incurs an energy and/or materials cost. Having a carbon footprint requirement precludes complex pathways and is penalizing for the energy and materials that must be used to do each chemical conversion. For example, the transformation of AS feedstocks to hydrocarbons often requires a high level of hydrogen to remove oxygen as gasoline hydrocarbons are not the typical output of a natural biomass growth or fermentation process apart from isolated examples like isoprene.

Prior to 2026, the fuel supplier had at their disposal an almost endless supply of fossil hydrocarbon components allowing for a wealth of blend options and supply security. Given the feedstock requirements and limits on conversion chemistry, the choice of AS components in 2026 will be limited. An emphasis on ethanol as a blend biocomponent or precursor to blend biocomponents is likely. Ethanol has a simple and well-established production: It has been produced for thousands of years by humans through fermentation and requires no conversion chemistry after the initial fermentation step. This simple two-carbon molecule also has many desirable attributes like knock-resistance, evaporative cooling, and flame speed that meet many of the high-performance requirements in motorsports fuel. For instance, ethanol is used as an important blend component in NASCAR and IndyCar fuels already.

Commercially practiced routes to bio-derived hydrocarbon fractions of gasoline are limited to the conversion processes of alcohols to hydrocarbons via processes with descriptive names like ethanol-to-gasoline (ETG) and methanol-to-gasoline (MTG).

CONCLUSIONS

An opportunity to understand F1 fuel formulations was highlighted by looking at the progression of fuels and engine specifications over the history of the sport. F1 fuels were optimized within the constraints of the fuel regulation period to compliment the characteristics of a specific engine regulation period which changed over time. The earlier F1 fuel specs were short and concise. New content was added over time that was mainly aimed at reinforcing relevancy of F1 fuels to commercial premium fuels as it became apparent that simply stating fuels needed to be like commercial fuels was inadequate. The challenge stems from the fact that "commercial fuel" is inherently ambiguous and encompasses a wide array of hydrocarbon mixtures. Over time the F1 specification incorporated more quantitative limits on composition and properties.

Specification limits were introduced to eliminate the use of high performance components that didn't find broad commercial use for gasoline, i.e. di-olefins, styrene, ring strained naphthenes, nitrogen based additives and disproportionate levels of olefins and aromatics. The drivers, including additional performance provided by these components was discussed and highlights some of the deviation in fuel compositions compared to commercial fuels used over the sports history.

Transitioning toward AS fuels in 2026 brings both challenges and opportunities. The extensive amount of text that will be added to the specification is directed to compliance of all components meeting second generation feedstock requirements with a GHG benefit of the final products more than 65% over petrol.

AKNOWLEDGEMENTS

We are especially grateful for the significant contributions from our colleagues James Turner, Vincent Pereme, and Peter Tibbetts from the Federation Internationale de l'Automobile who have provided the F1 specifications throughout the years and answered some of the probing questions relating to the motivation behind some of the fuel specification changes which was instrumental in making this paper possible.

REFERENCES

---

[1] Joe Saward, Why has FISA been abolished?, September 1, 1993, Grandprix.com

[2] FIA Formula One World Championship, Regulations, https://www.fia.com/regulation/category/110 (accessible July 2024)

[3] https://www.jomenvisst.de/fia/ (accessible July 2024)

[4] In 2026 there will be the addition of supplementary documents that contain ~100 pages, >24,000 words to cover the Advanced Sustainable requirements

[5] Richard Van Basshuysen, Fred Schaefer, Internal Combustion Engine Handbook, 2nd English Edition, SAE International, page 842

[6] ExxonMobil Heritage Booklet, 1000 Formula One Races, page 6

[7] W. Warnecke, B. Poulet, J. Landschof, A. Dreyer, J. Müller-Belau, G. Lovett, Innovation from Track to Road: The Role Fuels Can Play in Motorsport, 37th International Vienna Motor Symposium, 2016, page 108

[8] 50 Years of Technical Development in Formula One (autosport.com), https://atlasf1.autosport.com/evolution/1950s.html (accessible July 2024)

[9] With respect to the engines there were mainly two types of engines, with the smaller engine permitted to use supercharging forced induction. In the early part of the 60s, forced induction was banned but returned in the latter

part of the decade with the smaller size engines. There were multiple other changes to the engine regulations during this time, but in general, they maintained two options where a small engine was allowed to use forced air induction, and a large engine used normal aspiration that ranged in size from 0.75 L (although 1.5 L was more common) supercharged engines to 4.5 L atmospheric

[10] With respect to the engine the size difference between the two engines narrowed slightly, with 1.5 L with a compressor or 3.0 L without a compressor

[11] FISA still allowed the use aviation fuel in countries where no suitable fuel is commercially available

[12] Gasoline naming and RON characterization according to DIN EN 228

[13] Peter Wright, Tony Matthews, Formula 1 Technology, SAE International, 2001, page 18

[14] Values taken from Worldwide Specialty Chemicals Phillips 66, Reference Data for Hydrocarbons and Organosulfur Chemicals, Revised June 1998

[15] Yutaka Otobe, Osamu Goto, Hideyo Miyano, Michio Kawamoto, Akio Aoki and Tohru Ogawa, Honda formula One Turbo-charged V-6 1.5L Engine, SAE 890877

[16] Rocket engines are different than the Otto-cycle internal combustion engines used in racing, and important combustion qualities of fuels used for rocket engines are not characterized by knock resistance. MIL-DTL-25576 covers two kerosene propellant fuels for rocket engines referred to as RP-1 and RP-2 that have a maximum aromatics level of 5 v%. In fact, performance of rockets is highly sensitive to fuel weight which makes aromatics-containing fuels less appealing than fully saturated fuels which have a higher specific energy (energy released during combustion per unit mass of fuel)

[17] An exception would be JP10 which has a higher heating value of 39 MJ/L but still lower than a high aromatic gasoline fuel. The octane is not known for JP10 and its volatility is much too low to operate in a high revving gasoline engine

[18] See reference 5, "the secret of the first super fuel of the world", page 833

[19] Nitrogen m% limit moved around in the early years. Remained at 1 m% from 1979 to 1990, lowered in 1991 and 1992 to 0.2 m%, and increased back to 1 m% 1993. The limit declined from there to 0.2 m% again in 1994 where it remained until 2004 and was lowered to 500 ppm in 2005 where it remains today

[20] L. M. Gibbs, How Gasoline Has Changed, SAE 932828

[21] It takes 11 components at ~4.9 m% to be over 50 m% with a 12th component completing the blend

[22] Chevron, Motor Gasolines Technical Review, 2009, page 2

[23] It takes 7 components at ~4.9 m% to be over 30 m% with an 8th component completing the blend

[24] Francis A. Carey, Organic Chemistry 2nd Ed. McGraw-Hill, 1992, page 90

[25] Bryce, S., Lindsay, R., Galliard, I., and Glover, A., Fuels Development for Formula One, SAE Technical Paper 942540, 1994

[26] Farrell, J., Johnston, R., and Androulakis, I., Molecular Structure Effects On Laminar Burning Velocities At Elevated Temperature And Pressure, SAE Technical Paper 2004-01-2936, 2004

[27] Xin Meng, Erjiang Hu, Xiaotian Li, Zuohua Huang, Experimental and kinetic study on laminar flame speeds of styrene and ethylbenzene, Fuel 185 (2016), page 916–924

[28] Ian Bamsey, King of speed, Race Engine Technology Issue 73, page 20

[29] The term "Biological origin" refers to the source of the oxygenates used in fuel. To determine this, an isotope ratio test was conducted during the certification process. This test measured the ratios of carbon-13 to carbon-12 (13C/12C) and deuterium to hydrogen-1 (2H/1H) in the samples. FIA collected and analyzed empirical data from known sources, which was used to establish thresholds that could distinguish between oxygenates derived from conventional petroleum and those derived from crops like corn, wheat, sugar beet, and sugar cane. The FIA and FOFAP (the group of F1 fuel suppliers) agreed on these thresholds. During certification, a sample of the oxygenate had to be submitted for stable isotope analysis. This analysis classified the samples in a manner similar to the method described by Keiko ISHIDA-FUJII, Shingo GOTO, Ryu UEMURA, Keita YAMADA, Michikatsu SATO, Naohiro YOSHIDA, Botanical and Geographical Origin Identification of Industrial Ethanol by Stable Isotope Analyses of C, H, and O, Bioscience, Biotechnology, and Biochemistry, Volume 69, Issue 11, 1 January 2005, Pages 2193–2199, https://doi.org/10.1271/bbb.69.2193

[30] Cartoon rendition of the figure in 1.) Derek Seward, Race Car Design, 2017, Bloomsbury Publishing, page 236. 2.) The College of Petroleum and energy studies Oxford England, Gasoline technology and economics – Specifications, Quality, Blending components and Market developments, Section 2, Figure 3

[31] Heywood, John B. 2018. Internal Combustion Engine Fundamentals. 2nd ed. New York: McGraw-Hill Education, page 178

[32] Lorenzo Sassi, Ioannis Kitsopanidis, Guy Lovett, Evolution in F1 Engine Technology: Pursuing Performance From Todays Power Unit Through Efficiency, 2016 Vienna Engine Symposium

[33] Steve Rendle, Formula 1 Technology, The Engineering Explained, Evro Publishing Limited, 2023, page 174

[34] Definition can be found in 2022 Formula 1 technical regulations, Issue 13, 16 August 2022, See reference 2 for specification database link

[35] Jordan N. Metz and David J. Abdallah, Elucidating the influence of structural classes on azeotropic behavior in oxygenated fuels, 18th IASH 2024, Louisville, Kentucky, USA September 8, 2024

[36] 2026 Formula 1 technical regulations, Issue 8, 24 June 2024, section 16.2.8 defines Advanced Sustainable component. Specification states "With regard to fuel, the detailed requirements of this Article are intended to ensure the use of Advanced Sustainable (AS) fuels comprising solely AS components, that are composed of certified compounds and refinery streams and fuel additives and to prohibit the use of specific power-boosting chemical compounds." See reference 2 for specification database link

IASH 2024, the 18TH INTERNATIONAL CONFERENCE ON STABILITY, HANDLING AND USE OF LIQUID FUELS
Louisville, Kentucky, USA
September 8 – 12, 2024

REDUCED EMISSIONS USING DIESEL DETERGENT ADDITIVES

Mary Dery*, Gavin Medine, Eric Payne

Innospec Fuel Specialties, 200 Executive Dr. Downingtown,Pennsylvania 19335 USA

ABSTRACT

In 2022, the EPA adopted the Clean Trucks Plan which mandated the reduction of emissions and harmful air pollutants from heavy duty (HD) trucks. Reduced emissions that form smog and soot will apply to HD engines and vehicles beginning in model year 2027. The transportation industry is making changes to comply with these new regulations. Diesel technology has evolved to reduce the impact of petroleum-based combustion on the environment. This includes the incorporation of aftertreatment systems to trap or reduce particulate and gaseous emissions like NOx, CO and hydrocarbons. Burning fuel mixtures that include renewable and biodiesel further reduce a fleet's carbon footprint. Diesel performance additives are known to clean up deposits in a diesel engine that prevent optimal combustion of the fuel. Diesel performance additives also help to reduce emissions and enable the use of more biodiesel and renewable fuels. This presentation will review some of the recent findings and positive impacts of diesel detergents in field trials from around the globe.

*Corresponding author: Mary Dery, mary.dery@innospecinc.com

IASH 2024, the 18TH INTERNATIONAL CONFERENCE ON STABILITY, HANDLING AND USE OF LIQUID FUELS
Louisville, Kentucky, USA
September 8 – 12, 2024

FUNDAMENTAL UNDERSTANDING OF GASOLINE PIONA IN RELATION WITH KEY PROPERTIES

Zainab Saihati*, Radwan Bakor

Saudi Aramco, 2296 Research and Innovation Center, Dhahran 31311, Saudi Arabia

ABSTRACT

The world's increasing demand for liquid fuels has highlighted the need for a deeper understanding of fuels fundamental chemistries to optimize properties and aid us for the transition to liquid low carbon fuels. This abstract discusses the importance of understanding the various types of gasoline chemistries to aid in properties prediction and their effects in order to enhance fuel quality and performance. Delving into the chemical composition trends of gasoline fuels and their properties will enable us to build the understanding of such complex chemistries and potentially find innovative ways to synthesize low carbon fuels to reduce environmental impacts.

The following paper will be looking at the relationship between the different components and chemical groups using Gas Chromatography techniques to identify correlations with key properties such as RON, density, RVP, and distillation. The correlations can be a potential predictive tool for gasoline properties and further the understanding of properties and how they could be enhanced by the use blend components selection.

*Corresponding author: Zainab Saihati, zainab.saihati@aramco.com

IASH 2024, the 18TH INTERNATIONAL CONFERENCE ON STABILITY, HANDLING AND USE OF LIQUID FUELS
Louisville, Kentucky, USA
September 8 – 12, 2024

SUSTAINABILITY IN THE MARINE MARKET

Joseph L Stark*

Innospec Fuel Specialties LLC, 200 Executive Dr, Newark, DE 19702

ABSTRACT

The maritime industry has seen turbulent waters over the past several years in their quest to reduce greenhouse gas (GHG) emissions from international shipping in order to reach net zero climate neutrality by 2050. In 2020 the International Maritime Organization (IMO) mandated the reduction of sulfur levels in marine fuels used for international shipping to 0.5%, which effectively created a new fuel known as very low sulfur fuel oil (VLSFO).

Two ISO working groups (WG 17) have recently allowed biodiesel to be included as a blend component in bunker fuels up to 30% and will allow 100% biofuel in marine vessels by the end of 2024. The primary guidance for producers and blenders to create a quality, compliant bunker fuel is ISO 8217. The inclusion of biofuel increases the already wide range of streams that could be used as blend components to make bunker fuel for international shipping. With the inclusion of sustainable fuels such as biodiesel and renewable diesel, the potential for fuel quality issues continue to increase.

Processing these new bio-based fuels onboard ships can be difficult for ship operators as the energy content, density and viscosity of each fuel can vary greatly. In addition to the fuel property differences, the inherent stability, compatibility with other bunker fuels, cold flow characteristics and overall operational efficacy of the fuel can create additional challenges for fuel suppliers and ship operators that need to be managed.

Understanding the underlying composition, stability and compatibility of bunker blend components and finished blends such as VLSFO and marine diesel oil (MDO) fuels is critical in being able to identify potential handling and operational risks and in developing and implementing a successful mitigation strategy when an issue does arise.

*Corresponding author: Joseph L Stark, Joseph.Stark@innospecinc.com

IASH 2024, the 18TH INTERNATIONAL CONFERENCE ON STABILITY, HANDLING AND USE OF LIQUID FUELS
Louisville, Kentucky, USA
September 8 – 12, 2024

(Technical Poster)

COMPARISON OF DERIVED AND INDICATED CETANE NUMBER FOR AVIATION TURBINE FUELS AND CORRELATION WITH LEAN BLOW OUT

Jon Luecke*, Robert McCormick

National Renewable Energy Laboratory (NREL), 15013 Denver West Parkway MS 1634, Golden, CO 80401 USA

ABSTRACT

The introduction of sustainable aviation fuel (SAF) into aviation turbine fuels requires approval as described in ASTM D4054 to ensure compatibility with preexisting aviation hardware. Derived cetane number (DCN), measured using an Ignition Quality Tester (IQT) as described in ASTM D6890, is currently required to be in the range of 35—60 to meet the D4054 fit-for-purpose requirements due to previously developed correlation of DCN with measured phi for lean blow out. Additional ASTM methods for determining cetane number (CN) exist, the latest being ASTM D8183, which uses a device called the Advanced Fuel Ignition Delay Analyzer (AFIDA) to measure indicated cetane number (ICN). This method requires 40mL of fuel versus 100mL for DCN, a significant advantage for early stage SAF fuel development. Here we present DCN and ICN results for set of fuels that includes NJFCP A-1, A-2, A-3, C-1, C-2, C-3, C-4; additional commercial jet fuels and blends with SAF, and several pure compounds, with ICN ranging from 28—65. ICN values were measured to be 2 units on average higher than DCN values for CN>40, whereas ICN measurements for CN<35 were determined to be 2 units on average lower than DCN. A crossover equivalence in the region of cetane numbers from 36-38 is indicated. This dataset provides guidance for inclusion of D8183 ICN data into the D4054 specification as an alternative, lower volume method for measuring CN. Additionally, an approach that reduces the fuel requirement of D8183 from 40mL to 25mL is described.

*Corresponding author: Jon Luecke, hannes.luedtke@dlr.de

AN HPLC-ESI-QTOF METHOD TO ANALYZE POLAR HETEROATOMIC SPECIES IN TURBINE AVIATION FUEL VIA HYDROPHILIC INTERACTION CHROMATOGRAPHY THROUGH STATISTICAL ANALYSIS OF MASS SPECTRAL DATA

Amanda M. Arts[1], Zachary J. West[1], and Paul J. Wrzesinski[2]

[1]University of Dayton Research Institute, 300 College Park Dr, Dayton, Ohio 45469-0043, United States, Amanda.Arts@udri.udayton.edu, Zachary.West@udri.udayton.edu
[2]Air Force Research Laboratory, 1790 Loop Rd, Wright-Patterson AFB, Ohio 45433-7131, United States, paul.wrzesinski@us.af.mil

ABSTRACT

Aviation turbine fuel is a complex mixture containing of thousands of compounds. Although compounds containing oxygen, nitrogen, and sulfur functional groups are each found at low levels (<0.1% by mass) in fuels, their presence can generate significant effects on fuel properties. To combat the shortcomings of previous analytical approaches, a new analytical method using hydrophilic interaction liquid chromatography (HILIC) coupled with electrospray ionization and quadrupole time-of-flight mass spectrometry (ESI-QTOF) for the identification of heteroatomic, polar compounds in aviation turbine fuel has been developed. The HILIC method is a powerful separation technique that possesses many advantages; these include a minimal sample preparation requiring no extraction step ensuring no loss of compounds of interest, and the ability to acquire high-fidelity compound data for chemometric analysis of heteroatomic species in aviation turbine fuel.

This new method was used to analyze ESI-QTOF data from a set of aviation turbine fuels to allow for identification of significant heteroatom molecules through statistical analysis techniques. Each fuel sample contains dozens of chemical classes, and thousands of individual compounds. These mathematical tools offer a look inside the chemically complex composition of fuel to better interpret, describe, and understand these large datasets to help optimize the production and use of current, as well as future fuels. This effort developed a chemometric analysis workflow for linking identified polar compounds to key performance parameters, such as thermal oxidative stability. As proof of concept, connections were made using heteroatomic mass spectral data of 89 unique fuel samples to relate their chemical information to measured QCM autoxidative deposit. A Fold Change (FC) analysis, which is used to describe abundance ratios or differences between conditions, was conducted (with cutoff of >=2 on averaged normalized abundance data) to determine what nitrogen compounds could be associated with QCM deposit formation. The Down and Up regulated features that were extracted from the FC were then further down selected to analyze connections between known groups of 10 distinct nitrogen classes and thermal oxidative stability. Three main groupings of different trend cases emerged: 1) increase in deposit regardless of carbon number (DBE>/=4), 2) decrease in deposit at low carbon numbers, and an increase in deposit at high carbon numbers (DBE=1-3), and 3) increase in deposit for certain compounds, while others show no change (DBE=0). These findings were insightful, confirming or expounding upon previous work by discovering new information, as well as uncovering additional areas of further research.

KEYWORDS

HILIC, Heteroatoms, Aviation turbine fuel, Statistical analysis, Thermal stability

INTRODUCTION AND BACKGROUND

The composition of aviation turbine fuel is a complex mixture of hydrocarbons [1], which strongly influence the bulk fuel properties. Additionally, there are important properties that relate to the trace species found in fuel. These trace species consist of sulfur, oxygen, and nitrogen-containing organic molecules that are naturally occurring artifacts of the refining process. Even though the concentration of these S, O, N- heteroatomic species in aviation turbine fuel are low, around <0.1% by mass, their compositional information is critical knowledge to the understanding of aviation fuel thermal stability[1-4] as their presence in fuel impacts the formation of carbonaceous deposits. In addition, polar heteroatoms have impacts on synthetic aviation turbine fuels (SATF) due to the fact that most SATFs do not contain heteroatomic compounds, and the absence of these compounds has been linked to seal swell and lubricity issues[5].

Studies focused on predictive modeling of jet fuel thermal stability via chemical kinetic autoxidation mechanisms have yielded insightful information regarding the importance of certain heteroatomic species and their link to deposit formation. Previous work[6, 7] has shown that the oxidation and deposition characteristics of a variety of jet fuels could be positively correlated to measured concentrations of trace species classes, especially phenols, and the polar nitrogen classes of indoles and carbazoles. However, previous studies have provided conflicting information regarding the importance of other polar nitrogen species, such as pyridines, anilines, and quinolines[8-11]. There has been a proven link between sulfur species and deposit formation[12-14], but little is known about the specific species classes of sulfur species that play a role in the mechanism itself.

Further, the study conducted by Zabarnick *et al.*[3] expanded upon the role of heteroatomic species in the autoxidation mechanism and suggested that the inclusion of certain nitrogen species may be needed for adequate prediction of fuel deposition. The testing conducted encompassed individual compounds from several nitrogen classes including indoles, quinolines, anilines, pyridines, carbazoles, and indolines. The results demonstrated that nitrogen species interact with sulfides during jet fuel autoxidation to encourage deposit formation in the jet fuel thermal oxidation tester (JFTOT®) and the quartz crystal microbalance (QCM), both of which are commonly utilized for the study of thermal oxidative stability of jet fuel samples and have been described in detail elsewhere[7, 15-18].

Inspired by Zabarnick *et al.*, Christison *et al.*[19] built upon previous work using pyrolysis gas chromatography/mass spectrometry (Py-GC/MS)[20, 21] to analyze QCM deposit of hydrocracked jet fuel component systematically spiked with various combinations of nitrogen (indoles, amines, pyridines, quinolines), sulfur (sulfides and thiols), and phenolic compounds that had been previously correlated with thermal oxidative deposit formation. They uncovered three main compound classes that were identified in the analysis of thermal oxidative deposit: alkylphenols, quinolines, and substituted pyridines.

Several techniques have been used to identify and/or quantify polar species in turbine aviation fuels, including single use techniques such as two-dimensional gas chromatography (GC×GC) with solid phase extraction (SPE) pre-separation[22, 23], direct flow injection analysis using electrospray ionization and mass spectrometry (ESI-MS)[24], high-performance liquid chromatography (HPLC) with ESI-MS time-of-flight (TOF) using a standard C8 column [21, 25], or coupled techniques such as normal-phase (NP) HPLC with UV-vis absorption paired with SPE followed by HPLC fractionation and GC-MS detection[4].Some of these methods that utilize SPE are not ideal for identification and quantitation of polars as they are time-consuming with many steps and may result in the unintentional loss of some compounds of interest during the extraction process. While the direct-flow techniques minimize sample preparation, they are limited in instrument mass resolution therefore inhibiting high confidence compound identification. For the ESI-TOF methods, no observations have been made using any alternative column chemistries, which may offer more benefit for polars analysis to more accurately target and separate heteroatoms when compared to other established techniques.

A non-standard type of chromatography, called hydrophilic interaction chromatography (HILIC), was chosen as an alternative to more traditional types of chromatography. HILIC is becoming increasingly popular to analyze polar compounds due to its use of a polar stationary phase and a polar organic-aqueous mixture mobile phase solvent, with water being the strong eluting solvent and the majority of the solvent being organic. HILIC possesses many advantages including the capacity to separate anions, cations, and polar neutrals in a single run, short analysis times, and better MS performance compared to other HPLC techniques as it supports efficient evaporation of the solvent, thus enhancing sensitivity and altering ion suppression[26].

The present researchers have newly developed a method for exploring the unique combination of HILIC with LC-ESI-QTOF to analyze trace heteroatomic polars in aviation turbine fuel[27]. The method development focused on optimization of several method parameters known to be of importance regarding separation efficiency, retention time, and repeatability including column configuration, mobile phase composition, solvent gradient, re-equilibration time, injection volume, dilution factor, and sample solvent. The robustness of the method was confirmed in the development stage, and the ability to acquire high-fidelity compound data for chemometric analysis of heteroatomic species in aviation turbine fuel was proven.

Statistical tools for the analysis of aviation turbine fuel samples have been employed in varying capacities dating back to the 1950s[28], primarily for the purpose of developing predictive models that correlate the fuel chemical composition to the fuel physiochemical properties. Most studies have focused on of the use of one-[29-35] and two-dimensional[35-38] gas chromatography, spectroscopy[30-32, 39-46], and HPLC[40, 47] measurement techniques for the determination of chemical composition. These techniques have difficulty to accurately determine the identity of a singular compound in a complex mixture. As a result of this limitation, these techniques either analyzed known singular compounds of interest to relate to a particular compound class, or when analyzing an unknown compound, assign it to a group or compound class based of similarity to other analytes.

Statistical analysis of mass spectral data, specifically from LC-MS, can be used to accurately identify both unknown and known individual compounds to determine differences between samples to gain new insights into complex chemical processes. Christison *et. al* has conducted the only known research that has employed the use of statistical analysis of LC-MS jet fuel data, which utilized a TOF mass spectrometer with Agilent's Mass Profiler Professional (MPP) chemometric software[20, 21]. They provided a method to study the molecular origin of jet fuel thermal oxidative stability issues as measured via the JFTOT® by analyzing a small set of samples before and after thermal stressing. They offered understanding into the identities of molecules that correlate with poor thermal stability results and linked these to the soluble macromolecular oxidatively reactive species (SMORS) model for distillate fuel instability.

The goal of this current effort was to expound upon the work of Zabarnick *et al.* and Christison *et al.* utilizing a large dataset of jet fuel samples in combination with Agilent's MPP chemometric software to create a workflow for statistical analysis of mass spectral data from trace heteroatomic polar compounds generated via the newly developed HILIC method. It was desired to link targeted polar compounds of interest to key performance parameters, such as thermal oxidative stability. As proof of concept, connections were made using mass spectral data of a large dataset of unique fuel samples to relate their chemical information to measured QCM autoxidative deposit. Certain nitrogen-containing species classes were found to have varying effects on thermal stability, and will be discussed herein, confirming known information as well as offering new insight that can be used to link fuel composition to performance.

MATERIAL AND METHODS

An Agilent 1200 series liquid chromatograph (LC) in series with an Agilent 6546 Q-TOF was used for analyses. The HILIC column used was a bare silica gel Agilent InfinityLab Poroshell 120 HILIC, 2.1 mm inner diameter, 100 mm length, 2.7 μm particle size, narrow bore LC column with 120 Å pore size. The column temperature was 30 °C to create the recommended slightly-above ambient temperature condition. An Agilent Jet Stream (AJS) ESI source in positive mode was used with the following conditions: gas temperature of 250 °C, drying gas flow rate of 10 L/min, a nebulizing pressure of 35 psi, sheath gas temperature of 275 °C, sheath gas flow rate of 11 mL/min, a capillary voltage of 3500 V, a fragmentor voltage of 75 V, a skimmer voltage of 65 V, and an octupole radio frequency (RF) of 750 V. All solvents used were Optima-grade from Fisher Scientific, and acid modifiers used were also LC-MS grade and included formic acid (Fisher Scientific) and acetic acid (Fluke).

The values for the method parameters are reported in Table 1. The flow rate for all analyses was 0.40 mL/min. Two different solvents were explored as part of the optimization effort, designated as A solvent and B solvent. The A solvent for the mobile phase was water with methanol (MeOH) at 0.125 vol% and formic acid (FA) at 0.1 wt%. The B solvent for the mobile phase was acetonitrile (ACN) with 0.1 wt% FA. Samples were run with the HILIC analytical column plus a guard column (GC). Fuel samples were diluted 1:100 prior to injection in a combination of 80/20 vol% ACN and isopropanol (IPA). Sample volume injected was 5 μL. A gradient method was used to elute the analyzed samples with an isocratic hold at 10/90 vol% water/ACN for one minute, a ramp to 40/60 vol% water/ACN for four minutes followed by an isocratic hold for one to six minutes. A two-minute post-time was always used.

Table 1. Method Parameters

Column	Analytical Column and Guard Column
Mobile Phase	H_2O / ACN, FA 0.1 wt%
Gradient Elution	Gradient (strong/weak), 10/90 vol% to 40/60 vol%
Re-equilibration Time	Nine-minute run time with two-minute post-time
Injection Volume	5 μL
Dilution Factor	1:100
Sample Solvent	80/20 vol% ACN/IPA

A total of 89 unique jet fuels, a combination of Jet A, Jet A-1, F-24, and JP-8, were chosen for this work. Fuel property information can be found in Table 2. The samples were analyzed in triplicate in randomized order across three total sequential runs, each acquisition period spanning around 30 hours. Three separate runs were used as there were limitations due to the maximum mobile phase and calibrant check solution volumes that could be supplied. Sample solvent blanks and one fuel sample that was chose to be used as a quality control (QC) sample were utilized before, during, and after the fuel runs at periodic intervals. This allowed for the determination of drift in retention times and reproducibility of the data across the sequence.

Table 2. Fuel Dataset Properties

	Number of Samples			
Fuel Type	**Jet A**	**Jet A-1**	**F-24**	**JP-8**

	3	6	57	23
QCM deposit at 15hrs (μg/cm²)	**Low (<2.5)**	**Mid (2.5-6)**	**High (>6)**	
	41	35	13	
Total Speciated Polars (mg/L)	**Low (<300)**	**Mid (300-750)**	**High (>750)**	
	48	37	4	
Viscosity at -40° C (mm²/s)	**Low (<8.0)**	**Mid (8-11.9)**	**High (>=12.0)**	
	40	43	6	

Ranges estimated using historical information and relevant reports[48, 49].

Data preprocessing is especially critical prior to data being imported into MPP for statistical analysis. The first step in data preprocessing is extracting targeted "features" for each fuel sample. A "feature" refers to a compound in the Agilent MassHunter software with a unique m/z and retention time that is extracted using a deconvolution algorithm. Targeted feature extraction requires m/z specified by the user via a compound database, and user defined thresholds for mass and retention time windows, peak height, and absolute counts.

The targeted compound database used for feature extraction used in this work was comprised of 162 unique m/z values from ten nitrogen containing species classes (N-species) commonly present in aviation turbine fuel, and three nitrogen containing lubrication additives (tricresyl phosphate (TCP), dioctyl-diphenyl amine (DODPA), and phenyl-α-napthyl amine (PANA)). The N-species classes are characterized by double bond equivalence (DBE) number for molecules containing one N, and are listed in Table 3. Each N-species class contains a homologous series of compounds with the same DBE (or dehydrogenation number, z) but different carbon numbers (n), i.e., $C_nH_{2n-z}N$; therefore, multiple isomers of the same empirical formula exist for each carbon number and DBE class.

The N-species were the primary focus of this study, but additional heteroatomic compounds that also could be found in aviation turbine fuel were added to the database for later use. These 927 compounds included sulfur oxides, sulfur dioxides, disulfur dioxides, nitric oxides, nitrogen dioxides, and various N-O-S compounds for a range of carbon numbers (6 or 8 to 22) and a range of dehydrogenation numbers (+2 to -12 or +3 to -11).

Table 3. Targeted Nitrogen Species Classes

Empirical Formula	Carbon Number (n) Range	Example Classes	Double Bond Equivalent (DBE)	Example Structures
$C_nH_{2n+3}N$	6-20	Aliphatic amines	0	

$C_nH_{2n+1}N$	6-20	Piperidines, Pyrrolidines	1	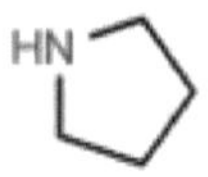
$C_nH_{2n-1}N$	6-20	Dihydropyrroles (DHP)	2	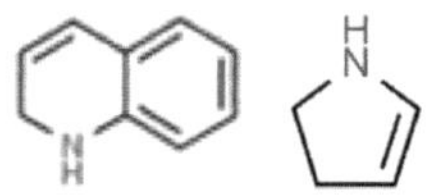
$C_nH_{2n-3}N$	6-20	Pyrroles	3	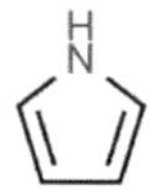
$C_nH_{2n-5}N$	6-20	Pyridines, Anilines	4	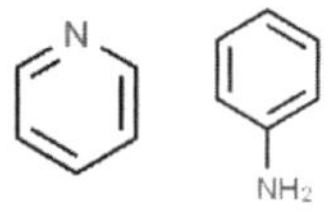
$C_nH_{2n-7}N$	8-20	Indolines, Tetrahydroquinolines (THQ)	5	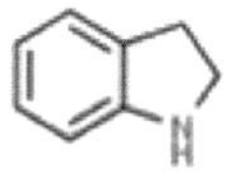
$C_nH_{2n-9}N$	8-20	Indoles, Dihydroquinolines (DHQ)	6	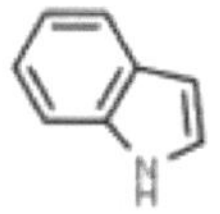
$C_nH_{2n-11}N$	9-20	Quinolines	7	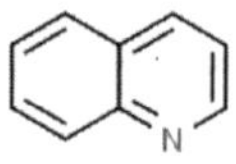
$C_nH_{2n-13}N$	11-20	Phenylpyridines	8	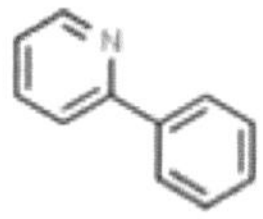
$C_nH_{2n-15}N$	12-20	Carbazoles	9	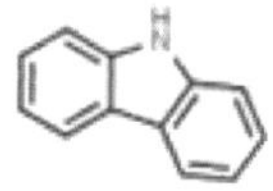

Once the targeted feature extraction was complete, compound hits were reviewed for accuracy, as well as chromatograms and spectra extracted. The extracted features information associated for each fuel sample is then exported for batch data processing using an in-house Python script. This batch data processing performs critical operations on all fuel data files to align the features by combining isomers and assigning exact mass to the targeted heteroatom species of interest. Once the data has been aligned, it is exported for use in MPP.

Upon import in MPP, data must be grouped and normalized before undergoing statistical analysis. Experiment grouping is where external metadata associated with a sample is added, which is a critical operation as it impacts interpretations of the data. Metadata that was added at this step for each sample included ID number and QCM deposit at 15 hours in $\mu g/cm^2$, although virtually any metadata associated with a fuel could be added to compare against the compositional data. Samples within the same experimental groups are considered replicates, and each new group is considered a condition to operate on. Normalization involves the scaling of data to distribute it more appropriately in a bell curve and makes it easier to visualize graphically. Many statistics tests operate under the assumption that the data is normally distributed, and since MS abundance data can be distributed over a wide range of values, MPP automatically performs a normalization that transforms peak intensity data on a Log2 scale.

RESULTS AND DISCUSSION

The trace heteroatomic polar species content of 89 unique aviation turbine fuels was analyzed using HPLC-ESI-QTOF with a HILIC column phase. Resulting individual fuel mass spectral data files were preprocessed by first extracting targeted features of interest and then by aligning those extracted features via batch processing. The parsed data files were then imported into MPP for chemometric analysis.

To determine if the data extracted and aligned was of quality that would be sufficient for statistical analysis and to determine if additional normalization was needed, a quality control (QC) verification was conducted. All 20 QC samples from across the three runs were analyzed to ensure that the QC samples showed no statistical difference from one another regardless of filtering applied. Originally, 817 features were reported as being present in at least one of the samples with an abundance cutoff of >5000 counts, and an ion cutoff of >/=2. A low-pass filter was then applied to the features list, based upon frequency, that would retain a feature only if it was present in 100% of samples to reduce the noise in the dataset, and this condensed the number of features to 432. Finally, to ensure that high-abundant, high-frequency compounds were not skewing the data, a high-pass filter on abundance to exclude the top 8.0% of features was applied, reducing the features list to 410. Upon each step in the filtering process, violin plots were extracted to look for consistency in the dataset. The violin plot in Figure 1 is a visual representation of the distribution of the numerical data depicting summary statistics and the density of each variable for each QC sample in the dataset after all filtering had been applied. The violin plot confirms that the QC samples are not statistically significantly different from one another, regardless of run or day, instilling confidence that the data is of good quality and that the process for data import is appropriate.

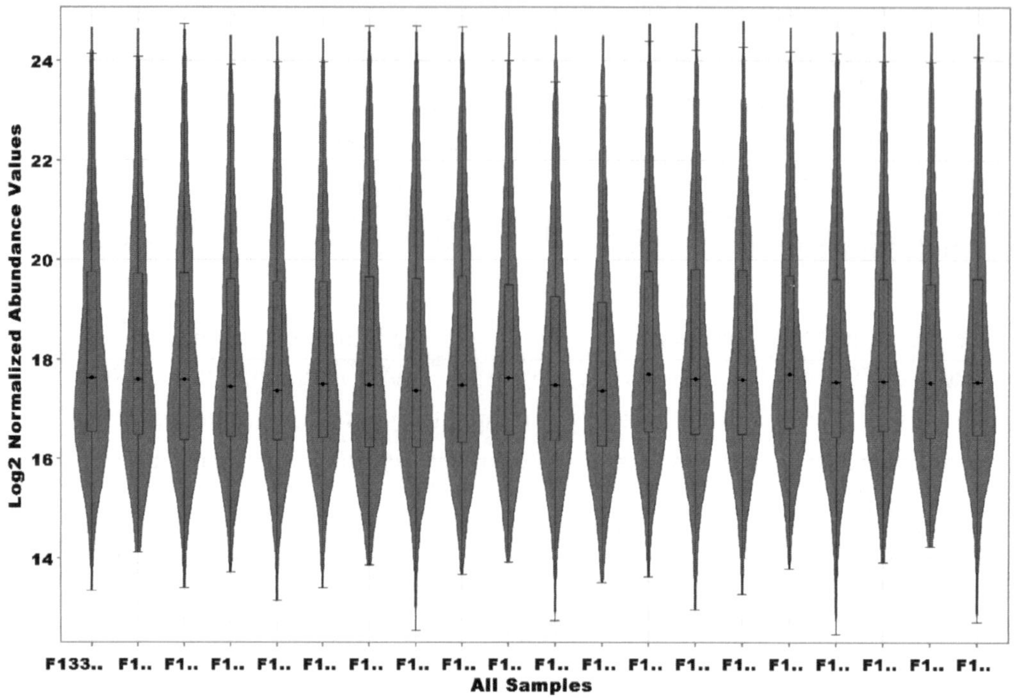

Figure 1. Violin plot of the 20 QC samples in the dataset after high-pass filtering.

A total of 273 samples representative of 89 unique fuels with designated ID numbers, including three replicates of the QC sample treated as samples from each run (nine total), were analyzed. Originally, 881 features were reported as being present in at least one of the samples with and abundance cutoff of >5000 counts, and an ion cutoff of >/=2, with 131 of these being representatives of the N-species. A low-pass filter was then applied to the features list based upon frequency that would retain a feature only if it was present in two out of the three replicates to reduce the noise in the dataset, this condensed the number of features to 599, where 111 were N-species compounds. A Principal Component Analysis (PCA)[50] plot of the dataset colored by ID number is shown in Figure 2. Examples of the three replicates of a single ID can be shown in the circled groupings. PCA is a mathematical method of compressing complex data into a few variables and is used in an unsupervised fashion to assess data quality, however it is important to note that PCA does not provide predictive capability or regression. The PCA plot confirms that the dataset is diverse, that samples of the same ID number group together appropriately, and that there are no obvious outliers.

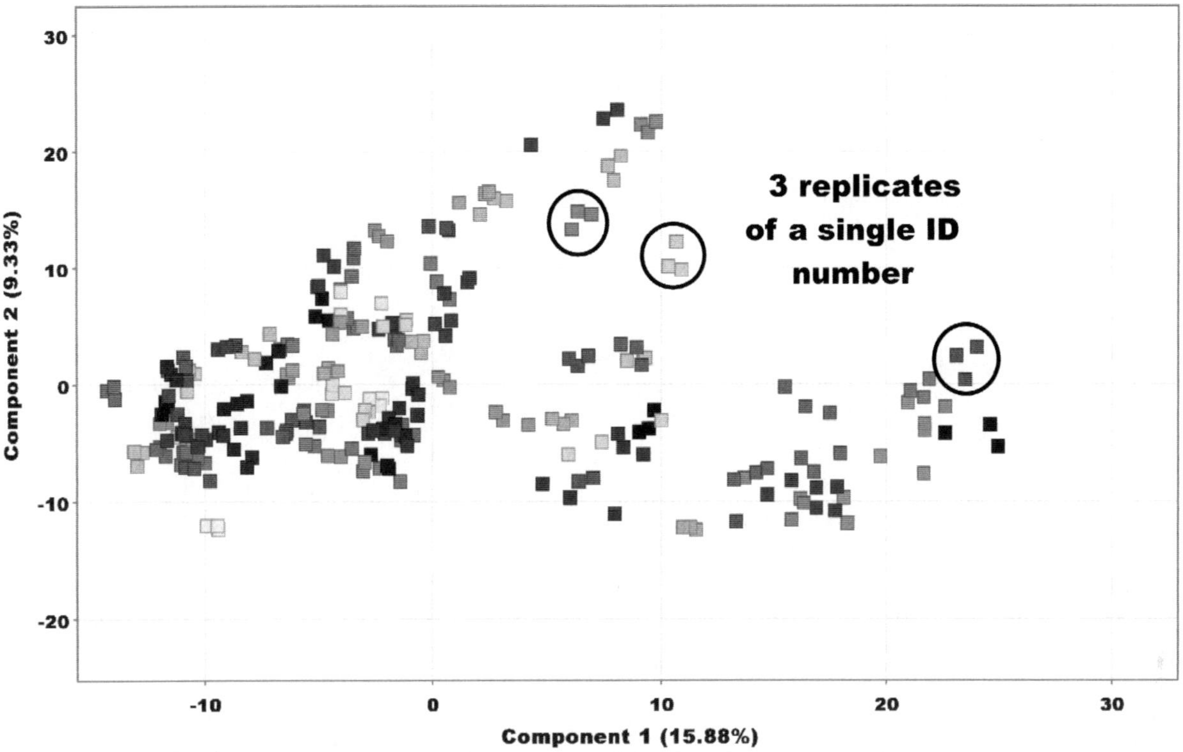

Figure 2. PCA plot of the 273 fuel samples representing the 89 ID numbers; markers are colored by sample ID, and superimposed circles identify three replicates of a single ID number.

Figure 3 shows the Kendrick Mass Defect (KMD)[51] plot for the N-species classes present in the dataset after filtering colored by retention time, where the Kendrick mass defect is plotted as a function of nominal Kendrick mass for ions observed in a mass spectrum. Families of homologous compounds are easily recognized in the KMD plot as they are positioned along a horizontal line in the plot, as denoted on the right side of the figure. This is a result of organic ions belonging to a homologous series having identical Kendrick mass defects but different nominal Kendrick masses. From bottom to top, the degree of unsaturation increases. The KMD plot shows the complex chemical nature of the polar compounds present in the fuel dataset and confirms the feature extraction worked appropriately, as it highlights the homologous series that were targeted. Further, it helps to describe the nature of elution of certain compounds, with the orange/red coloration representing relatively more polar features that eluted later in the chromatogram, with these include the majority of the N-species classes of interest.

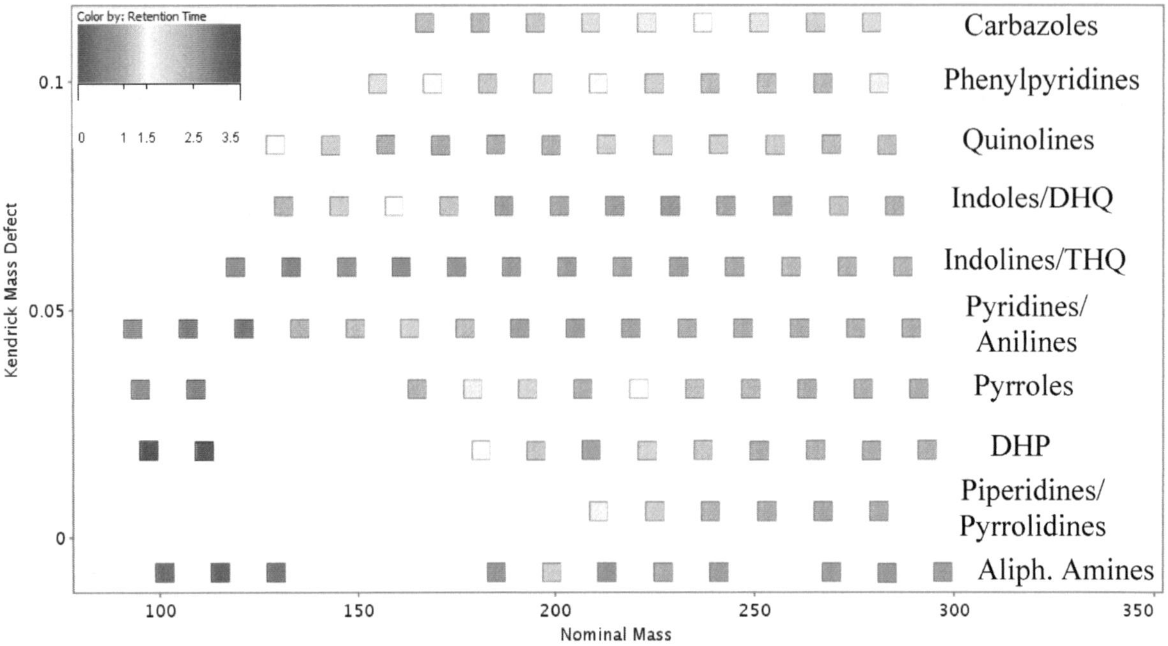

Figure 3. KMD plot for the nitrogen species classes present in the 273 fuel samples colored by retention time.

There is a desire to link chemical composition to performance property metadata. To validate this approach, previously measured QCM deposition data was imported as an associated experimental grouping as ranges. These ranges were chosen as a means to visualize trends and are consistent with those described in Table 2. The total range of QCM deposit for all fuels in the dataset was 0.7-18.0 $\mu g/cm^2$. A Fold Change (FC) analysis, which is used to describe abundance ratios or differences between conditions, was conducted to determine what compounds could be associated with QCM deposit formation. The normalized intensity data for a given N-species compound for each fuel in the denoted ranges was averaged for the FC. The FC cut-off on normalized data was chosen to be >= 2.0 on the Low vs. Mid and Mid vs. High pairings. After the FC analysis was complete, 94 out of the original 111 N-species features remained as flagging as Down or Up regulated. A profile plot colored by compound representing the FC>=2.0 for the chosen pairings is shown in Figure 4. The FC can be difficult to uncouple in this way due to the large number or compounds, therefore upon completion of the FC>=2.0 the regulation data was taken offline for further analysis. The Down and Up regulated features that were extracted were then further separated by N-species class (see Figures 5-7) for ease of visualization.

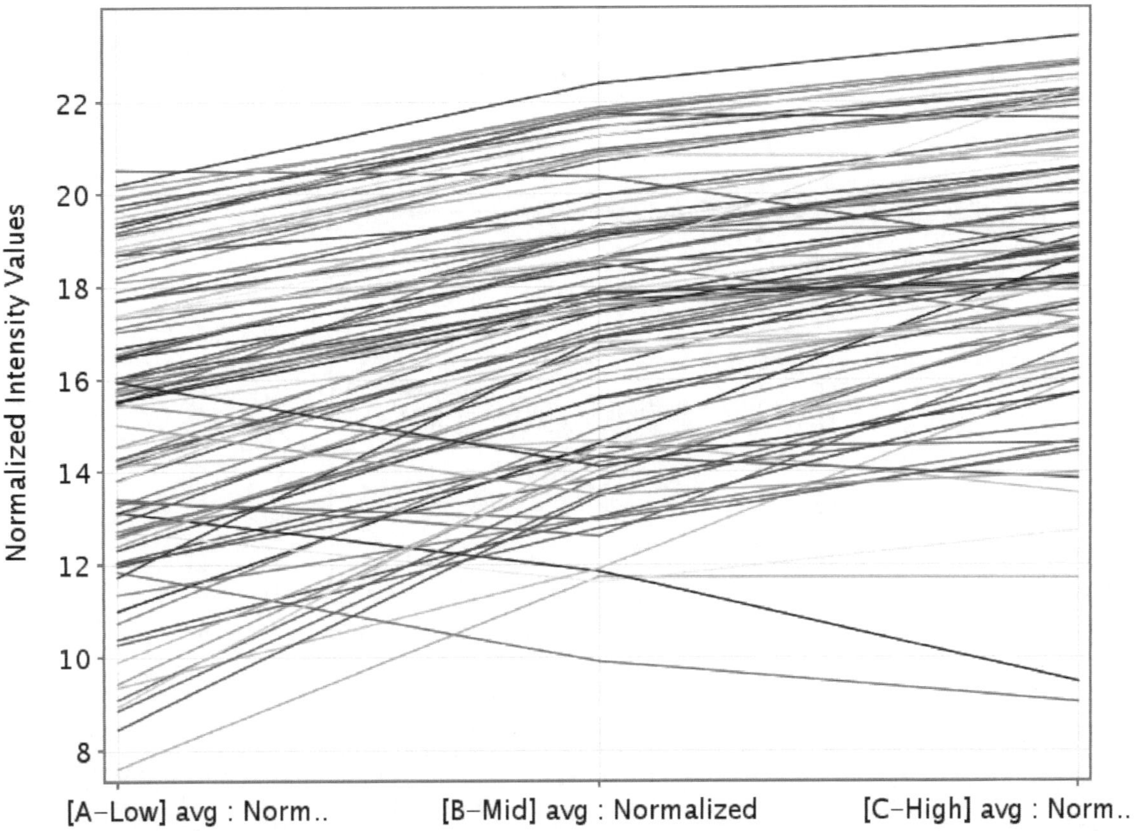

Figure 4. Profile plot for Low/Mid and Mid/High QCM deposit FC>=2.0 data of N-species.

Three main trends emerged within the N-species: 1) increase in deposit regardless of carbon number, 2) decrease in deposit at low carbon numbers, and an increase in deposit at high carbon numbers, and 3) increase in deposit for specific compounds. Figures 5-7 show representative plots for each of these three cases, with phenylpyridines representing case 1, pyrroles representing case 2, and aliphatic amines representing case 3. The visualization schemes for the three figures are representative of their respective regulation, with Up regulated values from Low to High corresponding to a red-colored solid line (signifying correlation to increase in deposit), a green-colored dashed line corresponding to a Down regulation (signifying correlation to a decrease in deposit), and a yellow-colored dotted line corresponding to no distinct regulation. It can be seen in Figure 5 that for the phenylpyridines an increase in their abundance value, regardless of their carbon number, resulted in an increase in deposit from the Low to Mid to High QCM cases represented. The profile plot for the pyrroles, Figure 6, highlights an increase in their abundance value had varying effects on deposit formation from the Low to Mid to High QCM cases represented, with the low carbon number of n=6 showing a decrease in deposit, the carbon numbers of n=11 to 15 showing no observed change, and the carbon numbers of n=16 to 20 showed increase in deposit. Figure 7 shows that for certain compounds in the aliphatic amines class, n=15 and 16, there was no observed change, while for other carbon numbers, n=12, 14, and 20, an increase in their abundance values led to an increase in QCM deposit. It can be seen that there is no clear pattern observed regarding which carbon numbers may or may not increase QCM deposit.

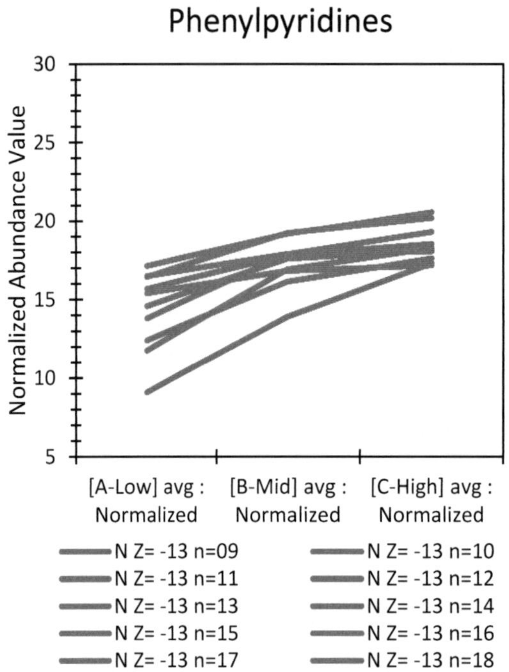

Figure 5. Representative profile plot of phenylpyridines for case one.

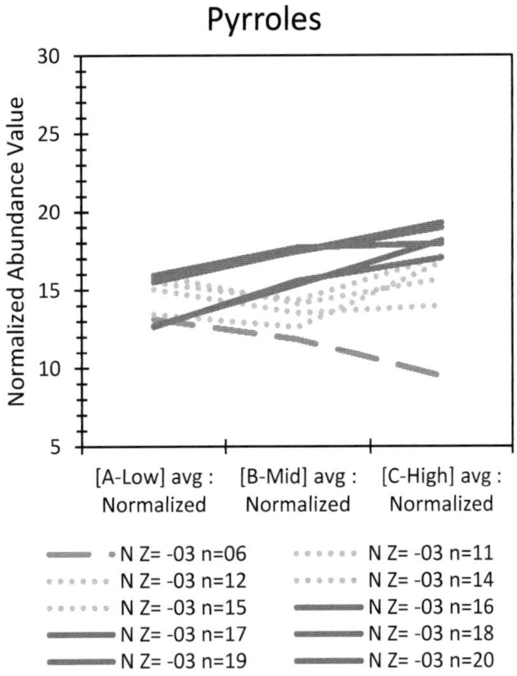

Figure 6. Representative profile plot of pyrroles for case two.

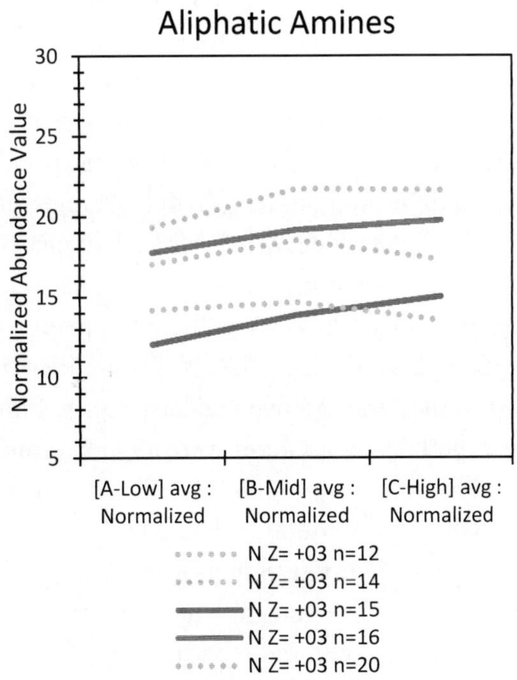

Figure 7. Representative profile plot of aliphatic amines for case three.

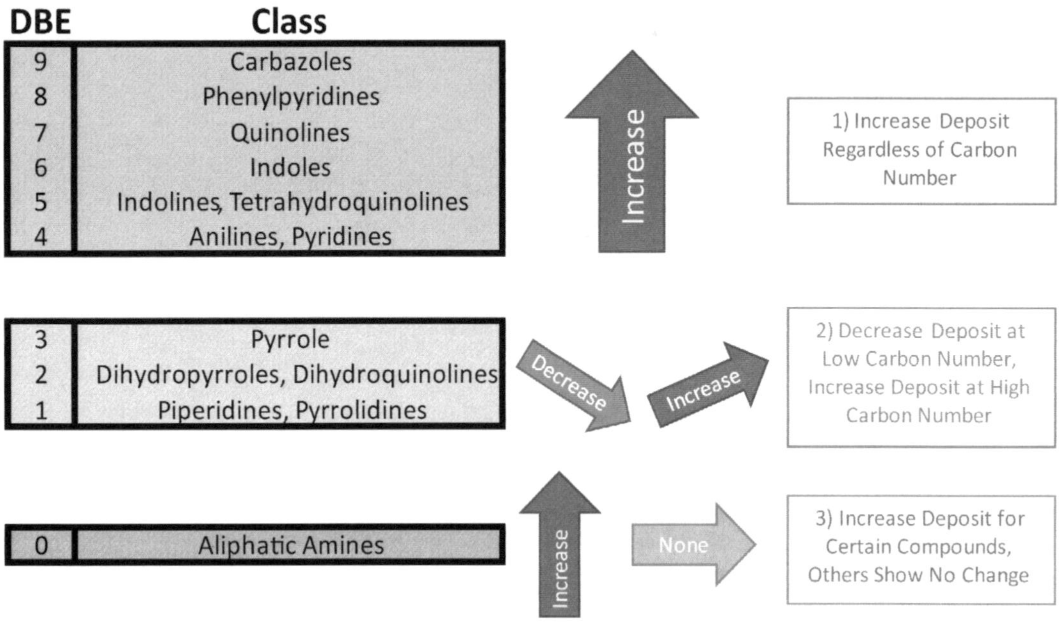

Figure 8. Case trends that emerged within an N-species class related to QCM deposit formation for the measured fuel dataset.

Figure 8 summarizes the observations of the FC study for how compounds relate to QCM data. This is a more inclusive approach in that the entire compound class is being measured, as opposed to previous approaches which analyze a single compound to represent an entire class of compounds. The 10 N-species classes are categorized using the three case trends previously described and are further organized using DBE number. Anilines/pyridines, indolines/tetrahydroquinolines, indoles, quinolines, phenylpyridines, and carbazoles compounds all show relation to an increase deposit regardless of carbon number. In regard to polar nitrogen species, it is well known that indoles and carbazoles impact deposit formation, and this study confirms their importance. Previously, there has been conflicting evidence regarding the importance of other polar nitrogen classes like anilines/pyridines and quinolines, this information helps to confirm their importance in deposit formation. Further, phenylpyridines and indolines/tetrahydroquinolines were all also found to increase deposit formation, leading to newfound information. An additional trend uncovered is that any polar nitrogen species class with a DBE of greater than or equal to 4 are linked to an increase in QCM deposit formation. This higher fidelity information regarding dehydrogenation and carbon number has never been obtained or realized before, and this knowledge gained helps to answer questions emerged from previous contradictory information.

Piperidines/pyrrolidines, dihydropyrroles/dihydroquinolines, and pyrroles decrease deposition at low carbon numbers (<\=6) and increase deposition at high carbon numbers (>/=16). Trends associated with carbon number have not before been examined, which could explain some of the previously conflicting information regarding whether or not certain classes correlate to increased deposit formation. Additionally, these three polar nitrogen species classes represent a DBE of 1, 2, and 3, respectively, further confirming that there is relationship between double bond equivalence and autoxidative deposit. Finally, where nitrogen classes are concerned, the species with the lowest DBE of 0, aliphatic amines, were shown to have increase in deposit for certain compound(s) in the group, while others show no change.

CONCLUSIONS

A method for the direct analysis of trace heteroatomic compounds in jet fuel was developed by combining the advanced techniques of HILIC chromatography and ESI-QTOF in the same analysis. This new method was then used to analyze ESI-QTOF data from a set of aviation turbine fuels to allow for identification of heteroatom molecules and subsequent statistical analysis. These mathematical tools offer an insight into the chemically complex molecular origins of fuel to better interpret, describe, and understand these large datasets to help optimize the production and use of current, as well as future fuels.

This effort developed a workflow for linking identified polar compounds to key performance parameters, such as thermal oxidative stability. As proof of concept, connections were made using a fold change analysis of mass spectral data from 89 unique fuel samples to relate their chemical information to measured QCM autoxidative deposit. Overall, the FC analysis relating QCM deposit levels to N-species classes proved insightful in building upon knowledge of heteroatoms and their relationships with the autoxidation mechanism. This study validates prior work and gains understanding of previous unknowns in relation to certain N-species. The trends uncovered linking the polar nitrogen species to the DBE phenomenon and the linkage to carbon number for certain classes are vital information that can be further utilized for future fuel applications.

ACKNOWLEDGEMENTS

The authors thank Shane Kosir for help with programming the parser code in Python.

This material is based on research sponsored by AFRL/RQTF under agreement number FA8650 16-2-2605. The U.S. Government is authorized to reproduce and distribute reprints for Governmental purposes notwithstanding any copyright notation thereon. The views and conclusions contained herein are those of the authors and should not be interpreted as necessarily representing the official policies or endorsements, either expressed or implied, of AFRL/RQTF or the U.S. Government.

REFERENCES

1. Hazlett, R. N., *Thermal oxidation stability of aviation turbine fuels*. Astm International: 1991; Vol. 1.

2. Zabarnick, S.; West, Z. J.; Arts, A.; Griesenbrock, M.; Wrzesinski, P., Studies of the impact of fuel deoxygenation on the formation of autoxidative deposits. *Energy & Fuels* **2020,** *34* (11), 13814-13821.

3. Zabarnick, S.; West, Z. J.; Shafer, L. M.; Mueller, S. S.; Striebich, R. C.; Wrzesinski, P. J., Studies of the role of heteroatomic species in jet fuel thermal stability: model fuel mixtures and real fuels. *Energy & Fuels* **2019,** *33* (9), 8557-8565.

4. Balster, L. M.; Zabarnick, S.; Striebich, R. C.; Shafer, L. M.; West, Z. J., Analysis of polar species in jet fuel and determination of their role in autoxidative deposit formation. *Energy & fuels* **2006,** *20* (6), 2564-2571.

5. Link, D. D.; Baltrus, J. P.; Zandhuis, P.; Hreha, D. C., Extraction, separation, and identification of polar oxygen species in jet fuel. *Energy & fuels* **2005,** *19* (4), 1693-1698.

6. Kuprowicz, N. J.; Zabarnick, S.; West, Z. J.; Ervin, J. S., Use of measured species class concentrations with chemical kinetic modeling for the prediction of autoxidation and deposition of jet fuels. *Energy & fuels* **2007,** *21* (2), 530-544.

7. Sander, Z. H.; West, Z. J.; Ervin, J. S.; Zabarnick, S., Experimental and modeling studies of heat transfer, fluid dynamics, and autoxidation chemistry in the jet fuel thermal oxidation tester (JFTOT). *Energy & Fuels* **2015,** *29* (11), 7036-7047.

8. Taylor, W. F.; Frankenfeld, J. W., Deposit formation from deoxygenated hydrocarbons. 3. Effects of trace nitrogen and oxygen compounds. *Industrial & Engineering Chemistry Product Research and Development* **1978,** *17* (1), 86-90.

9. Frankenfeld, J. W.; Taylor, W. F.; Brinkman, D. W., Storage stability of synfuels from oil shale. 1. General features of sediment formation in model fuel systems. *Industrial & engineering chemistry product research and development* **1983,** *22* (4), 608-614.

10. Taylor, S. E., An alternative view of the thermal oxidative stability of jet fuels. *ACS Division of Petroleum Chemistry, Inc. Preprints* **2002,** *47* (3), 165-169.

11. Frankenfeld, J. W.; Taylor, W. F.; Brinkman, D. W., Storage stability of synfuels from oil shale. 2. Effects of nitrogen compound type and the influence of other nonhydrocarbons on sediment formation in model fuel systems. *Industrial & engineering chemistry product research and development* **1983,** *22* (4), 615-621.

12. Kauffman, R., The effects of different sulfur compounds on jet fuel oxidation and deposition. **1997**.

13. Taylor, W.; Wallace, T., Kinetics of deposit formation from hydrocarbons. Effect of trace sulfur compounds. *Industrial & Engineering Chemistry Product Research and Development* **1968,** *7* (3), 198-202.

14. Taylor, W. F., Deposit formation from deoxygenated hydrocarbons. II. Effect of trace sulfur compounds. *Industrial & Engineering Chemistry Product Research and Development* **1976,** *15* (1), 64-68.
15. International, A., Standard Test Method for Thermal Oxidation Stability of Aviation Turbine Fuels. ASTM D3241-20c, 2023.
16. Zabarnick, S.; Mick, M., Inhibition of jet fuel oxidation by addition of hydroperoxide-decomposing species. *Industrial & engineering chemistry research* **1999,** *38* (9), 3557-3563.
17. Zabarnick, S., Studies of jet fuel thermal stability and oxidation using a quartz crystal microbalance and pressure measurements. *Industrial & engineering chemistry research* **1994,** *33* (5), 1348-1354.
18. International, A., Standard Practice for Thermal Oxidative Stability Measurement via Quartz Crystal Microbalance. ASTM D7739-11(2020), 2022.
19. Christison, K. M.; Boursalian, G. B.; Franz, A. H., Fundamental study of jet fuel oxidative deposit formation. *Fuel* **2024,** *358*, 130343.
20. Christison, K. M. Exploring the Molecular Origin of Jet Fuel Thermal Oxidative Deposition through Statistical Analysis of Mass Spectral Data and Pyrolysis Gas Chromatography/Mass Spectrometry of Deposits. University of the Pacific, 2019.
21. Christison, K. M.; Lorenz, R. M.; Xue, L.; Sparkman, O. D., Exploring the molecular origin of jet fuel thermal oxidative instability through statistical analysis of mass spectral data. *Energy & Fuels* **2019,** *33* (2), 830-836.
22. Shafer, L.; Striebich, R.; Zabarnick, S.; West, Z. In *Identification and quantitation of polar species in liquid transportation fuels by GC× GC*, International Conference on Stability, Handling, and Use of Liquid Fuels, 2015.
23. Striebich, R. C.; Contreras, J.; Balster, L. M.; West, Z.; Shafer, L. M.; Zabarnick, S., Identification of polar species in aviation fuels using multidimensional gas chromatography-time of flight mass spectrometry. *Energy & fuels* **2009,** *23* (11), 5474-5482.
24. Rostad, C. E., Screening of polar components of petroleum products by electrospray ionization mass spectrometry. *Energy & fuels* **2005,** *19* (3), 992-997.
25. Adams, R. K.; Zabarnick, S.; West, Z. J.; Striebich, R. C.; Johnson, D. W., Chemical analysis of jet fuel polar, heteroatomic species via high-performance liquid chromatography with electrospray ionization–mass spectrometric detection. *Energy & fuels* **2013,** *27* (5), 2390-2398.
26. Agilent Technologies, I. Hydrophilic Interaction Chromatography Method Development and Troubleshooting 2018. https://www.agilent.com/cs/library/technicaloverviews/public/5991-9271EN_HILIC_method_development_TechOverview.pdf.
27. Arts, A. M.; Wrzesinski, P. J.; West, Z. J., An HPLC-ESI-QTOF Method to Analyze Polar Heteroatomic Species in Aviation Turbine Fuel via Hydrophilic Interaction Chromatography. *Journal of Chromatography A* **2024**, 464754.
28. Nelson, W., *Oil & Gas J.* **1955,** *31:269.*
29. Benavides, A.; Benjumea, P.; Cortés, F. B.; Ruiz, M. A., Chemical composition and low-temperature fluidity properties of jet fuels. *Processes* **2021,** *9* (7), 1184.

30. Cookson, D. J.; Lloyd, C. P.; Smith, B. E., Investigation of the chemical basis of kerosene (jet fuel) specification properties. *Energy & Fuels* **1987,** *1* (5), 438-447.
31. Cookson, D. J.; Smith, B. E.; Shaw, I. M., Determination of hydrocarbon compound class abundances in kerosine and diesel fuels: Chromatographic and nuclear magnetic resonance methods. *Fuel* **1987,** *66* (6), 758-765.
32. Cookson, D. J.; Iliopoulos, P.; Smith, B. E., Composition-property relations for jet and diesel fuels of variable boiling range. *Fuel* **1995,** *74* (1), 70-78.
33. Petrovic, K.; Vitorovic, D., A new method for the estimation of the freezing point of jet fuels based on the n-paraffin content. **1973**.
34. Cramer, J. A.; Hammond, M. H.; Myers, K. M.; Loegel, T. N.; Morris, R. E., Novel data abstraction strategy utilizing gas chromatography–mass spectrometry data for fuel property modeling. *Energy & fuels* **2014,** *28* (3), 1781-1791.
35. Vozka, P.; Modereger, B. A.; Park, A. C.; Zhang, W. T. J.; Trice, R. W.; Kenttämaa, H. I.; Kilaz, G., Jet fuel density via GC× GC-FID. *Fuel* **2019,** *235*, 1052-1060.
36. Shi, X.; Li, H.; Song, Z.; Zhang, X.; Liu, G., Quantitative composition-property relationship of aviation hydrocarbon fuel based on comprehensive two-dimensional gas chromatography with mass spectrometry and flame ionization detector. *Fuel* **2017,** *200*, 395-406.
37. Vozka, P.; Mo, H.; Šimáček, P.; Kilaz, G., Middle distillates hydrogen content via GC× GC-FID. *Talanta* **2018,** *186*, 140-146.
38. Flora, G.; Karimzadeh, F.; Kahandawala, M. S.; DeWitt, M. J.; Corporan, E., Prediction of Hydrocarbons Ignition Performances Using Machine Learning Modeling. *Available at SSRN 4664979*.
39. Cookson, D. J.; Smith, B. E., Calculation of jet and diesel fuel properties using carbon-13 NMR spectroscopy. *Energy & fuels* **1990,** *4* (2), 152-156.
40. Cookson, D. J.; Smith, B. E., Observed and predicted properties of jet and diesel fuels formulated from coal liquefaction and Fischer-Tropsch feedstocks. *Energy & fuels* **1992,** *6* (5), 581-585.
41. Cooper, J. B.; Larkin, C. M.; Schmitigal, J.; Morris, R. E.; Abdelkader, M. F., Rapid analysis of jet fuel using a handheld near-infrared (NIR) analyzer. *Applied Spectroscopy* **2011,** *65* (2), 187-192.
42. Cramer, J. A.; Morris, R. E.; Rose-Pehrsson, S. L., Use of genetic algorithms to improve partial least squares fuel property and synthetic fuel modeling from near-infrared spectra. *Energy & Fuels* **2010,** *24* (10), 5560-5572.
43. Morris, R. E.; Hammond, M. H.; Cramer, J. A.; Johnson, K. J.; Giordano, B. C.; Kramer, K. E.; Rose-Pehrsson, S. L., Rapid fuel quality surveillance through chemometric modeling of near-infrared spectra. *Energy & Fuels* **2009,** *23* (3), 1610-1618.
44. Fodor, G.; Mason, R.; Hutzler, S., Estimation of middle distillate fuel properties by FT-IR. *Applied Spectroscopy* **1999,** *53* (10), 1292-1298.

45. Gómez-Carracedo, M.; Andrade, J.; Calviño, M.; Prada, D.; Fernández, E.; Muniategui, S., Generation and mid-IR measurement of a gas-phase to predict security parameters of aviation jet fuel. *Talanta* **2003,** *60* (5), 1051-1062.

46. Wang, Y.; Ding, Y.; Wei, W.; Cao, Y.; Davidson, D. F.; Hanson, R. K., On estimating physical and chemical properties of hydrocarbon fuels using mid-infrared FTIR spectra and regularized linear models. *Fuel* **2019,** *255*, 115715.

47. Caswell, K. A.; Glass, T. E.; Swann, M.; Dorn, H. C., Rapid prediction of various physical properties for middle distillate fuels utilizing directly coupled liquid chromatography/proton nuclear magnetic resonance. *Analytical Chemistry* **1989,** *61* (3), 206-211.

48. COORDINATING RESEARCH COUNCIL, I. *STUDIES OF SCANNING BROOKFIELD VISCOMETRY AS A REPLACEMENT FOR FREEZING POINT IN AVIATION FUEL SPECIFICATIONS*; October 2013, 2013.

49. COORDINATING RESEARCH COUNCIL, I. *Fuel and Water Characterization in Support of the CRC Panel on Engine Component Deposits*; CRC Report No. AV-25-16; May 2021, 2021.

50. Wold, S.; Esbensen, K.; Geladi, P., Principal component analysis. *Chemometrics and intelligent laboratory systems* **1987,** *2* (1-3), 37-52.

51. Kendrick, E., A mass scale based on CH2= 14.0000 for high resolution mass spectrometry of organic compounds. *Analytical chemistry* **1963,** *35* (13), 2146-2154.

IASH 2024, the 18TH INTERNATIONAL CONFERENCE ON STABILITY, HANDLING AND USE OF LIQUID FUELS
Louisville, Kentucky, USA
September 8 – 12, 2024

INCREASING THE LEVEL OF DETAIL OF COMPOSITIONAL ANALYSES WITH GCXGC-FID/ QMS FOR PRESCREENING OF SUSTAINABLE AVIATION FUELS

Hannes Lüdtke*, Florian Pütz, Uwe Bauder, Markus Köhler, Andreas Huber, Patrick Oßwald

Institute of Combustion Technology, German Aerospace Centre

ABSTRACT

To evaluate sustainable aviation fuels (SAF) and other complex samples, non-targeted comprehensive analysis is required. The commonly selected method for this purpose is the two-dimensional gas chromatography (GCxGC) as it provides high peak capacities and generates detailed compositional information through multiple separation mechanisms. The obtained results are the basis for probabilistic property models that are able to estimate a number of key properties without the need for additional experimental measurements. Applications for this Prescreening methodology is preliminary feedback prior to approval processes and the possibility for support during the optimization and design of novel SAFs production routs. However, the quality of predictions is dependent on the degree of detail provided by the compositional data. GCxGC results are usually categorized for carbon number and group type. Since the different isomers are not accounted for, but can differ significantly in their properties, uncertainties arise when predicting the properties of a fuel.

The DLR GCxGC system is equipped with a dual detector setup comprising of a quadrupole mass spectrometer (qMS) and a flame ionization detector (FID). This ensures the capability of simultaneous identification of uncommon fuel compounds while maintaining a sensitive and universal quantification method. Nevertheless, MS proves to be not always sufficient for complete structural isomeric differentiation. In an attempt to improve the detail of classification through identification of smaller subgroups, an investigation on the correlation of structurally influencing parameters to the separation behavior was carried out. The impact of higher compositional detail on the prediction of different properties is discussed.

*Corresponding author: Hannes Lüdtke, hannes.luedtke@dlr.de

IASH 2024, the 18TH INTERNATIONAL CONFERENCE ON STABILITY, HANDLING AND USE OF LIQUID FUELS
Louisville, Kentucky, USA
September 8 – 12, 2024

SGMM - A NOVEL MODEL UTILIZING COMPOSITIONAL SUBGROUP INFORMATION FOR INCREASING ACCURACY OF FUEL PROPERTY PREDICTION IN PRESCREENING

Florian Pütz*, Hannes Lüdtke, Astrid Ramirez, Uwe Bauder, Patrick Le Clercq, Andreas Huber

DLR (German Aerospace Center), Pfaffenwaldring 38-40 Stuttgart, 70569 Germany

ABSTRACT

The prescreening process developed by Heyne, Rauch, Le Clercq, and Colket has become an important element in supporting development of new sustainable aviation fuel production pathways and increases the likelihood of successful ASTM approval. Prescreening involves the determination of the fuel composition at a molecular level using two-dimensional gas chromatography (GCxGC). Based on the composition, probabilistic machine learning models can then predict the properties of the fuel. However, a major limitation results from the inability of GCxGC to precisely resolve molecules on the level of structural isomers, which leads to serious disadvantages when predicting fuel properties. This limitation is particularly relevant for properties that differ significantly due to the isomeric structure of the molecules, such as freezing point and viscosity. Since these properties are strongly isomer dependent, the assumption that all known isomers can be present in the fuel has a particularly negative impact on both the inaccuracy and uncertainty of fuel property predictions.

To address this challenge, the German Aerospace Center (DLR) is developing a method to identify subgroups within GCxGC bins. Building on this, we introduce the novel SubGroupMeanMatrix (SGMM) model that integrates these subgroups into the property prescreening framework to improve prediction accuracy and reduce the prediction uncertainty. In this study, we explain the methodology behind this new model and evaluate its impact on the accuracy of property prediction for selected representative aviation fuels as well as limitations of the new model.

*Corresponding author: Florian Pütz, florian.puetz@dlr.de

IASH 2024, the 18TH INTERNATIONAL CONFERENCE ON STABILITY, HANDLING AND USE OF LIQUID FUELS
Louisville, Kentucky, USA
September 8 – 12, 2024

DEVELOPMENT AND EVALUATION OF RUGGED PORTABLE SENSOR FOR IN-FIELD MEASUREMENT OF CORROSION INHIBITOR/LUBRICITY IMPROVER ADDITIVE

N. Scott Bergey*[1], Corrianna R. Boucher[2], Eric M. Johnson[2], Lena E. Flynn[1], John J. Buffin[1], Terrence W. Dickerson[1], Richard A. Kamin[1]

[1]*U.S. Navy, 22229 Elmer Rd. Bldg. 2360 Patuxent River, Maryland 20670 USA*
[2]*Peraton*

ABSTRACT

The US Navy has developed and tested an in-line field additization cart that injects the three required F-24 additives into unadditized D1655 fuel. Fuel system icing inhibitor and static dissipator additive both have readily field-portable measurement techniques available to validate that the in-line injection correctly blended the additive to its specified concentration as required in AFLP-3747. Since there is currently no portable measurement technique to measure corrosion inhibitor/lubricity improver (CI/LI) additive, the US Navy has spearheaded development of a simple, portable, and inexpensive field kit to reliably quantify CI/LI. This work showcases the final prototype kit and the processes used to optimize and simplify the measurement technology to maintain viability across challenging environmental conditions.

*Corresponding author: N. Scott Bergey, scott.bergey1@navy.mil

IASH 2024, the 18TH INTERNATIONAL CONFERENCE ON STABILITY, HANDLING AND USE OF LIQUID FUELS
Louisville, Kentucky USA
8 – 12 September 2024

PREDICTIVE MODELLING OF LIQUID DENSITY AND SURFACE TENSION FOR SYNTHETIC AVIATION FUELS USING NUCLEAR MAGNETIC RESONANCE ATOM TYPE CONTRIBUTIONS

Robert P. Parker, Mark Kelly, Tiarnán Watson-Murphy, Stephen Dooley

School of Physics, Trinity College Dublin, Ireland
parkerro@tcd.ie,

ABSTRACT

Prescreening of Synthetic ("Sustainable") Aviation Fuel (SAFs) is becoming regarded as an essential technique to secure the early-stage development of SAF candidates. Prescreening refers to the study of small volumes of SAF at the early stages of product development for the purpose of assessing the compatibility of the SAF candidate with the American Society for Testing and Materials (ASTM D4054) evaluation procedure for new aviation turbine fuels. The basis of the prescreening process relies on the availability of numerical models that can predict the fit-for-purpose properties of the ASTM evaluation procedure without the volumes of the SAF candidate needed to perform all the necessary measurements being available.

In this work, such models are developed that predict two important fit-for-purpose ASTM D4054 properties, the temperature dependent liquid density and temperature dependent surface tension of complex hydrocarbons mixtures. The basis of the method is the determination of atom type compositions by a ^{1}H ^{13}C Heteronuclear Single Quantum Coherence (HSQC) Nuclear Magnetic Resonance (NMR) spectroscopy methodology.

The dependence of each of, temperature dependent liquid density, and temperature dependent surface tension, on the range of atom types that comprise typical aviation fuel hydrocarbons is learned by the multiple linear regression modelling to a database of pure component and pure component mixture data gathered from the literature. A data base of 1,241 entries is gathered for density, and a data base of 1,260 entries gathered for surface. Each database is divided in an 80:20 ratio for model training and model testing. Extensive iterations of model training and testing objectively identify the same seven atom types as being most influential to each property. The collection of seven atom types, in addition to a temperature dependent term, is shown to be sufficient to determine each property quite precisely.

The resulting models are then coupled with the ^{1}H ^{13}C HSQC NMR determined atom type composition of two real fossil derived aviation fuels, four real synthetic aviation fuels and four reference gasoline fuels. Impressively, the overall end-to-end methodology shows high performance in predicting the unseen liquid density of these real fuels to an absolute error range of 0.00 – 5.59 %, and the unseen surface tension to an absolute error range of 0.29 - 4.41%. The methodology is also shown to be capable of predictively differentiating between the relative magnitude of each property over this set of real fuels. It is concluded that the method holds utility for prescreening applications.

Keywords: Synthetic aviation fuels, ASTM D 4054, Prescreening, Fuel properties, Density, Surface tension, NMR, Atom types.

1. INTRODUCTION

Synthetic ("Sustainable") Aviation Fuel (SAFs), produced using carbon neutral feedstocks, are important in mitigating the environmental impact of aviation, reducing greenhouse emissions, and enabling long-term sustainability in the industry. Introducing new aviation fuels and fuel additives in accordance with ASTM D4054-23 [1] (American Society for Testing and Materials) poses challenges for companies due to the number of fuel property measurements and the high volume of fuel required by the evaluation process, presenting a high economic barrier to entry and a deterrent for investment.

Prescreening is a novel technique where the properties of the candidate SAF can be predicted using statistical regression or other mathematical models based on some form of compositional information. Heyne et al. [2] have summarized six key properties to which prescreening should be first applied to aid in eventual fuel qualification. These include density at 15°C and surface tension at 22°C. Density is the ratio of mass to volume; this is a critical property for fuel energy density, atomisation and other important behaviours.

Surface tension is the energy required to increase the surface area of a liquid. Cohesive forces hold molecules together in a drop, while adhesive forces arise between the liquid and the interface [3]. Surface tensions crucially affects the droplet size and atomization dynamics of fuels in gas turbine combustor, in turn impacting the operability and efficiency of the engine [2].

Ideally, prescreening methods use millilitres of the fuel, allowing for reduced expenditure in initial SAF development. Heteronuclear Single Quantum Coherence (HSQC) Nuclear Magnetic Resonance (NMR) spectroscopy has potential in this regard. HSQC-NMR can identify and quantify the atom types, which are the common fundamental sub-structures comprising molecules or mixtures of molecules. Atom types are localised structures within a molecule characterised by a specific electronic environment. With regard to analysis of fuels, HSQC-NMR has the advantage over chromatographic techniques that it does not require separation, identification, and subsequent quantification of each and every molecule comprising the fuel. The fuel can be examined in situ, with sub-parts per million detection precision available if desired.

Group contribution methods are techniques that estimate the properties of pure components through the summation of contributions from specific molecular groups or structures within the compound [4-7]. Predictive models for density that utilise group contribution methods already exist in the literature [8-13]. Skander et al. [11] conducted a study which used molecular structures and a correction term based on the relative structure location to CH_3 groups and carbon atom placement in ring structures for the estimation of physical properties of hydrocarbons, including liquid density at 20°C. The resulting model predicted liquid density of all hydrocarbon classes with an Average Absolute Deviation (AAD) of 1.3%. Wakeham et al. [13] utilised carbon atoms/bonds and molecular energies within hydrocarbons to estimate the liquid density and critical properties with an Average Relative Error (ARE) of 1.17% for liquid density at 20°C.

The group contribution methods of all prior works available have been concerned with the estimation of liquid density of pure hydrocarbons and not their mixtures, nor the very complex mixtures that would be typical of SAF candidates.

Similarly, group contribution methods for the estimation of surface tension are available in the literature [9, 14-19]. Gharagheizi et al. [15] conducted a study where the surface tension for ionic liquids were computed based on their chemical structures. These structures were defined by the anions and cations within the ionic liquids. After evaluating 34 different chemical structures, they found that only 19 were effective for accurately estimating surface tension. The model's performance was characterised by an Average Absolute Relative Deviation (AARD) of 3.20. This study of Gharagheizi et al. represents the first instance in literature where surface tension estimation of a liquid relied solely on group contributions and temperature.

Building upon this approach, Shahsavari et al. [19] predicted the surface tension of ionic liquids using a group contribution method based on the amount of each constituent element present, rather than chemical structures. The resulting model displayed an AARD of 4.16%. These studies demonstrate that estimating surface tension through group contributions is feasible, however, its application is limited to known samples with discrete quantifiable chemical structures.

To date, group contribution methods have not been applied to the prediction of density or surface tension of fuel-like mixtures, presumably for the reason of the difficulty in assigning molecular group fractions to mixtures of such complexity. In this regard, we have previously developed the ^{1}H ^{13}C HSQC NMR methodology for the purpose of atom type analysis of fuels [20-22]. It has been shown that this methodology can identify and precisely quantify ten discrete atom types most prevalent in complex hydrocarbon fuels [22].

In this study, multilinear regression (MLR) models, based on atom type group contributions, are developed for liquid density and surface tension for the specific use case of aviation fuel complex mixtures.

2. METHODOLOGY

The basis of the methodology is the assumption that fuel properties are principally dependent on the molecular structure of a fuel which can be measured by NMR, and that this relationship can be learned by mathematical analysis of an appropriate database of molecular structures for each property (i.e., in this study, temperature dependent liquid density and surface tension).

There are three technical conditions that must be met when employing MLR for this purpose: (1) The atom types which most significantly contribute to surface tension and density must be identifiable through the ^{1}H ^{13}C HSQC NMR spectrum of the real fuel requiring analysis. (2) There must be property data for pure components and/or pure component mixtures of sufficient quality and quantity to train the numerical model. (3) The resulting operating equation must be concise and only contain statistically significant parameters. This approach enhances our understanding of the importance of, and interactions between, atom types, the molecules they comprise, and the "physical" properties they define.

2.1 Atom Types by ^1H ^{13}C HSQC NMR Spectroscopy

The atom types utilized in this study are presented in Figure 1. The method of Ure et al. has been adapted to additionally enable the identification and quantification of aromatic-ring junction carbon atom types [23]. In order to validate the full procedure, a set of real fuels with unknown atom type compositions were analysed. HSQC, quantitative ^1H, and qualitative ^{13}C NMR spectra were collected using a Bruker Avance III 400 NMR spectrometer with the parameters suggested by Ure et al. [22]. Following Ure et al., the H/C ratio of each fuel was determined using an Exeter Analytical CE 440 Elemental Analyser and used to deduce the quantity of quaternary paraffinic carbon present. The *^1H ^{13}C HSQC* spectrum was used to assign spectral regions for each atom type independently for each sample. These regions were then integrated in the quantitative ^1H spectrum to determine the atom type composition in mol% hydrogen, which was then used to calculate the inferred mol% carbon composition using the extended method of Watson-Murphy et al. [23]. As an example of the result, a typical acquired spectrum is given in Figure 2 for POSF 7629 Sasol Iso-Paraffinic Kerosene (IPK) derived from Coal [24], and its atom type assignment in Table 1.

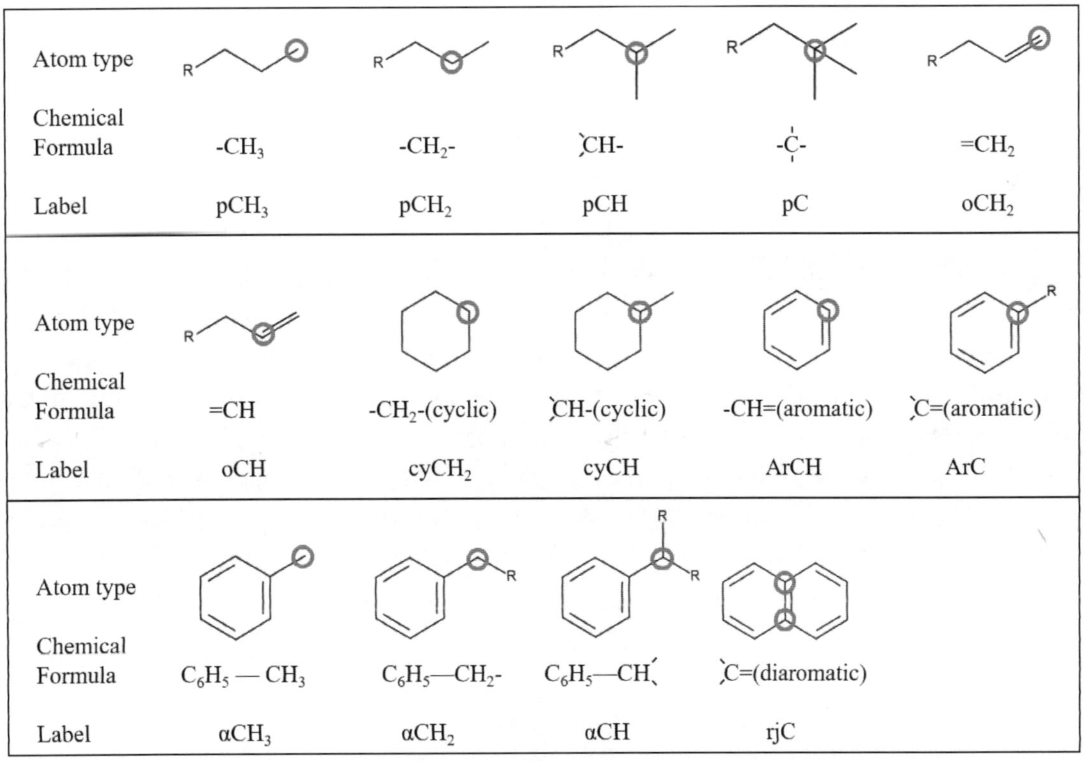

Figure 1. Jet aviation fuel atom type formula classification and structure, with shorthand nomenclature.

Additionally, this methodology was applied to the real aviation and gasoline fuels available in our laboratory: POSF 5729 Shell Fischer-Tropsch Synthetic Paraffinic Kerosene derived from natural gas [24], POSF 7720 Hydrotreated Esters And Fatty Acid (HEFA) derived from Camelina oil [24], POSF 10151 Gevo Alcohol-to-Jet (ATJ) derived from iso-butanol, and the CRC Face Gasolines A, F, C, I which are each a different compositional design [25].

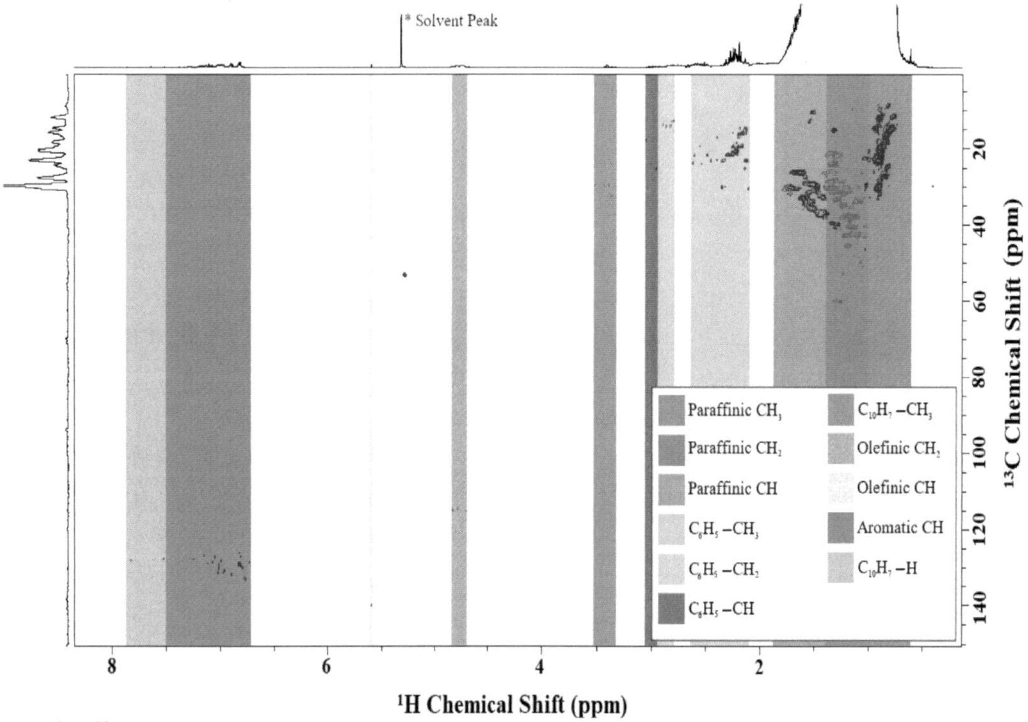

Figure 2. *^1H ^{13}C HSQC NMR* spectrum of POSF 7629 Sasol IPK derived from coal.

2.2 *Data, Regression & Models*

To train the MLR model, density and surface tension data of experimental measurements was gathered by critical examination of suitable literature for both pure components and their mixtures. It contains 2,501 entries [26-46], generally comprised of: normal paraffins, iso-paraffins, cycloparaffins, aromatics and binary mixtures of these classes. For model generation a discrete database is utilised for each property. Each entry in each database is expressed by its atom type composition, drawing from the palette of Figure 1, alongside the associated temperature at which the liquid density or surface tension was measured. MLR models were produced with operating equations of the form,

$$y = a_0 + \sum_{n=1}^{N} a_n x_n \qquad [1]$$

where α_0 is the intercept, α_n are the coefficients learned for each independent variable, x_n are the atom type mole fractions or temperature, and the dependent variable, y, is liquid density or surface tension. Individual models for temperature dependent liquid density and temperature dependent surface tension were then trained using the Linear Regression method from the SciKit Learn Python package [47].

The 1,241 member density and 1,260 member surface tension datasets were both divided in an 80:20 ratio of training to testing data. The training datasets train the MLR models, and the testing datasets validate the performance of the models, ensuring they have been reliably trained. The independent variables chosen to describe each property were systemically evaluated using the p-value of each coefficient. The p-value is a statistical indicator that measures the probability of the observed data occurring by chance under the assumption that there is no actual dependence effect. A low p-value (<0.05) for a coefficient implies a significant impact of a particular atom type over the range of fractions the atom type occurs in the training data set, on liquid density or surface tension in this context. Each model was developed by the backward selection methodology, where the independent variables with a p-value greater than 0.05 were singularly removed and the model was retrained with fewer atom types until the coefficient of determination (R^2) degraded significantly, at which point the specific atom type was retained. This process ensures only the variables with a statistically confirmed influence are considered, leading to more accurate, more concise, and more reliable predictive models. The models presented in Figures 3 and 4, and Table 2, are those that satisfy the criteria regarding p-value, whilst achieving the highest coefficient of determination of all models generated.

2.3 Model Validation

True validation data i.e., density or surface tension measurements for real or even realistic (e.g., model fuels, surrogates) complex liquid fuels, is sparse. A further challenge in demonstrating validation arises from the novelty of the atom type NMR methodology, but this is also a limitation of other prescreening approaches such as GCxGC [48]. Atom type assignments, particularly as obtained through the reliable demonstrated Ure et al. method of this study, are not available in the existing literature. As a result, studies of this nature are constrained by the availability of fuel samples within the laboratory, the measurement capability of the laboratory, and/or the availability of data on the property of interest from the literature. For example, for the real fuel samples (gasoline, kerosene, diesel) available at Trinity College Dublin, liquid density or surface tension data is exclusively found in Edwards et al. [49, 50] for (synthetic) aviation fuels, and that of the CRC Fuels for Advanced Combustion Engines (FACE) reports for gasoline-type fuels [25]. The confluence of these logistical limitations, which are inherent to any other property prediction methodology, results in the fuel selections as listed above, for study of the properties density and surface tension.

3 RESULTS AND DISCUSSION

Table 1. Operating equation declaring learned atom types, learned coefficients (α) and associated p-values for temperature dependent density and surface tension models. p-values <0.05 indicate high statistical significance. Atom type mole fraction (X).

Equation	$\alpha_0 + \alpha_1 T + \alpha_2 x_{pCH3} + \alpha_3 x_{pCH} + \alpha_4 x_{pC} + \alpha_5 x_{cyCH2} + \alpha_6 x_{cyCH} + \alpha_7 x_{ArCH} + a_8 x_{rjC}$									
Models		α_0	α_1	α_2	α_3	α_4	α_5	α_6	α_7	α_8
Density, ρ [kg/m³]	coefficients	1101.0	-0.9	-556.9	539.2	995.8	-69.9	430.1	28.3	771.4
	p-values	<0.01	<0.01	<0.01	<0.01	<0.01	<0.01	<0.01	<0.01	<0.01
Surface Tension, σ [mN/m]	coefficients	61.9	-0.1	-36.7	30.2	58.2	-6.3	14.4	-2.8	49.9
	p-values	<0.01	<0.01	<0.01	<0.01	<0.01	<0.01	<0.01	<0.01	<0.01

The temperature dependent liquid density (ρ) and temperature dependent surface tension (σ) regression model coefficients with the respective p-values are provided in Table 1 (aka "the models").

The first and compelling result is that the best performing models for both liquid density and surface tension contain the same set of atom types as learned independent variables. This implies that the same atom types are principally influential for these physical properties. Secondly, and strikingly, the p-values of all the variables are below 0.01, demonstrating the strong influence of each parameter in the operating equation. This observation was specifically interrogated and confirmed to be true.

The predictive performance of each model was then assessed by evaluation against two discrete databases; (1) The training and testing dataset discussed previously comprised of pure components and pure component mixtures from the literature; and (2) The reference real fuels available at Trinity College Dublin as outline above.

3.1.1 Atom Type Composition of Real Fuels

To evaluate model performance against (2), atom type data is needed. The results of the *^{1}H ^{13}C HSQC NMR* assignment are provided in Table 2 and graphically in Figure 3, which we propose as a useful way to simplify complex fuel compositions, allowing visual inspection of "what a fuel looks like" to the eye. The conventional fossil aviation fuels such as Jet-A POSF 10325 and JP-5 POSF 10289 have similar atom type distributions with a large quantity of paraffinic atom types, and a notable fraction of aromatic and cycloparaffinic types. The synthetic aviation fuels have a

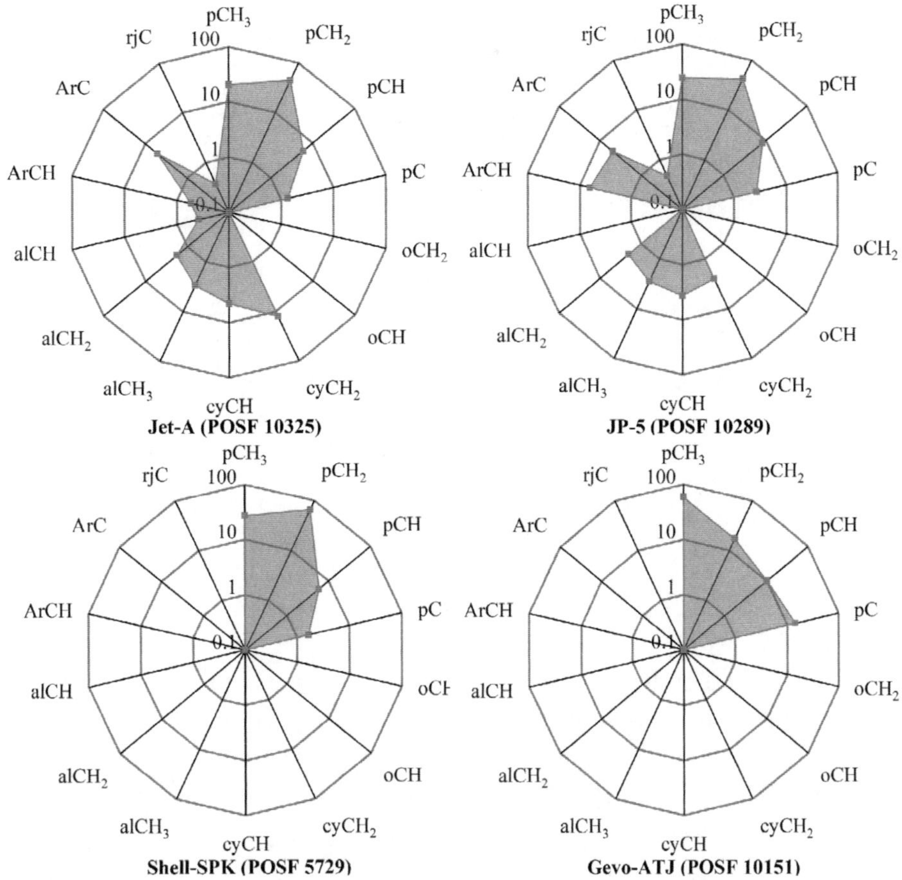

Figure 3. Atom type mole fraction on a log-radar plot for select fossil aviation fuels and synthetic aviation fuels, showing considerable compositional differences.

similarly large quantity of paraffinic atom types, but do not contain as high levels of aromatics, and strikingly no cycloparaffins for Shell-SPK POSF 5729 and Gevo-ATJ POSF 10151.

For the FACE gasolines, the compositional spectrum is mostly consistent with the detailed hydrocarbon analysis performed by gas chromatography in the literature [25]. Each gasoline has a significant amount of n-paraffins and i-paraffins and each unique fuel design structure is captured.

3.1.2 Performance to Validation Dataset #1: 80:20 Split Training Set

Table 2. Atom type compositions determined using HSQC and quantitative ^{1}H NMR spectroscopy for real synthetic and fossil aviation fuels and reference FACE gasolines.

Sample	POSF	pCH_3	pCH_2	pCH	pC	oCH_2	oCH	$cyCH_2$	cyCH	αCH_3	αCH_2	αCH	ArCH	ArC	rjC
Jet A	10325	20.34	44.09	5.97	1.33	-	-	12.54	4.54	2.97	1.79	0.39	0.54	5.14	0.37
JP-5	10289	23.99	42.21	8.38	2.69	-	-	2.5	3.69	2.79	2.05	-	6.4	4.84	0.47
FACE A	-	53.11	23.87	15.32	5.77	-	-	0.82	0.41	0.10	-	-	0.51	0.10	-
FACE F	-	40.44	19.22	10.13	6.79	1.41	1.41	10.74		2.16	0.19		5.18	2.34	-
FACE C	-	52.27	26.13	13.61	4.80	-	-	-	-	1.03	0.31	-	0.52	1.34	-
FACE I	-	44.46	33.61	12.24	4.65	1.01	1.01	1.82	0.71	-	-	-	0.51	-	-
Shell-SPK	5729	27.11	65.30	5.83	1.64	-	-	-	-	0.03	-	-	0.06	0.03	-
HRJ-Camelina	7720	29.64	60.30	8.71	1.17	-	-	-	-	0.03	0.01	-	0.09	0.05	-
Gevo-ATJ	10151	58.46	17.02	10.09	14.3	0.05	-	-	-	0.01	-	0.02	0.02	0.03	-
Sasol-IPK	7629	48.68	27.17	17.57	5.69	0.02	0.01	-	-	0.21	0.07	0.03	0.22	0.32	0.01

The predictive performance of the models using 80% ("seen") training data and 20% ("unseen") testing data is shown in Figure 4 and 5 for density and surface tension respectively.

$$\rho = \alpha_0 + \alpha_1 T + \alpha_2 x_{pCH3} + \alpha_3 x_{pCH} + \alpha_4 x_{pC} + \alpha_5 x_{cyCH2} + \alpha_6 x_{cyCH} + \alpha_7 x_{ArCH} + a_8 x_{rjC}$$

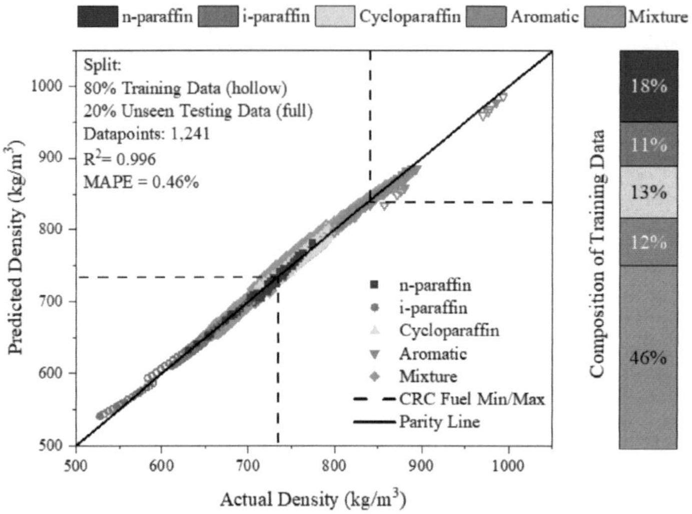

Figure 4. Temperature dependent density, predicted density versus actual measured density with an 80:20 split in training (seen data, hollow symbols) and testing (unseen data, full symbols) data respectively.

$$\sigma = \alpha_0 + \alpha_1 T + \alpha_2 x_{pCH3} + \alpha_3 x_{pCH} + \alpha_4 x_{pC} + \alpha_5 x_{cyCH2} + \alpha_6 x_{cyCH} + \alpha_7 x_{ArCH} + a_8 x_{rjC}$$

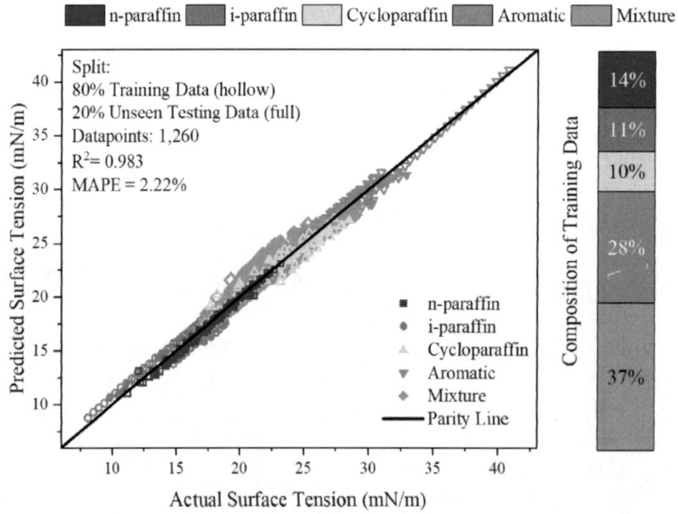

Figure 5. Temperature dependent surface tension, predicted surface tension versus actual surface tension with an 80:20 split in training (unseen data, hollow symbols) and testing (seen data, full symbols) data respectively.

For density, the model predicts unseen data with a mean average percentage error of 0.47%, closely similar to its ability to replicate the training data (0.3%). This performance is considered excellent. Similarly, the surface tension model exhibits impressive replication of unseen data, showing a mean average percentage error of 2.22%, comparable to the replication achieved for the surface tension training data (1.7%).

3.1.3 Performance to Validation Dataset #2: Reference Real Aviation Fuels and FACE Gasolines

The performance of the density model when predicting the density of reference real aviation fuels [50] and FACE [25] gasolines is summarised in Table 3 and Figure 6. It is important to appreciate that these tests are entirely unseen, for these fuels, the model only has knowledge of the NMR determined atom types. The comparisons thus present a meaningful demonstration of the effectiveness of the entire methodology, metrology, and underpinning fundamental concepts.

Table 3. Performance analysis of machine learned density model produced in this work for FACE gasolines at 15 °C [25][a] and synthetic aviation fuels at 16 °C [50][b] from literature. The predictions were made using atom types quantified by Watson-Murphy et al.

Sample	POSF	Actual Density [kg/m^3]	Model Density [kg/m^3]	Absolute Error in Density	Model Surface Tension [mN/m]
Jet A	10325	803[b]	805	0.21 %	24.9
JP-5	10289	826[b]	825	0.23 %	26.1
FACE A	-	685[a]	675	1.52 %	17.1
FACE F	-	707[a]	723	2.29 %	19.9
FACE C	-	691[a]	678	1.89 %	17.3
FACE I	-	697[a]	714	2.56 %	19.6
Shell-SPK	5729	737[b]	740	0.39 %	21.6
HRJ-Camelina	7720	757[b]	757	0.00 %	22.7
Gevo-ATJ	10151	761[b]	718	5.59 %	19.1
Sasol-IPK	7629	760[b]	728	4.23 %	20.5

The density model demonstrates a strong predictive capability for the density of the reference fuels, with an average error of 1.96% for densities at 15 and 16°C, Table 3.

The predictive performance of the density model against fossil jet aviation fuels is analysed in Figure 6. The density model is evaluated using the only temperature dependent experimental data available, that from Edwards et al. for Jet-A POSF 10325 and JP-5 POSF 10289 [49] showing an average error of 0.44%, and a maximum error of 1.18% at 65°C for JP-5 POSF 10289. Importantly, Figure 6 shows that the overall methodology would predict both fuels to be inside the limits presented by the ASTM D1655 standard specification for aviation turbine fuels [51]. Such a verification is the purpose of prescreening.

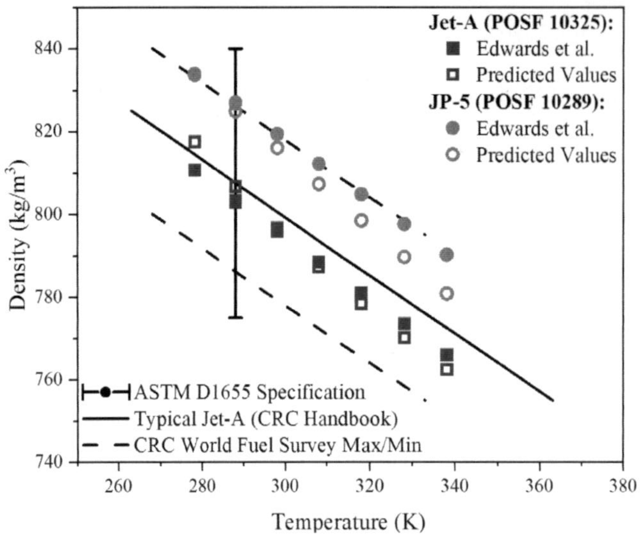

Figure 6. Liquid density versus temperature, literature experiment (full symbols) and unseen model predictions of this study (hollow symbols). The ASTM fit-for-purpose properties are included as lines for reference. The ASTM D1655 specification is used as a qualification (limit) metric in the initial ASTM D 4054 evaluation.

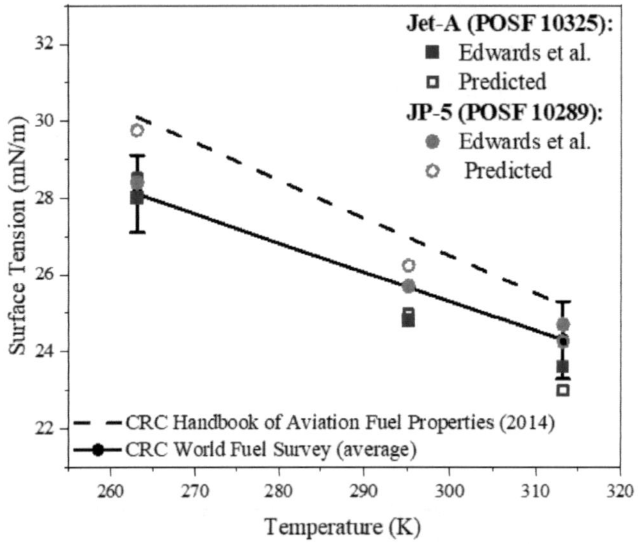

Figure 7. Surface tension versus temperature, literature experiment (full symbols) and unseen model predictions of this study (hollow symbols). For reference, the ASTM fit-for-purpose properties are included as lines. There are no specification limits for surface tension in the ASTM D4054 evaluation.

With regard to surface tension, much less experimental data is available for true unseen testing. Of the fuels analysed by *^{1}H ^{13}C HSQC NMR*, only Jet-A POSF 10325 and JP-5 POSF 10289 have measurements reported. The surface tension predictions for the other reference fuels are supplied in Table 3, showing the aviation fuels to be in the range 17-22 mN/m at 22°C. This is slightly lower than the range specified in the CRC world fuel survey, and within the range to conform with ASTM D4054, Standard Practice for Evaluation of New Aviation Turbine Fuels and Fuel Additives.

Moreover, Figure 7 presents direct comparison of entirely unseen model predicted temperature dependent surface tension to the measurements of Edwards et al. for Jet-A POSF 10325 and JP-5 POSF 10289. The model demonstrates a strong predictive capability, showing an average error of 2.3%. More specifically, an error of 0.29-2.99% for Jet-A POSF 10325 and 1.68-4.41% for JP-5 POSF 10289 demonstrates the potential of the overall technique.

3.2 Compositional Space of the Training Data

Figures 6 and 7 indicate that while both models do achieve a useful predictive capability, the models do not perfectly capture the temperature-dependent density and temperature dependent surface tension behaviours of the JP-5 in particular. Of curiosity, is that the quality of the model prediction deviates with fuel identity for both density and surface tension. Though uncertainty in the experimental measurement should not be discounted as a source of this discrepancy, it is interesting to analyse the models in this regard.

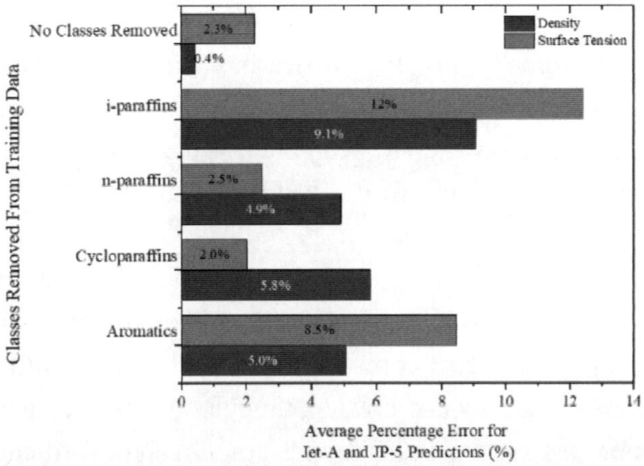

Figure 8. Average percentage error in the model prediction of temperature dependent density and temperature dependent surface tension of Jet-A POSF 10325 and JP-5 POSF 10289, relative to experiment, for different descriptions of the compositional space. The y-axis denotes the singular absence of each molecular class from the training data.

Here, the surface tension model in particular, shows less accuracy, which can be indicative of an inherently greater complexity in this phenomena compared to density (Figure 4 vs, Figure 5). Despite these shortcomings, the predictions for both properties are sufficiently accurate to classify the analysed fuels as “fit-for-purpose”, thus fulfilling the primary objective of this study.

Obviously, the reference real fuels against which the predictive ability of the models is ultimately analysed, are highly complex fluids comprised of several hundred to a few thousand individual molecules. This immediately suggests a need for a more diverse set of complex mixtures in the training data. Inherent to the methodology set out in this study, is the mathematical fact, that the models produced cannot be more accurate than the precision to which they replicate the data with which they are trained. In the ideal scenario, a model would be trained to a training data set that comprehensively and evenly covers the compositional space of the fluids to be analysed. In practice, this is difficult owing to the vastness of the potential compositional space offered by the wide variety of conventional and synthetic aviation fuels, and to the time and cost requirement in making such experimental determinations for reference pure components mixtures.

From Figure 4, the large variabilities in the composition space occupied by the real reference fuels becomes obvious, particularly the lower prevalence of aromatic and cycloparaffinic atom types in the synthetic aviation fuels.

Both factors provoke analysis of how important coverage of the compositional space actually is. To inform this, the influence of the compositional space on the performance of the models is studied. Model training by multiple linear regression was performed again, but with individual molecular classes removed from the training set, and the performance of the resulting model assessed, all other variables held constant. Figure 8 shows the results against density and surface tension data for Jet-A POSF 10325 and JP-5 POSF 10289

In particular, for both density and surface tension, the removal of isomerised paraffins and aromatics have a considerable degrading effect on the predictive ability, emphasizing their importance.

With respect to density, the removal of any hydrocarbon class leads to a considerable, but more consistent increase in the model error, of the order of 5%. The extent of degradation for surface tension is more variable according to which molecular class training data are removed. This feature further points to the more complex nature of the surface tension physical phenomenon, over that of liquid density.

The analysis evidences the importance of having a comprehensive coverage in molecular class/atom type training data, and coverage in as even number density as is practical, in order to accurately predict these properties of real complex fuels, by this, and similar methodologies.

Furthermore, and of fundamental scientific value, though density and surface tension are usually labelled as "physical" properties as no covalent bonds are broken or formed, the analysis presented here makes clear that these phenomena are chemical (i.e., dependent on molecular structure) in nature.

4. CONCLUSION

Predictive models for liquid density and surface tension have been created and their applicability to prescreening of synthetic ("sustainable") aviation fuel candidates has been investigated. These models, produced through multiple linear regression, utilise atom type mole fractions that identified and quantified by a ^{1}H ^{13}C Heteronuclear Single Quantum Coherence (HSQC) Nuclear Magnetic Resonance (NMR) spectroscopy methodology. A 1,241 member and 1,260 member database of density and surface tension measurements was constructed from the literature.

This data was used to train and test simple numerical models by which temperature dependent density, temperature dependent surface tension can be described on the basis of atom types.

The liquid density model displays an excellent predictive performance of unseen pure components and pure component mixtures with an average error of 0.47%.

This model also predicts the density of complex liquid fuels such as real fossil and synthetic aviation fuels, and gasolines to within 5.59 %.

The surface tension model similarly has an excellent predictive performance showing an average error of 2.22% to unseen pure components and pure component mixtures and predicting the surface tension of real aviation fuels to within 4.41%.

The models presented in this work can be improved by including more complex mixtures in the dataset and improving the coverage of the training data set in compositional space. However, doing so this poses a significant challenge, as it necessitates extensive experimental efforts and the procurement of suitable fuel samples for the dataset.

ACKNOWLEDGEMENTS:

This publication has emanated from research conducted with philanthropic financial support of the Ryanair Sustainable Aviation Research Center at Trinity College Dublin, the European Union through the European Research Council, Mod-L-T, action number 101002649 and the European Regional Development Fund under the AMBER award. RP is funded by a SURE undergraduate research fellowship from Trinity College Dublin. The authors would like to acknowledge Dr John O'Brien and Dr Manuel Ruether, who provided support for NMR.

REFERENCES:

1. Committee D. ASTM D4054: Standard Practice for Evaluation of New Aviation Turbine Fuels and Fuel Additives. Conshohocken, PA: ASTM International2023.
2. Heyne J, Rauch B, Le Clercq P, Colket M. Sustainable aviation fuel prescreening tools and procedures. Fuel. 2021;290:120004.
3. Petrucci RH. General chemistry : principles and modern applications. 10th ed. Toronto, Ont.: Pearson Canada Toronto, Ont.; 2011.
4. Ambrose D. Correlation and estimation of vapour-liquid critical properties : I. critical temperatures of organic compounds / by D. Ambrose. National Physical Laboratory . Division of Chemical S, editor. Teddington: Teddington : National Physical Laboratory; 1978.
5. Constantinou L, Gani R. New group contribution method for estimating properties of pure compounds. AIChE Journal. 1994;40(10):1697-710.
6. Joback KG, Reid RC. Estimation of Pure-Component Properties From Group-Contributions. Chemical Engineering Communications. 1982.
7. Lydersen AL, Station EE, Greenkorn RA, Hougen OA. Estimation of Critical Properties of Organic Compounds by the Method of Group Contibutions: University of Wisconsin; 1955.
8. Carreón-Calderón B, Uribe-Vargas V, Ramírez-Jaramillo E, Ramírez-de-Santiago M. Thermodynamic Characterization of Undefined Petroleum Fractions using Group Contribution Methods. Industrial & Engineering Chemistry Research. 2012;51(43):14188-98.
9. Mishra S, Bukkarapu KR, Krishnasamy A. A composition based approach to predict density, viscosity and surface tension of biodiesel fuels. Fuel. 2021;285:119056.

10. Paduszyński K. Extensive Databases and Group Contribution QSPRs of Ionic Liquids Properties. 1. Density. Industrial & Engineering Chemistry Research. 2019;58(13):5322-38.
11. Skander N, Chitour CE. A New Group-Contribution Method for the Estimation of Physical Properties of Hydrocarbons. Oil & Gas Science and Technology - Revue d'IFP Energies nouvelles. 2002;57(4):369-76.
12. Stefanis E, Constantinou L, Tsivintzelis I, Panayiotou C. New Group-Contribution Method for Predicting Temperature-Dependent Properties of Pure Organic Compounds. International Journal of Thermophysics. 2005;26(5):1369-88.
13. Wakeham WA, Cholakov GS, Stateva RP. Liquid Density and Critical Properties of Hydrocarbons Estimated from Molecular Structure. Journal of Chemical & Engineering Data. 2002;47(3):559-70.
14. Deng J, Zhang Y, Jia G. Surface tension of binary and ternary mixtures mapping with ASP and UNIFAC models based on machine learning. Physics of Fluids. 2023;35(6).
15. Gharagheizi F, Ilani-Kashkouli P, Mohammadi AH. Group contribution model for estimation of surface tension of ionic liquids. Chemical Engineering Science. 2012;78:204-8.
16. Mulero A, Cachadiña I, Cardona LF, Valderrama JO. Pressure–Surface Tension–Temperature Equation of State for n-Alkanes. Industrial & Engineering Chemistry Research. 2022;61(9):3457-73.
17. Naef R, Acree WE. Calculation of the Surface Tension of Ordinary Organic and Ionic Liquids by Means of a Generally Applicable Computer Algorithm Based on the Group-Additivity Method. Molecules. 2018;23(5):1224.
18. Randová A, Bartovská L. Group contribution method: Surface tension of linear and branched alkanes. Fluid Phase Equilibria. 2016;429:166-76.
19. Shahsavari S, Mesbah M, Soroush E, Farhangian H, Alizadeh S, Soltanali S. A simple group contribution correlation for modeling the surface tension of pure ionic liquids. Journal of Molecular Liquids. 2018;265:292-8.
20. Abdul Jameel AG, Naser N, Emwas A-H, Dooley S, Sarathy SM. Predicting Fuel Ignition Quality Using 1H NMR Spectroscopy and Multiple Linear Regression. Energy & Fuels. 2016;30(11):9819-35.
21. Dussan K, Won SH, Ure AD, Dryer FL, Dooley S. Chemical functional group descriptor for ignition propensity of large hydrocarbon liquid fuels. Proceedings of the Combustion Institute. 2019;37(4):5083-93.
22. Ure AD, O'Brien JE, Dooley S. Quantitative NMR Spectroscopy for the Analysis of Fuels: A Case Study of FACE Gasoline F. Energy and Fuels. 2019.
23. Watson-Murphy T, Ure AD, Ghaani MR, Dooley S, editors. Machine Learning and HSQC-NMR for viscosity prediction of sustainable aviation fuel. American Chemical Society; 2023; Indianapolis, United States of America,.
24. Corporan E, Edwards T, Shafer L, DeWitt MJ, Klingshirn C, Zabarnick S, et al. Chemical, Thermal Stability, Seal Swell, and Emissions Studies of Alternative Jet Fuels. Energy & Fuels. 2011;25(3):955-66.
25. CRC. FACE GASOLINES AND BLENDS WITH ETHANOL: DETAILED CHARACTERIZATION OF PHYSICAL AND CHEMICAL PROPERTIES. COORDINATING RESEARCH COUNCIL, INC.; 2014. Contract No.: CRC Report No. AVFL-24
26. Abdussalam AA, Ivaniš GR, Radović IR, Kijevčanin ML. Densities and derived thermodynamic properties for the (n-heptane+n-octane), (n-heptane+ethanol) and (n-

octane+ethanol) systems at high pressures. The Journal of Chemical Thermodynamics. 2016;100:89-99.
27. Aguila-Hernandez J, Hernandez I, Trejo A. Temperature dependence of the surface tension for binary mixtures of n-butanenitrile+ n-alkanes. International journal of thermophysics. 1995;16:45-52.
28. Aliaj F, Gjevori A, Llozana A, Syla N, Zenuni G, Arbneshi T, et al. Thermophysical Properties of the 2-Methylpropan-1-ol + Cyclohexane + Benzene Ternary System and Its Binary Subsystems Within the Temperature Range (293.15–333.15) K and Under Ambient Pressure. International Journal of Thermophysics. 2023;44(7):113.
29. Aralaguppi MI, Jadar CV, Aminabhavi TM. Density, Refractive Index, Viscosity, and Speed of Sound in Binary Mixtures of Cyclohexanone with Hexane, Heptane, Octane, Nonane, Decane, Dodecane, and 2,2,4-Trimethylpentane. Journal of Chemical & Engineering Data. 1999;44(3):435-40.
30. González B, Domínguez I, González EJ, Domínguez Á. Density, Speed of Sound, and Refractive Index of the Binary Systems Cyclohexane (1) or Methylcyclohexane (1) or Cyclo-octane (1) with Benzene (2), Toluene (2), and Ethylbenzene (2) at Two Temperatures. Journal of Chemical & Engineering Data. 2010;55(2):1003-11.
31. Grigoryev BA, Nemzer BV, Kurumov DS, Sengers JV. Surface tension of normal pentane, hexane, heptane, and octane. International Journal of Thermophysics. 1992;13(3):453-64.
32. Hales JL, Townsend R. Liquid densities from 293 to 490 K of nine aromatic hydrocarbons. The Journal of Chemical Thermodynamics. 1972;4(5):763-72.
33. Jasper JJ, Kring EV. The Isobaric Surface Tensions and Thermodynamic Properties of the Surfaces of a Series of n-Alkanes, C5 to C18, 1-Alkenes, C6 to C16, and of n-Decylcyclopentane, n-Decylcyclohexane and n-Dcylbenzene. The Journal of Physical Chemistry. 1955;59(10):1019-21.
34. Kahl H, Wadewitz T, Winkelmann J. Surface Tension of Pure Liquids and Binary Liquid Mixtures. Journal of Chemical & Engineering Data. 2003;48(3):580-6.
35. Kang K, Wang X. Liquid densities for n-decane + p-xylene mixtures from 293.15 K to 363.15 K at pressures up to 60 MPa. Fluid Phase Equilibria. 2018;458:142-52.
36. Klein T, Yan S, Cui J, Magee JW, Kroenlein K, Rausch MH, et al. Liquid Viscosity and Surface Tension of n-Hexane, n-Octane, n-Decane, and n-Hexadecane up to 573 K by Surface Light Scattering. Journal of Chemical & Engineering Data. 2019;64(9):4116-31.
37. López-Lázaro JdlS, Iglesias-Silva GA, Estrada-Baltazar A, Barajas-Fernández J. Density and Surface Tension of Binary Mixtures of 2,2,4-Trimethylpentane + n-Heptane, 2,2,4-Trimethylpentane + n-Octane, Ethyl Acetate + Benzene, and Butanenitrile + Benzene from (293.15 to 323.15) K. Journal of Chemical & Engineering Data. 2015;60(6):1823-34.
38. Mark M, Jolene S. Certification Report for SRM 2214: Density of Iso-octane for Extended Ranges of Temperature and Pressure. Special Publication (NIST SP), National Institute of Standards and Technology, Gaithersburg, MD; 2016.
39. Mohsen-Nia M. Measurement and modelling of surface tensions of systems containing n-hexadecane, n-heptane and n-pentane. Physics and Chemistry of Liquids. 2011;49(5):608-14.
40. Regueira T, Pantelide G, Yan W, Stenby EH. Density and phase equilibrium of the binary system methane + n-decane under high temperatures and pressures. Fluid Phase Equilibria. 2016;428:48-61.

41. Regueira T, Yan W, Stenby EH. Densities of the Binary Systems n-Hexane + n-Decane and n-Hexane + n-Hexadecane Up to 60 MPa and 463 K. Journal of Chemical & Engineering Data. 2015;60(12):3631-45.
42. Sanmamed YA, Dopazo-Paz A, González-Salgado D, Troncoso J, Romaní L. An accurate calibration method for high pressure vibrating tube densimeters in the density interval (700 to 1600)kg·m−3. The Journal of Chemical Thermodynamics. 2009;41(9):1060-8.
43. Vijande J, Pineiro MM, García J, Valencia JL, Legido JL. Density and surface tension variation with temperature for heptane+ 1-alkanol. Journal of Chemical & Engineering Data. 2006;51(5):1778-82.
44. Vladimir Diky CDMAYSABRDCJWMYPAFKSATEWLM. NIST Standard Reference Database 103b. Thermodynamic Data for Pure Compounds. NIST/Thermodynamics Research Center (TRC).
45. Wohlfarth C, Wohlfarth B. Binary Mixtures: References: Datasheet from Landolt-Börnstein - Group IV Physical Chemistry · Volume 16: "Surface Tension of Pure Liquids and Binary Liquid Mixtures" in SpringerMaterials (https://doi.org/10.1007/10560191_5). Springer-Verlag Berlin Heidelberg.
46. Yang J, Wu J, Bi S. Surface Tension Measurements by Pendant Drop Method of 10 Pure Long-Chain Alkanes and Alcohols for Temperatures up to 573.15 K. Journal of Chemical & Engineering Data. 2021;66(6):2615-28.
47. Fabian Pedregosa, Gael Varoquaux, Alexandre Gramfort, Vincent Michel, Bertrand Thirion, Olivier Grisel, et al. Scikit-learn: Machine Learning in Python. Journal of Machine Learning Research. 2011;12.
48. Heyne J, Bell D, Feldhausen J, Yang Z, Boehm R. Towards fuel composition and properties from Two-dimensional gas chromatography with flame ionization and vacuum ultraviolet spectroscopy. Fuel. 2022;312:122709.
49. Edwards JT. Reference Jet Fuels for Combustion Testing. 55th AIAA Aerospace Sciences Meeting.
50. Edwards JT. Jet Fuel Properties. Air Force Research Laboratory, Wright-Patterson AFB, OH; 2020 2020.
51. Committee D. ASTM D1655: Standard Specification for Aviation Turbine Fuels. Conshohocken, PA: ASTM International2022.

IASH 2024, the 18TH INTERNATIONAL CONFERENCE ON STABILITY, HANDLING AND USE OF LIQUID FUELS
Louisville, Kentucky, USA
September 8 – 12, 2024

(Technical Poster)

PERFORMANCE AND STORAGE STABILITY OF RENEWABLE FUELS WITH STATIC DISSIPATIVE ADDITIVE

Asha Patel, Sankaran Murugesan*, Don Wolfe

Baker Hughes, 12645 West Airport Blvd, Sugarland, TX 77478 USA

ABSTRACT

Renewable fuels are becoming an increasing part of the fuels offering for domestic and commercial applications, and more recently an emphasis towards production of Sustainable Aviation Fuels (SAF). In the industry, different renewable feed stocks are used to make renewable fuels such as vegetable oils, used cooking oils and various biomasses using different catalytic conversion process to produce renewable fuels. The properties of renewable fuels are related to the feeds and processes used in their manufacture. Generally, in traditional distillate fuels, Static Dissipative Additives (SDA) are used to facilitate the safe dissipation of electrostatic charges through the liquid hydrocarbon, by reducing the accumulation of hazardous surface charges. However, in case of renewable fuels, the investigation of conductivity improver performance has not been explored. This presentation gives a comprehensive study of conductivity performance of SDA in three different renewable fuels in comparison with conventional ULSD. Detailed discussion of the SDA performance with regard to dosage variation, temperature and long-term fuel storage stability is considered.

*Corresponding author: Sankaran Murugesan, sankaran.murugesan@bakerhughes.com

IASH 2024, the 18TH INTERNATIONAL CONFERENCE ON STABILITY, HANDLING AND USE OF LIQUID FUELS
Louisville, Kentucky, USA
September 8 – 12, 2024

(Technical Poster)

EFFECT OF CHEMICAL COMPOSITION ON FREEZE POINT OF HYDROCARBON MIXTURES

David C. Bell*, Randall Boehm, Zhibin Yang, Joshua Heyne,

Washington State University, 2710 Crimson Way, Richland, Washington 99354 USA

ABSTRACT

Accurate property predictions for the freezing point of hydrocarbon fuels is an ongoing challenge in the aviation fuel community. Freezing point has historically been attributed to high concentrations of n-alkane species and high temperatures at the upper distillation curve values (T90 and FBP), but these models are strictly correlative methods based around experience with conventional fuels. For alternative fuels, the applicability of these methods is uncertain. Recent work has suggested that freeze point is determined by the concentration of a limiting species, where the balance has negligible impact. This work tests this hypothesis and improves a method for freeze point predictions for complex fuels. In this work, low concentrations of high freeze point components are added to low freeze point solvents to isolate the impact of several explanations of freeze point behavior. These include the impact of aromatics on solvency, the impact of mixture complexity, and the impact of structural similarity. The results of this study show that mixtures of fuels will result in freeze points lower than any of fuels on their own and demonstrate the basis for a widely applicable prediction method for alternative and conventional aviation fuels.

*Corresponding author: David C. Bell, david.c.bell@wsu.edu

IASH 2024, the 18TH INTERNATIONAL CONFERENCE ON STABILITY, HANDLING AND USE OF LIQUID FUELS
Louisville, Kentucky, USA
September 8 – 12, 2024

(Technical Poster)

DEFINING QUALIFICATION REQUIREMENTS FOR RENEWABLE DIESEL FUELS

David Evans*, Tori Guarnera, Rachelle Hewitt, Nathan Matheson

Defence Science and Technology Group, 506 Lorimer Street Fishermans Bend, 3207 Australia

ABSTRACT

The aviation industry maintains stringent guidelines for qualifying synthetic blending components, governed by ASTM D4054. Following approval by the ASTM D02.J subcommittee, the new Synthetic Blending Component (SBC) is integrated into ASTM D7566, a process closely overseen by Airframe and Engine OEMs, fuel producers, and industry experts. Conversely, the diesel industry lacks a comparable mechanism to assess SBC quality for diesel fuel. Although ASTM D975 permits non-conventional feedstock, it emphasizes the potential need for additional testing. Diesel engine OEMs generally accept SBCs, and sometimes neat synthetic diesel, provided they meet ASTM D975, EN950 or EN15940 performance standards.

The US Navy has established a rigorous qualification testing protocol outlined in Navy Standard Work Package 44FL-005, which aligns with ASTM D4054 and is considered the gold standard in technical requirements. With the renewable diesel production market expanding rapidly, there's an imperative to establish minimum testing standards to mitigate risks that is readily accessible by industry. This paper proposes minimum technical requirements based on the ASTM D4054 Fast Track approach to ensure performance and fit for purpose requirements are satisfied and introduces additional specifications for diesel fuel containing synthetic components.

*Corresponding author: David Evans, david.evans29@defence.gov.au

IASH 2024, the 18TH INTERNATIONAL CONFERENCE ON STABILITY, HANDLING AND USE OF LIQUID FUELS
Louisville, Kentucky, USA
September 8 – 12, 2024

INSIGHTS INTO SCALING UP HEFA-SPK PRODUCTION

Jesse Vilja, Kati Sandberg, Peter John

Neste Oyj, Keilaranta, P.O. Box 95, 00095 Neste FINLAND. jesse.vilja@neste.com

ABSTRACT

Achieving net-zero emissions in the aviation industry by 2050 necessitates the widespread adoption of sustainable aviation fuels (SAF), in addition to advancements in technology, operations and infrastructure[1]. This requires a rapid global scale-up of SAF production, a challenge addressed by Neste's recent investments. Neste's Singapore refinery expansion, completed in 2023 after an initial investment decision in 2018, utilizes the proprietary NEXBTL™ technology to produce SAF. Combined with the upcoming modification to Neste's Rotterdam refinery, enabling SAF production alongside Renewable Diesel, Neste will reach a world-leading SAF production capability of 1.5 Mt/a in 2024. In the NEXBTL™ process, diverse fatty acid and fatty acid ester feedstocks are converted through hydroprocessing, isomerization and fractionation into HEFA-SPK, a synthetic blend component (SBC). In the production and in the downstream supply chain, quality assurance principles set by industry standards such as EI/JIG 1530 and EI 1533 are followed. Neste's ASTM D7566 certified HEFA-SPK can in neat form and over its lifecycle reduce greenhouse gas emissions by up to 80 % compared to using fossil jet fuels (calculated with established life cycle assessment methodologies, such as CORSIA). Neste MY Sustainable Aviation Fuel also significantly reduces non-CO2 emissions and contrails compared to fossil jet fuel use[2]. Its blending with conventional jet fuel results in a drop-in jet fuel for existing aircraft engine and airport fuel infrastructure purposes.

Keywords: SAF, SBC, HEFA-SPK, aviation, quality

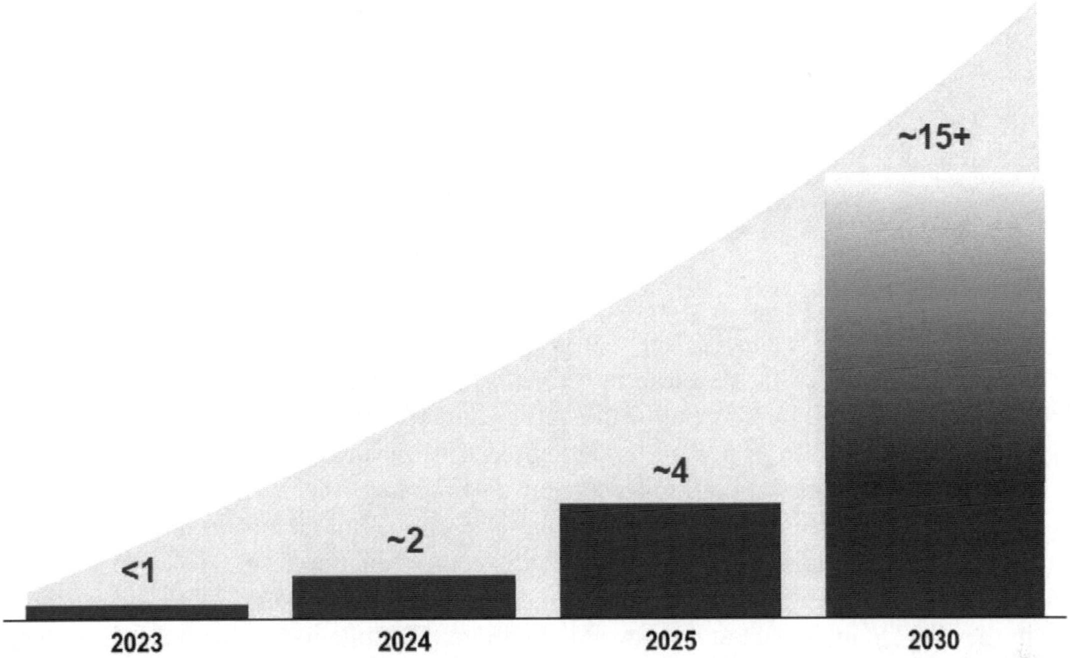

Figure 1. SAF demand (Mt) outlook until 2030[3].

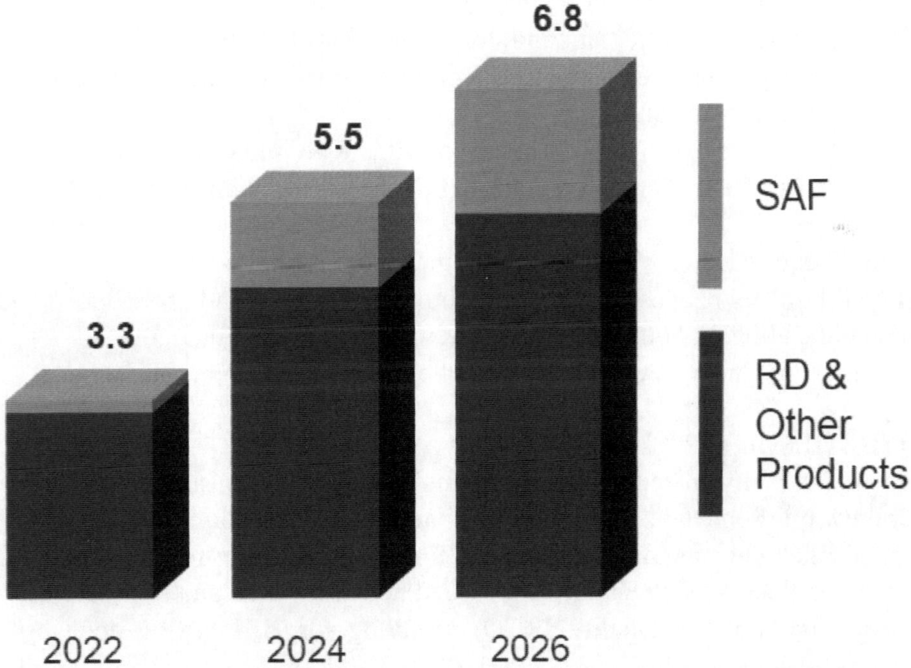

Figure 2. Neste's end-of-year, nameplate renewable and circular capacities (Mt/a), with product optionality. SAF: Sustainable Aviation Fuel, RD: Renewable Diesel.

DISCUSSION - Neste's HEFA-SPK production and supply chain setup with key quality aspects

Below listed are the main phases of the HEFA-SPK supply chain from the feedstock to the end-product. For each main phase, the governing standards or guidance materials are listed in parentheses.

Feedstock[4]

- Waste and residues consisted 92 % of the Neste total renewable feedstock pool of 4.3 Mt in 2023, with SAF produced 100 % from renewable waste and residue feedstock
- Sourced globally for the renewables refineries in Finland, the Netherlands and Singapore, and for our joint venture in California in the United States
- Efficient sourcing, transporting and flexible use of various mixes of raw materials to produce a wide range of high-quality renewable products incl. SAF
- Unique pretreatment capability to remove impurities, unlocking the use of even lower-quality raw materials while ensuring the high end-product quality
- Future raw material alternatives include acid oils, brown greases and novel vegetable oils in the short to mid-term, and renewable hydrogen, lignocellulosic waste and residues and microalgae in the long-term

NEXBTL™ Process (EI/JIG 1530[5], EI 1533[6], ASTM D7566[7])

- Utilized at Singapore, Rotterdam and Porvoo facilities for SBC production
- Hydroprocessing of pre-treated feedstocks into *n*-paraffins while removing impurities such as oxygen, nitrogen and sulfur
- Isomerization of the hydrocarbons into *i*-paraffins with improved cold properties
- Fractionation as the final step to produce HEFA-SPK compliant with ASTM D7566 Annex A2
- Dosage of required additives directly into the column rundown
- Quality follow-up between each major processing step, and from the dedicated SBC rundown line utilizing both online and physical sample analytics
- Management of Change evaluations as part of the refinery operations

Tank Farm (SBC) (EI/JIG 1530, EI 1533)

- Tanks are internally coated with cone down floors to a central sump which is equipped for water draw off. Some tanks are fitted with internal floating roofs
- One tank filled per time and relaxed for 30 mins after filling followed by upper-middle-lower check. Positive segregation is provided between tested and untested fuel tanks
- Refinery Certicifate of Quality (RCQ) sampling for HEFA-SPK done by nominated inspectors
- After 24h settling and RCQ passing, certified as an SBC, followed by draining, and release under a Release Certificate (RC) with a unique batch number for loading into a vessel/barge
- Daily water draining for certified batches
- Batch documentation (RCQ, Batch Composition Form) stored in an online database for traceability purposes

Vessel Loading (EI/JIG 1530, EI 1533, HM 50[8])

- Loading done via an SBC dedicated jetty, with loading line sample follow-up at regular intervals
- Vessel compartment first foot sampled by nominated surveyors for verification of product integrity
- On first-foot samples passing, the loading is resumed until completion, followed by collection of composite samples for vessel recertification testing (RT)
- Based on composite samples and loading line samples passing, a Vessel Release Certificate is created and the jetty is embarked from
- In case any of the previous samples results fail, vessel loading is halted and re-sampling is required
- Vessel RT samples are stored for retention purposes for 12 months

Blend Terminal Network (EI/JIG 1530, EI 1533, ASTM D7566, ASTM D1655[9] or DEF STAN 91-091[10])

- With the vessels typically considered as non-dedicated transport modes, the SBC cargo's integrity is verified via RT testing prior to discharge, during which samples are collected at regular intervals
- The discharge is done either into a dedicated SBC receipt tank or, with close consideration of the case-by-case max. blend ratio achievable, directly into a semi-synthetic jet fuel (SSJF) blending tank which already contains the conventional jet fuel blend component
- Conforming to the max. blend limit of 50 vol-% for HEFA-SPK, the SSJF batch is certified against ASTM D7566 Table 1 and its homogeneity is verified
- With Neste's blending hubs located in the EU, Singapore and in the US (CA, TX), the SSJF is finally released and distributed to the end-users

CONCLUSIONS

For the aviation industry to reach net-zero emissions by 2050, regulatory development and voluntary commitments are crucial in supporting the long-term demand outlook and thus in untapping the global potential of waste and residue oils and fats which amounts to ca. 40 Mton/a by 2030.

As the global SBC production gets scaled up and the supply chain complexity increases through the entry of both neat and blended SAFs, it is equally important that firm quality assurance principles will serve as the bedrock for sustained SAF growth without compromising the aviation safety.

However, with the medium-term influx of SBCs from currently approved and future production routes, and the rising share of SAF through regulation (e.g. ReFuelEU), the handling requirements should not discourage growth in an industry where economics can already be challenging vs. fossil-based solutions.

REFERENCES

1 Air Transport Action Group (ATAG): Waypoint 2050, 2nd edition, 2021.
2 Märkl et al.: Powering aircraft with 100 % sustainable aviation fuel reduces ice crystals in contrails, Atmos. Chem. Phys., 24, 3813–3837, https://doi.org/10.5194/acp-24-3813-2024, 2024.
3 Neste analysis based on IATA, IEA, Argus
4 Neste: Annual Report 2023, https://www.neste.com/investors/financials/annual-report, 2024.
5 Energy Institute & Joint Inspection Group: Quality assurance requirements for the manufacture, storage and distribution of aviation fuel to airports, 2nd edition, 2019.
6 Energy Institute: EI 1533 - Quality assurance requirements for semi-synthetic jet fuel and synthetic blending components (SBC), 1st edition, 2022.
7 ASTM International: D7566 Standard Specification for Aviation Turbine Fuel Containing Synthesized Hydrocarbons, Edition 24a, 2024.
8 Energy Institute: Guidelines for the preparation of tanks and lines for marine tank vessels carrying petroleum and refined products, 6th edition, 2024.
9 ASTM International: D1655 Standard Specification for Aviation Turbine Fuels, Edition 24, 2024.
10 Ministry of Defense: Defence Standard 91-091, Issue 17, 2024.

IASH 2024, the 18TH INTERNATIONAL CONFERENCE ON STABILITY, HANDLING AND USE OF LIQUID FUELS
Louisville, Kentucky, USA
September 8 – 12, 2024

(Technical Poster)

UPDATE TO THE DEVELOPMENT OF A SHIPBOARD TEST KIT FOR DISSOLVED COPPER IN A JET FUEL PROTOTYPE FIELD TEST KIT

Kristina Myers*[1], Thomas Loegel[2], Robert E. Morris[1], Jacob Guthrie[1]

[1]*Nova Research, Inc.*, 1900 Elkin St., Ste. 230, Alexandria, VA 22308 USA
[2]*Naval Research Laboratory, 4555 Overlook Ave SW Washington, DC, 20375 USA*

ABSTRACT

Recently, the Navy has been investigating functionalized filtration technologies to selectively remove copper from fuel. In order to utilize this system, there is a need to monitor copper content in the filter effluent. We have revisited the bathocuproine (BCP) colorimetric methodology [1]. The chemistry was further refined and some work was performed to extend the method applicability to both jet and diesel fuels. An upgraded and more capable data acquisition and analysis kit was also developed.

A range of copper concentrations were obtained by combining a copper-free JP-5 with that same fuel doped with copper from copper metal, as well as a set of copper-bearing fuels from Navy ships. The analyzer performed adequately at copper concentrations from 400 to 50 µg/kg, which was deemed adequate for the intended purpose of monitoring copper filter breakthrough.

However, it was found that fuels from a particular aircraft carrier displayed negative results due to decreased absorbance when BCP was added with respect to the reagent blank. Chemical characterizations of these anomalous JP-5 fuels were conducted in an effort to determine any anomalous species responsible for this behavior. Investigations into alternate reducing agents, protocols, and reagents were also conducted in order to develop a test method to accommodate these outlier fuels.

[1] Morris, R. E.; Collins, G. E.; Lu, Q.; Wei, J-F,; Serino, P. M.; Guo, Y. "Development of A Rapid, On-Site Method for The Determination of Copper in Jet Fuels" Presented at the 8th International Conference on Stability and Handling of Liquid Fuels, September 14-19, 2003.

*Corresponding author: Kristina Myers, kristina.m.myers9.ctr@us.navy.mil

IASH 2024, the 18TH INTERNATIONAL CONFERENCE ON STABILITY, HANDLING AND USE OF LIQUID FUELS
Louisville, Kentucky, USA
September 8 – 12, 2024

(Technical Poster)

CETANE IMPROVER EFFECTIVENESS IN SEMI-SYNTHETIC JET FUELS AND RELEVANCE OF CETANE NUMBER IN AERONAUTICAL COMBUSTION

Mikael Orain[1], Mickaël Sicard*[1], Amandine Herbaut[2], Nathalie Brassart[3], Mickaël Matrat[4]

[1]*ONERA, Chemin de la Huniere Palaiseau, 91123 France*
[2]*CETSEO*
[3]*IFPEN*

ABSTRACT

Safe operation of diesel engines requires a fuel with a cetane value over 40. However, kerosene may be used in land vehicles as the single fuel on the battlefield. Kerosene cetane value is yet lower than that of diesel. It is then possible to add a cetane improver (CI) to increase this value. Nowadays, the incorporation of alternative fuels (SBC) raises the question of the effectiveness of this additive. Moreover, some SBCs have very low cetane values. In order to study the behavior of two different CI, a fuel matrix, composed of JP-8/ATJ and JP-8/HEFA-ARO blends were additivated at different levels. The derived cetane number (DCN) results show that an increase of SBC content decreases the cetane boost. Comparison of the two CI shows that both have similar responses.

Although the cetane number was developed to characterize the behavior of a fuel in a diesel engine, recent studies have shown that it would be possible to use it to characterize a fuel in an aeronautical combustor. To support this hypothesis, droplet sizing and ignition tests were carried out on different SBCs for different temperatures and pressures.

The results show that there is no clear influence between drop sizing and the fuel. The ignition tests, performed in a model combustor, exhibit significant variations between the fuels. However, considering only the cetane value does not allow to explain the results. These are highly dependent on fuel temperature, and therefore almost certainly on parameters such as viscosity and surface tension.

*Corresponding author: Mickaël Sicard, mickael.sicard@onera.fr

IASH 2024, the 18TH INTERNATIONAL CONFERENCE ON STABILITY, HANDLING AND USE OF LIQUID FUELS
Louisville, Kentucky, USA
September 8 – 12, 2024

(Technical Poster)

HANDLING, MAINTENANCE AND SAFETY USE OF AVIATION FUEL - INFORMATIVE BULLETINS

Maor Saban, Liel Mazar, Yalfal Siyum, Haim Judah, Moshe Rabaev*,

Israeli Air Force

ABSTRACT

Optimal care and maintenance of aviation fuels and infrastructure is of crucial importance for ensuring the safety of passengers, personnel and the environment. In order to increase the awareness of fuel technicians, engineers and decision makers about safety issues concerning the maintenance and handling of fuel in the field, we have produced numerous informative bulletins. These are written in simple language without much jargon, making it possible to convey important and essential knowledge in an easy, digestible manner for the layperson. Topics covered in the bulletins include information about "Apple Jelly", microbial contamination, fuel filters, static electricity, aviation fuel additives, laboratory tests, critical fuel properties, fuel contaminations and real fuel-related aircraft safety events - emphasizing how a minor human mistake can have major safety implications. The bulletins have greatly raised the awareness of our organization's personnel in their handling of aviation fuel, and therefore we find it essential to share them with the global community.

*Corresponding author: Moshe Rabaev, rabaevm@gmail.com

IASH 2024, the 18TH INTERNATIONAL CONFERENCE ON STABILITY, HANDLING AND USE OF LIQUID FUELS
Louisville, Kentucky USA
8 – 12 September 2024

EXPERIMENTAL STUDY OF HYDROPEROXIDE CONTENT IN SURROGATE JET-FUELS USING HPLC

Ryma Benrabah[1], Boris Roux[1], Kanika Sood[1], Philippe Arnoux[1], René Fournet[1], Pierre-Alexandre Glaude[1], Baptiste Sirjean[1]

[1]Université de Lorraine, CNRS, LRGP, F-54000 Nancy, FR. *ryma.benrabah@univ-lorraine.fr*

ABSTRACT

Hydroperoxides are major products of the autoxidation of hydrocarbon fuels and play a central role in oxidation kinetics. The quantification of hydroperoxide is a key factor for the qualification of jet-fuel and is determined through a standardized method. A recent comparison of different iodometric methods of quantification showed that an HPLC chain coupled with a Post-reaction Column (HPLC-PcR) is highly accurate with a high reproducibility[1]. In this work, HPLC-PcR is used to detect and quantify the hydroperoxide content and analyze different oxidized surrogate fuels, by separating the oxygenated fractions. Surrogate fuels were oxidized in a PetroOxy and in a microreactor and then analyzed by HPLC-PcR. n-decane and cumene were used as surrogate jet-fuels and n-butanol as a surrogate biofuel. The oxidation of n-decane yields several decane hydroperoxides and other oxygenated compounds that were separated in the HPLC. The separated fractions of aged cumene in HPLC-PcR were analyzed by Gas Chromatography – Mass Spectrometry (GC-MS) demonstrating that the typical hydrocarbon oxidation pathways take place for cumene autoxidation, yielding only cumene hydroperoxide, 2-phenyl-2-propanol and acetophenone. On the contrary, for *n*-butanol, we demonstrate that the typical mechanism leading to ROOH does not prevail and that its autoxidation mainly yields H_2O_2 as peroxide content.

KEYWORDS: Hydroperoxide (ROOH), High performance Liquid Chromatography (HPLC)

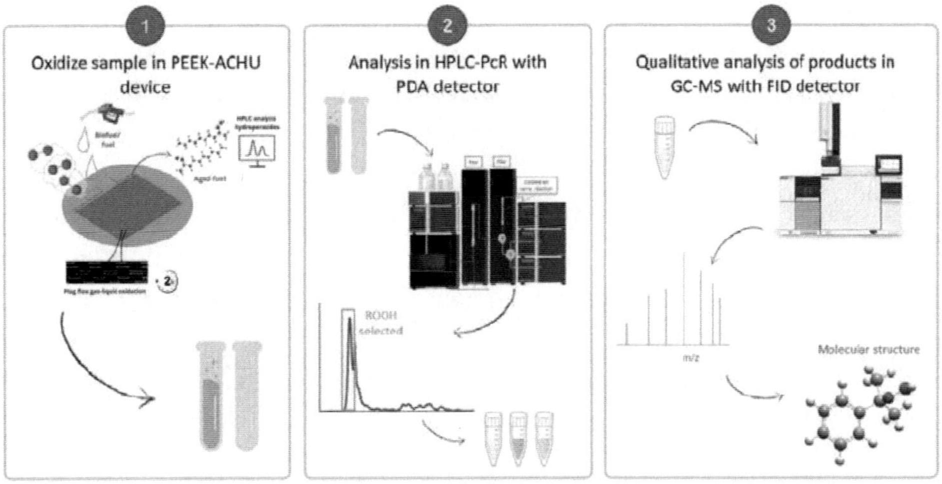

INTRODUCTION

As part of the new European and global standards on reducing carbon emissions in aviation, the use of Sustainable Aviation Fuels (SAF) is a solution for using fuels with a reduced carbon footprint compared to those derived from fossil resources [2,3]. SAF can be used as drop-in fuels, allowing for immediate implementation without the need to wait for a new fleet of aircrafts specifically designated for e-fuels such as hydrogen. However, these fuels must comply with several standards and physic-chemical tests that may be more stringent than those applied to traditional fuels, given their different production sources and formulations. One of the standard tests required to validate the use of these fuels involves assessing their oxidation stability. There are several standard tests to verify this property, such as PetroOxy or JFTOT tests, which measure oxidation stability by monitoring the evolution of different physical properties of the fuel [4]. Additionally, the oxidation stability of a fuel can also be gauged by quantifying the amount of hydroperoxides, for which a standard test method (ASTM D3703) is available [5]. Monitoring hydroperoxides concentration is a good indicator of the extent of the fuel auto-oxidation process, as hydroperoxides are the primary oxygenated products formed according to the simplified mechanism illustrated in Figure 1.

$$RH \longrightarrow R\cdot \qquad \textbf{Initiation}$$

$$R\cdot + O_2 \longrightarrow ROO\cdot \qquad \textbf{Propagation}$$

$$ROO\cdot + RH \longrightarrow ROOH + R\cdot$$

$$ROO\cdot + ROO\cdot \longrightarrow ROH + RCHO\ (ou\ RCOR') + O_2 \qquad \textbf{Termination}$$

Figure 1 : Simplified hydrocarbon autoxidation pathways 6.

This measurement not only provides quantitative data on fuels stability, but also offers insights on other aspects, such as the adverse effect of hydroperoxides on some elastomers [5]. Hydroperoxides quantification is also crucial in other applications. For example, they are measured in edible oil to assess the edibility of the product [7]. As a result, studies aimed at optimizing this measurement are widespread, but they often focus solely on the total ROOH content, without differentiating between the different types of hydroperoxides in the system. On the other hand, in the field of low-temperature gas-phase oxidation, major efforts have been made over the years to analyze hydroperoxides using conventional chromatography, coupled to mass spectrometry. However, these techniques have faced technical limitations [8,9]. In the 2010s, multiplexed synchrotron-photoionization mass spectrometry made remarkable progress in the detection and identification of these unstable compounds [10,11]. More recently, the use of ultra-high pressure liquid chromatography and high-resolution mass spectrometry (Orbitrap®), was shown to improve the characterization of low temperature oxidation products, including hydroperoxides [8,12].

Nevertheless, for in-depth kinetic studies of the oxidation mechanisms of liquid fuels or other molecules, it is necessary to know the total amount and the molecular structures of the hydroperoxides to deduce the oxidation pathways precisely.

A recent comparative study published by our team in 2024 enabled us to investigate and compare different iodometric methods for quantifying the total amount of hydroperoxides [1]. Iodometry is the basis of the method used in ASTM D370 and is historically recognized for its effectiveness in quantifying hydroperoxides. A reaction between iodide ions and hydroperoxides (ROOH) takes place in an acidic medium, as shown in reaction (1). The quantification of I_2 provides the hydroperoxide concentration in the mixture.

$$ROOH_{\,(organic)} + 2H^+_{\,(organic)} + 2I^-_{\,(organic)} \rightarrow I_{2\,(organic)} + ROH_{\,(organic)} + H_2O \qquad (1)$$

Our benchmark study compared different titrations methods, different titrators, endpoint detections methods, and UV-Vis absorption detection of iodine. We demonstrated that the HPLC with a post column reactor (PcR), where the iodometric reaction (1) occurs on the fractions exiting the separation column, along with UV-Vis absorption detection of I_3^-, gives the best results in terms of quantification limits, precision and repeatability. In addition, the chromatogram enables us to separate the different types of hydroperoxides formed in the separation column and collect these fractions for further analysis.

Equation (1) shows that hydroperoxides reduction also yields alcohols. As alcohols are more stable than hydroperoxides, their detection and characterization can be easily achieved via analytical techniques such as GC-MS. This technique is widely used in chemistry and other industrial plants due to its precision for stable molecules at high temperatures. However, alcohols can also be formed during other fuel oxidation processes, making it difficult to attribute the proportion of alcohols strictly produced in the iodometric reaction. In this context, indirect detection can be used. For example, a method to quantify hydroperoxides with GC-MS was developed by West et al. [13] using triphenylphosphine (TPP). In this method, TPP reduces the ROOHs, and the resulting quantity of triphenylphosphine oxide is measured by GC-MS. This technique allows an accurate quantification of the hydroperoxide content in liquid fuels.

In this work, we used the HPLC-PcR to quantify the most abundant hydroperoxides formed during the oxidation reaction of a surrogate jet-fuel components, *n*-decane and cumene, as well as a surrogate biofuel, *n*-butanol [14 15]. Additionally, the GC-MS technique is also used to compare the results obtained from both HPLC-PcR and HPLC methods for the determination of the composition of the oxidized surrogate molecules.

MATERIAL AND METHODS

Reactants

Anhydrous 1-butanol (99.8 %) was purchased from Sigma Aldrich. Cumene (99.9 % extra pure), 2-phenyl-2-propanol (99%), *n*-octane (98+%) and acetone (99+%) were purchased from Thermo Scientific and, *n*-decane (≥ 99 %) from Acros Chemicals. Potassium iodide (KI) anhydrous, sodium iodide (NaI) (>99%), and absolute ethanol were purchased from Carlo Erba. Butylated hydroxytoluene (BHT) (≥ 99 %), triphenylphosphine (TPP) (reagent ≥ 99 %) and acetic acid (≥ 99.8 %) were purchased from Sigma Aldrich. Acetophenone (for synthesis), cumene hydroperoxide (80% solution in cumene, for synthesis) came from Sigma-Aldrich. Hydrogen peroxide (30% in water) came from VWR Chemicals.

Reactors used

Two different reactors were used in this study; the first one is the PetroOxy from Anton Paar, which is a reference for the Rapid Small-Scale Oxidation Test according to ASTM D7545 methods. In this test, a 5 mL sample of liquid fuel is placed in the gold-plated cell, pressurized with pure O_2 at 7 bar, and heated to 413K. The pressure is recorded over time, and the induction period is defined as the time it takes to lose 10% of the maximum pressure reached in the cell. The O ring sealing of the cell was replaced between each test, and the reactor was thoroughly cleaned at the end of each experiment.

For the *n*-decane oxidation study, we used the PEEK-ACHU device described in our previous work [16] and is shown here in Figure 2. This reactor enables homogenous oxidation of liquid fuels and allows for sampling of aged fuel at the predetermined residence time chosen at the beginning of the experiment. This technique allowed us to study *n*-decane at low extents of reaction and to observe the ROOHs formed at different extents of reaction.

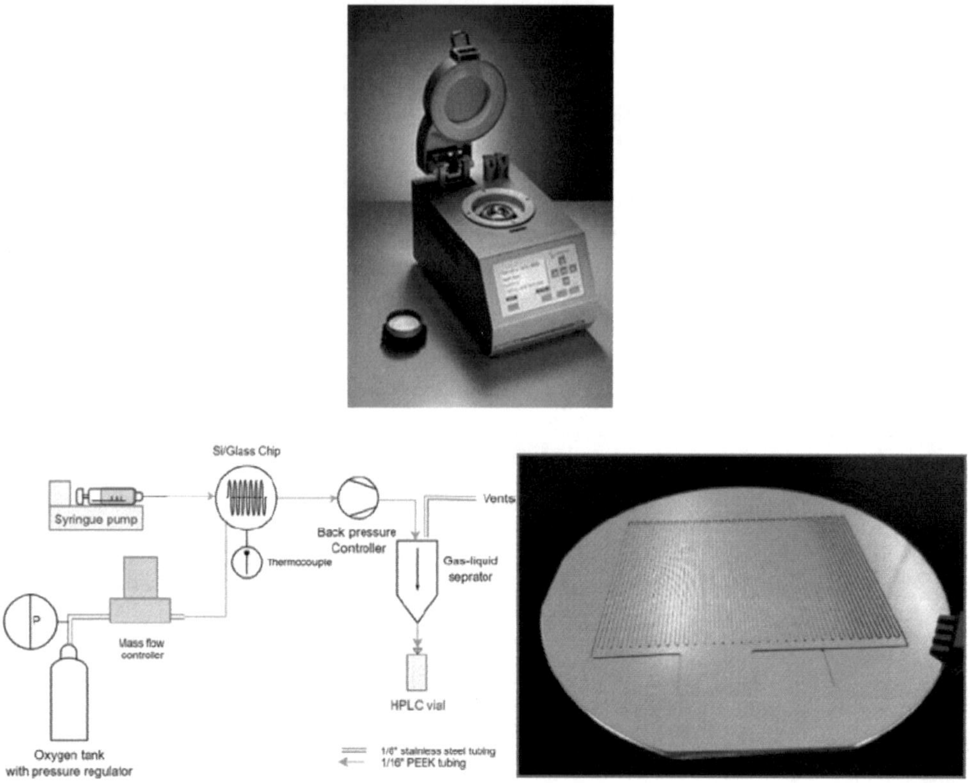

Figure 2: Photograph of PetroOxy device (up) | PEEK-ACHU rig setup overview (down-left) | Photograph of the microreactor 4-inches in Si/glass (down right).

HPLC-PcR for hydroperoxides quantification

The total hydroperoxide content was measured using the HPLC-PcR designed recently by our group. A detailed description of this analytical device can be found in the article recently published by our team [1]. This method enabled us to quantify hydroperoxides for concentrations ranging between $5x10^{-7}$ and $1x10^{-3}$ mol L^{-1}. The HPLC-PcR, schematically presented in Figure 3, was used at a flow of 1 mL per minute of solvent (isooctane:absolute ethanol 95:5) and 10 µL of the aged surrogate fuel was injected using the autosampler.

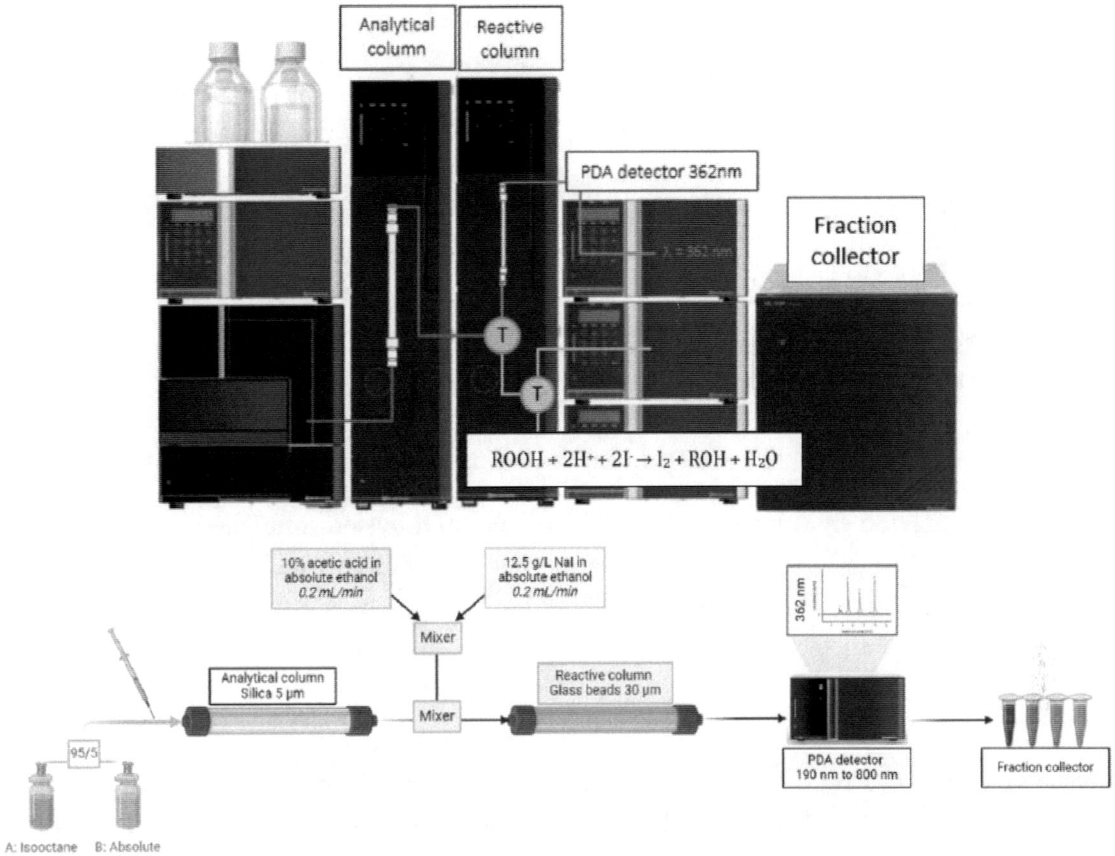

Figure 3: Representation of the HPLC-PcR to determine the total ROOH content of each sample. The final chromatogram combined with a calibration curve is used for the quantification of the hydroperoxide content of each sample [1].

The iodometric reaction takes place in the reactive column, that allows a precise quantification of the different hydroperoxides exiting the separation column. At the entrance of the reactive column, the fractions are mixed with a 10% acetic acid solution in absolute ethanol and a 1.5 g/mol solution of NaI in absolute ethanol. A chromatogram at 362 nm, capturing the absorbance of I_3^- for each sample.

HPLC

A traditional HPLC (Shimadzu Nexera X) was used to qualitative analysis of the liquid fuel surrogates. It utilizes a polar separative column in normal phase (silica, 5µm) with an eluent composed of a mixture of isooctane and absolute ethanol in a 95/5 proportion, at a flow of 1 mL per minute. 10 µL of the oxidized surrogate fuels was injected using an autosampler. This system is equipped with an autoinjector that provides dilution for samples that are too concentrated. We determined that this option has a reproducibility of 12%, calculated from our samples before and after the dilution in the autoinjector. Detection is achieved with a Photo-Diode Array (PDA) detector, capable of recording chromatograms across wavelengths ranging between 190 and 800 nm. A representation of the device is shown in Figure 4.

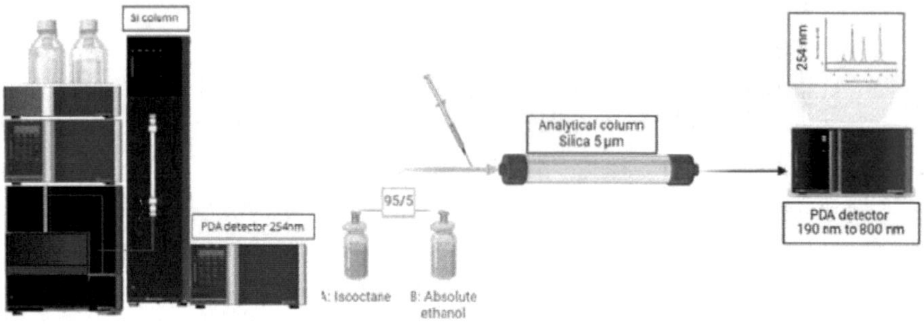

Figure 4: Representation of the HPLC set-up for the analysis and the detection of aromatics and hydrogen peroxide

For the experiments on traditional HPLC, the post column system remained connected to the system to rigorously compare the retention time between the two techniques. This approach ensured that the residence time in the system will be the same, while the two additional pumps supplied only absolute ethanol without any reactants.

GC-MS/FID

Analysis with GC-MS/FID (Flame Ionization Detector) were performed with an Agilent 7890A coupled to 5975C mass spectrometer and an FID. A scheme representing the overall methods used is shown below in Figure 5.

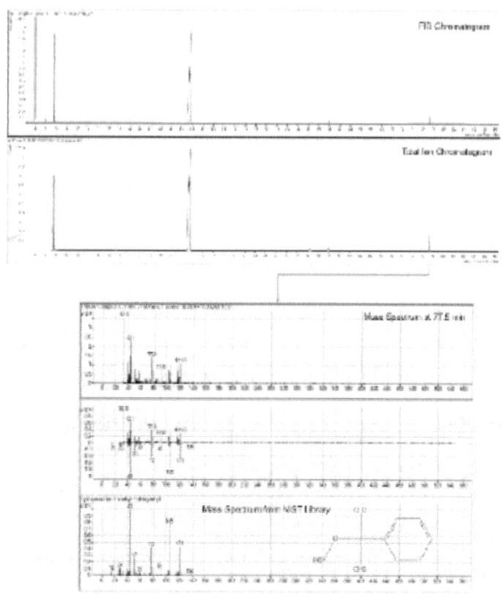

Figure 5: Representation of the GC/MS/FID used in this study together with some exemplary chromatograms

A 0.5g aliquot of the liquid sample was diluted in 25 mL of acetone, and 0.5g of the internal standard, *n*-octane. Subsequently, 1µL of this solution was injected into the GC apparatus. The sample passes through the HP-5ms J&W capillary column (30m x 0.25mm x 0.5µm) with a helium flow of 2 mL/min. The temperature program started with 40°C during 20 min, followed by a ramp of 5 °C/min to reach 320°C, where it was held constant for an additional 10 min.

Our system is equipped with 2 detectors: an FID for quantification and an MS for identification. A splitter was installed at the column outlet to divert the flow into 2 controlled flows, enabling both detectors to analyze the same analytes simultaneously. Quantification with the FID was achieved with an internal calibration based on the De St Laumer method [17]. The quantity of internal standard, *n*-octane, injected is known, as it is measured with a precision balance. The MS spectra were compared with those in the NIST library to identify the molecules present.

RESULTS AND DISCUSSIONS

n-Butanol oxidation and characterization of oxidation products

The oxidation of *n*-butanol was performed in the PetroOxy device, and its IP was 90.88 min at 413 K. The total hydroperoxides content was then measured in the HPLC-PcR and the associated chromatogram is shown in Figure 6.

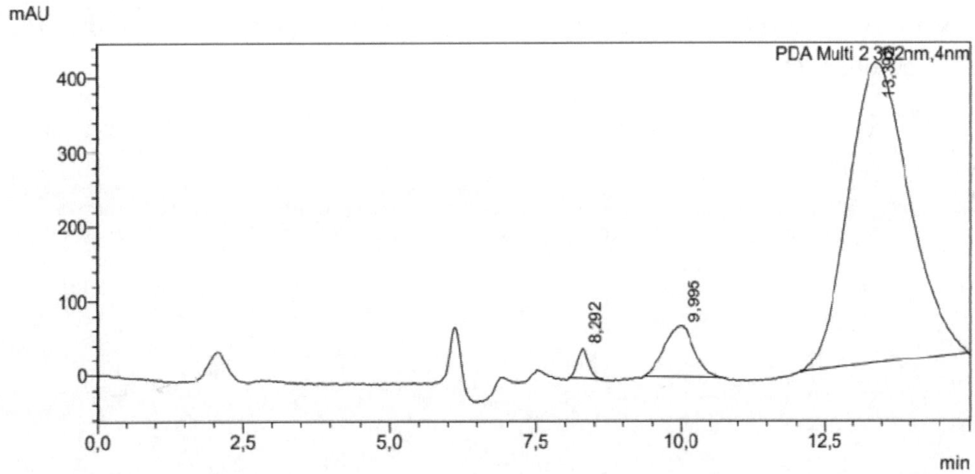

Figure 6: Chromatogram of hydroperoxides measured by HPLC-PcR after 90.88 min of oxidation of *n*-butanol in PetroOxy | sample diluted 15 times | injection volume: 10µL.

The chromatogram presented in Figure 6 shows five hydroperoxides peaks in oxidized *n*-butanol. In our previous work on alcohols autoxidation [18], we hypothesized, based on DFT theoretical calculations, that the most abundant hydroperoxide in the oxidation of *n*-butanol was H_2O_2. Indeed, the H-abstractions in the oxidation mechanism mainly occur at the bottom of the hydroxy group of *n*-butanol, yielding a predominant α- R•OH radical, that is subsequently oxidized by O_2 into butanal and HO_2. The reaction of the latter radical with *n*-butanol leads to the formation of

H_2O_2. This mechanism differs from the classical pathway observed in alkanes, where the initial R• radical formed from the RH fuel predominantly leads to the formation of ROOH, as shown in Figure 1. Based on this prior analysis, we suggest that the main peak at 13.4 min is H_2O_2. Other smaller peaks are isomers of hydroperoxybutan-1-ol and butaneperoxoic acid. The principal hydroperoxides produced during n-butanol autoxidation are shown Figure 6.

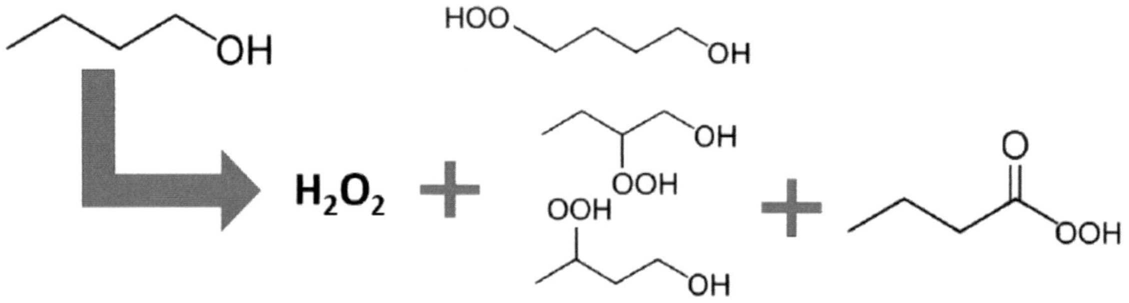

Figure 7: Hydroperoxides formed during the oxidation of *n*-butanol

By comparing the retention time of tert-butyl hydroperoxide in HPLC-PcR, it can be concluded that the peaks at 8.3 and 10 min in Figure 6 correspond to the two hydroperoxybutan-1-ol isomers represented in Figure 7. To confirm that the hydroperoxide observed at a retention time of 13.4 min is indeed H_2O_2, we injected hydrogen peroxide (30% in water) directly in the HPLC-PcR. The resulting chromatogram is shown in Figure 8.

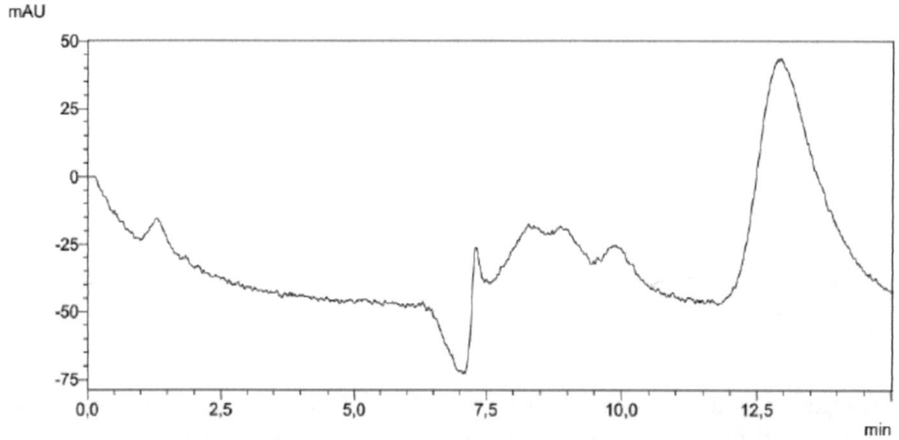

Figure 8: H_2O_2 chromatogram in HPLC-P-cR diluted 15 times in isooctane at 362 nm

As we can see, the ROOH peak of the mixture appears near 13 min. The presence of H_2O_2 in a highly polar water solvent, which is not soluble in isooctane, leads to perturbations in the baseline. To further confirm the H_2O_2 peak identification, we added 2 µL of 30% H_2O_2 in water to the oxidized *n*-butanol (1/15 dilution). The chromatogram obtained is presented in Figure 9 .

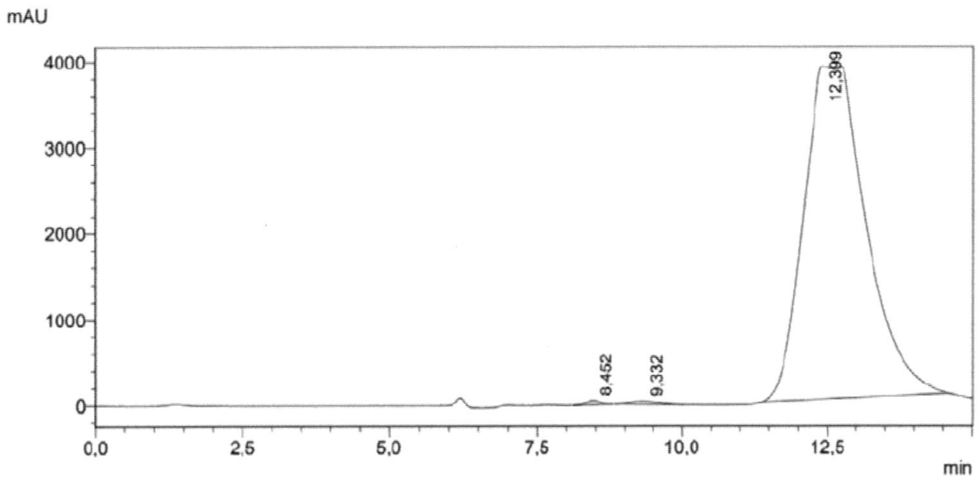

Figure 9: Chromatogram of oxidized *n*-butanol mixed with H_2O_2 (see text) in HPLC-PcR at 362 nm

The chromatogram in Figure 9 reveals a drastic increase in the peak area around 13.0 minutes when H_2O_2 was added to the oxidized n-butanol sample, providing groundbreaking evidence. This, for the first time in the literature, confirms the novel finding that the primary oxidation pathway of n-butanol deviates significantly from that of traditional alkanes. Rather than leading to ROOH as typically observed, the pathway predominantly yields butanal and the HO_2• radical, a crucial chain carrier. This observation fundamentally shifts our understanding of alcohol oxidation chemistry.

From the area of the peaks and the calibration curve established in a previous work [1], the total concentration of hydroperoxides detected is 4.66 10^{-2} mol/L, with 89.91% of them identified as H_2O_2.

Cumene oxidation and HPLC

The oxidation mechanism of cumene primarily involves the abstraction of the tertiary hydrogen, which leads to a resonance stabilized radical. Moreover, since it is an industrial process, the expected products of the oxidation are well known: cumene hydroperoxide, 2-phenyl-2-propanol and acetophenone [19], all of which are presented in Figure 10.

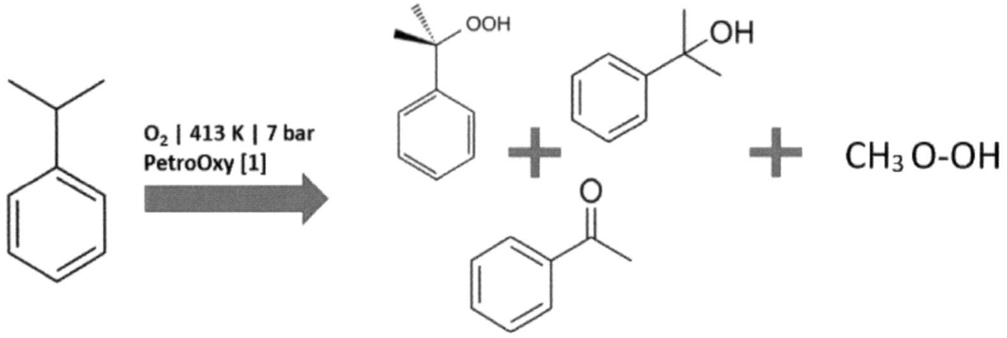

Figure 10: Autoxidation products of cumene

In the literature, the sole hydroperoxide considered in the autoxidation of cumene is cumene hydroperoxide. However, our analysis demonstrates that a second hydroperoxide, previously undetected, is formed in oxidized cumene, which we have identified as methyl hydroperoxide. This finding is discussed in detail later in this section.

To determine the accuracy of the HPLC with the PDA detector, which is highly effective for quantifying aromatics, oxidized cumene was directly injected in the HPLC without the use of the post-column reactor. This approach allows us to precisely measure the aromatic compounds in the oxidized cumene sample. Autoxidation of cumene was achieved in the PetroOxy, until the IP was reached at 18 min. The oxidized cumene was collected and a sample was used to determine the total ROOH concentration using the traditional iodometric titration with potentiometric detection of the endpoint, using TPP as titrant [20]. Another sample was analyzed with HPLC-PcR and the chromatogram measured at 254 nm can be found in Figure 11.

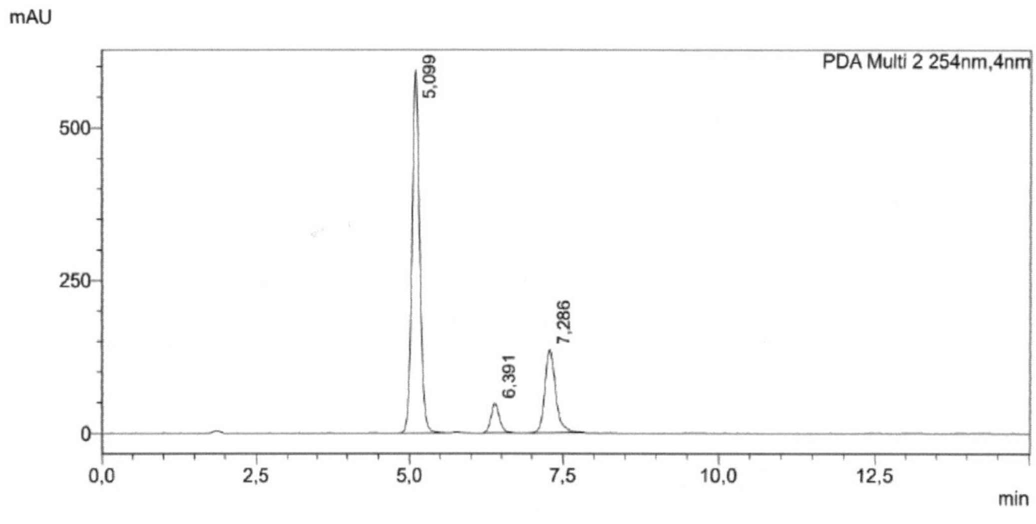

	Concentration (mol/L)	% molar
Cumene	6.91	96.4 %
Acetophenone	0.028	0.4 %
Cumene hydroperoxide	0.201	2.8 %
2-phenyl 2-propanol	0.036	0.5 %

Figure 11: HPLC chromatogram (254 nm) of cumene aged for 18 min in a PetroOxy. The table below the chromatogram gives the corresponding concentration from our calibration.

At 254 nm, the different oxidation products can be observed, as their aromatic group absorb strongly at this wavelength, enabling their identification and analysis. It may be recalled that the chromatograms of hydroperoxides are measured at 362 nm, by measuring the absorption of I_3^-. A preliminary study was conducted to identify the composition of the aged fuel. Each oxidation product was prepared at different concentrations and passed through the system. The retention time of each product was determined, and the quantification was conducted by correlating the area of the peaks with the concentration given by the Sigma-Aldrich on the pure component. The data obtained includes the retention times and the concentrations of each product. The table in the Figure 12 summarizes the retention times and the calibration coefficient α (to correlate the area of the peak with its concentration) of the main products of cumene oxidation: acetophenone, cumene hydroperoxide and 2-phenyl-2-propanol.

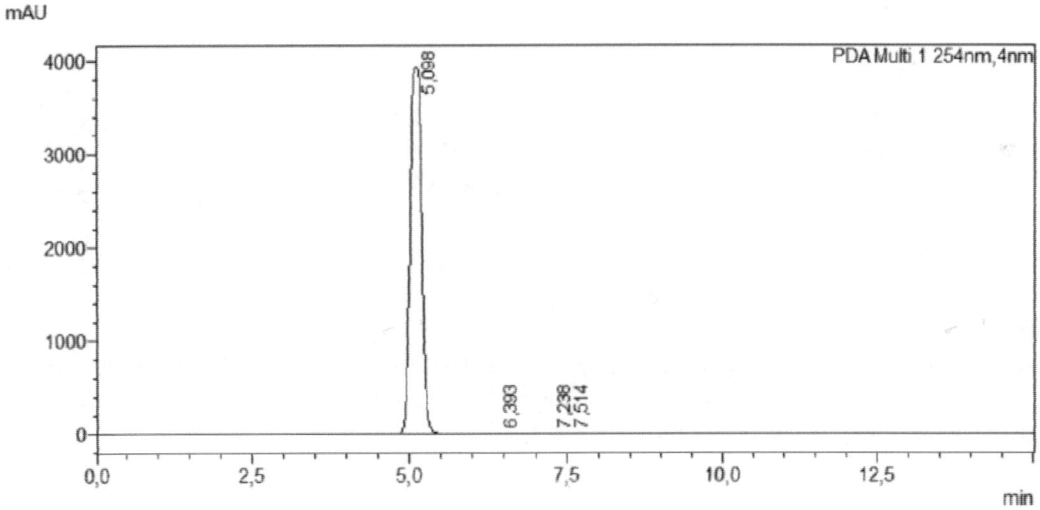

Molecule	Retention time (min)	Concentration= α *Area (x10⁷)
Cumene	5.099	7.75
Acetophenone	6.392	162
Cumene Hydroperoxide	7.238	3.84
2-phenyl-2-propanol	7.531	5.98

Figure 12: HPLC chromatogram of fresh cumene at 254 nm. Calculations realized to calculate the conversion of cumene unaged cumene.

Pure cumene exhibited a single peak at 5.099 minutes, confirming its purity at 99.9%. The initial concentration was determined using the area of the pure cumene chromatogram, and the difference of the peaks of pure and aged cumene was calculated. This study showed that for a residence time of 18 minutes, corresponding to the IP, the conversion of cumene is 3.6%. A similar analysis was conducted for acetophenone and 2-phenyl-2-propanol to determine their concentrations in the aged mixture, with the results shown in Figure 11.

A study at 362 nm was conducted in the HPLC-PcR to quantify and identify the ROOH formed in cumene autoxidation when the IP time is reached. The chromatogram is presented in Figure 13.

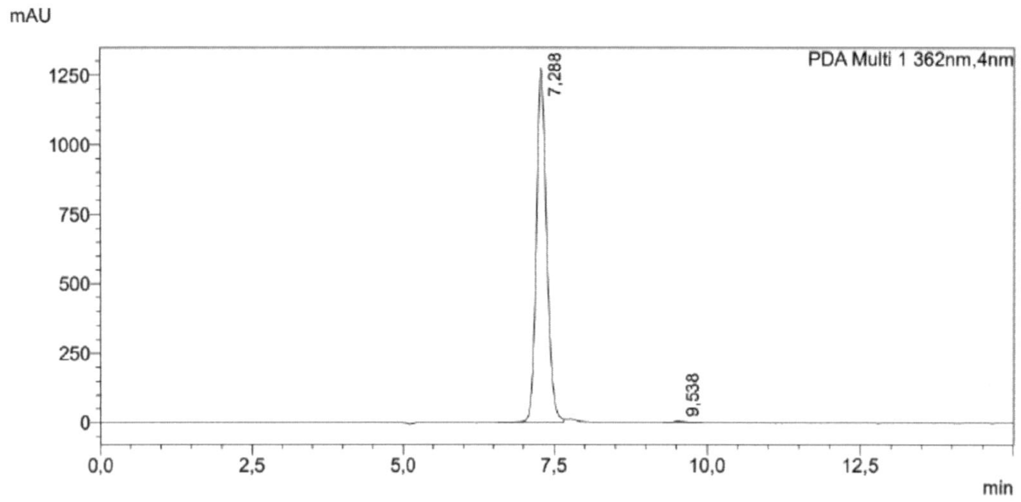

	[ROOH] fresh cumene (mol/L)	[ROOH] aged cumene (mol/L)
Iodometric titration	-	1.70 x 10^{-1} mol/L
HPLC-P-c-R	< 1x10^{-7} mol/L	2.03 x 10^{-1} mol/L
Deviation	-	16 %

Figure 13: HPLC-PcR chromatogram measured at 362 nm of oxidized cumene. Table: total hydroperoxides concentrations of fresh and aged cumene, with HPLC-PcR technique and iodometric titration.

Only two hydroperoxides are observed in oxidized cumene, with cumene hydroperoxide being the principal one. Another peroxide is detected at a retention time of 9.5 min. The total hydroperoxide quantification techniques were compared in the table below the chromatogram in Figure 13 and the small deviation found is mainly due to the dilution that was done for the analysis in HPLC-PcR, since the sample was too concentrated (~1 x 10^{-1} mol/L) for the detector. We already showed that dilutions with micropipettes contribute to increased uncertainty in the quantification of hydroperoxides [1]. The minor peak at 9.5 min in Figure 13, was attributed to CH_3OOH. Indeed, in the literature, it has been reported that there is a competition between the β-scission of cumyloxy radical $C_6H_5C(CH_3)_2O\bullet \rightarrow CH_3\bullet$ + acetophenone and its H-abstraction from cumene: $C_6H_5C(CH_3)_2O\bullet$ + cumene → 2-phenyl 2-propanol + cumyl radical [21]. If one assumes that all the acetophenone formed comes from the β-scission of the cumuloxy radical, then an equivalent amount of $CH_3\bullet$ is produced. The fate of $CH_3\bullet$ in the liquid is either to abstract an H-atom from cumene or to add onto O_2. Our CBS-QB3 calculations predicts a barrier at 0K of 8.5 kcal/mol for the H-abstraction from cumene by $CH_3\bullet$ [22]. Therefore, the addition onto O_2, which is a barrierless process yielding $CH_3OO\bullet$, is favoured from a potential energy point of view. Since liquid cumene is supposed to be saturated with O_2 in the PetroOxy experiments, $[O_2]$ should not be limiting the rate of the $CH_3\bullet + O_2 \leftrightarrows CH_3OO\bullet$ reaction. As most peroxy radicals in hydrocarbon autoxidation, most $CH_3OO\bullet$ will lead to the formation of CH_3OOH by H-abstraction. This represents a high limit, as some $CH_3\bullet$ may be consumed by termination or addition on unsaturated molecules but is consistent with the results observed in HPLC-PcR.

n-decane autoxidation: HPLC-PcR and GC-MS analysis

n-decane was oxidized in the PEEK-ACHU device for 30 min at 413 K, under a pressure of 20 bars of O_2, and the sample was recovered and collected in a 1.5 mL vial. We showed that hydroperoxides found in liquid *n*-decane oxidized in both PetroOxy and PEEK-ACH devices are equivalent. Notably here, the residence time of 30 min is 6 times shorter than the IP time of *n*-decane in PetroOxy, at 413K under 7 bar of O_2 (IP = 180 min). The sample was first analyzed with HPLC-PcR, after which several peaks were sampled using a fraction collector for further analysis by GC-MS. Figure 14 shows the chromatogram of hydroperoxides formed after 30 min of *n*-decane oxidation.

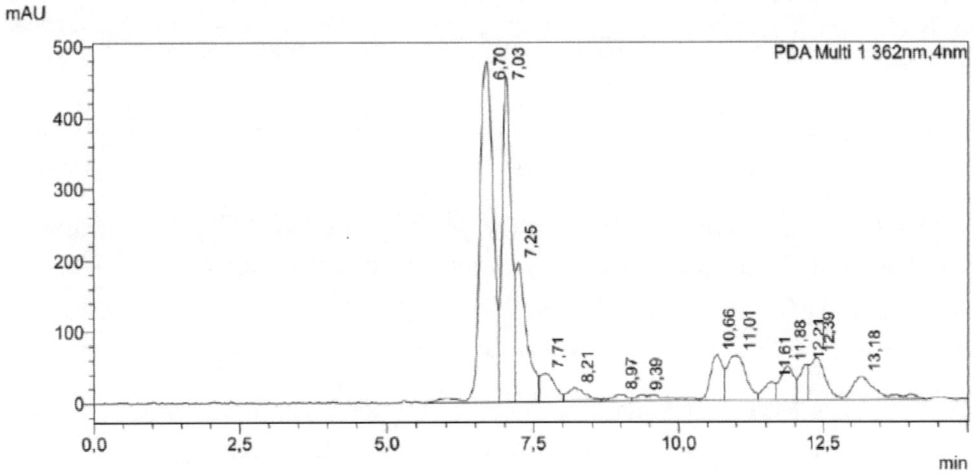

Figure 14: Chromatogram of aged *n*-decane in PEEK-ACHU rig for 30 min at 20 bar of O_2 and 413 K | each peak represents a different molecule of hydroperoxide.

A large number of peaks are obtained in the HPLC-PcR analysis of aged *n*-decane demonstrating the presence of several molecules bearing, at least, one hydroperoxide functional group. The chromatogram appears to be more complex than the ones obtained for 1-butanol (Figure 9) and cumene (Figure 13). Among the 14 peaks observed in Figure 14, it can be expected that the most abundant are the 5-decane hydroperoxide isomers (which elute between 6 to 8 minutes in the chromatogram of Figure 14). Based on comparisons with the smaller hydroperoxides identified in HPLC-PcR analysis for oxidized cumene and 1-butanol, one can expect that smaller hydroperoxides with characteristics similar to CH_3OOH will appear around 9 minute mark, while H_2O_2 will be detected around 12 minutes. Other peaks may be attributed to dihydroperoxides, peracids or hydroperoxide dimers, further contributing to the complexity of the chromatogram.

In the spirit of finding the most abundant hydroperoxide formed during the aging of the fuel, the fraction collector was used to collect samples from each peak for subsequent analysis via the GC-MS, as illustrated in Figure 15.

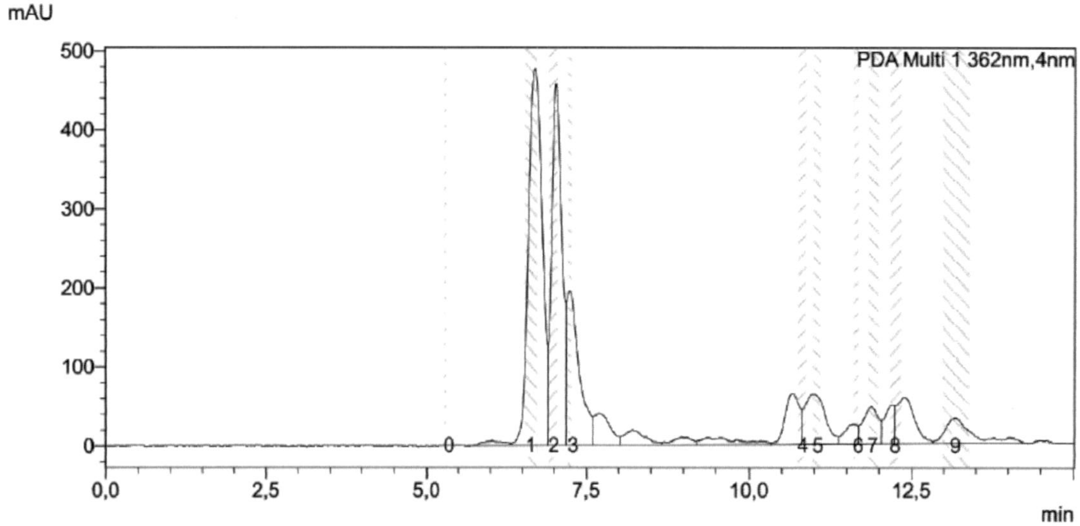

Figure 15: Chromatogram of the aged *n*-decane sample and in light blue: fraction recovered in the vial and the associated number

It should be recalled that the chromatogram obtained using HPLC-PcR corresponds to the absorbance of I_3^- at 362 nm, produced by the reduction of hydroperoxides by I^- ions, yielding alcohols, I_2, and water (reaction (1)). The GC-MS analysis of the collected fractions is therefore expected to give information solely into the alcohols formed within the post-column reactor.

The analysis of the collected fractions is a challenge that remains to be addressed. Indeed, each fraction collected contains only a very small quantity of a few hundred microliters, which is often insufficient for significant peak detection compared to the signal of other species analyzed. To explore the limit at which the signal of alcohol formed in the post-column reactor can be detected, we increased the number of experiments, allowing us to accumulate 3 accumulations of the fraction corresponding to peak #1 in Figure 15. The GC-MS spectrum is given in Figure 16.

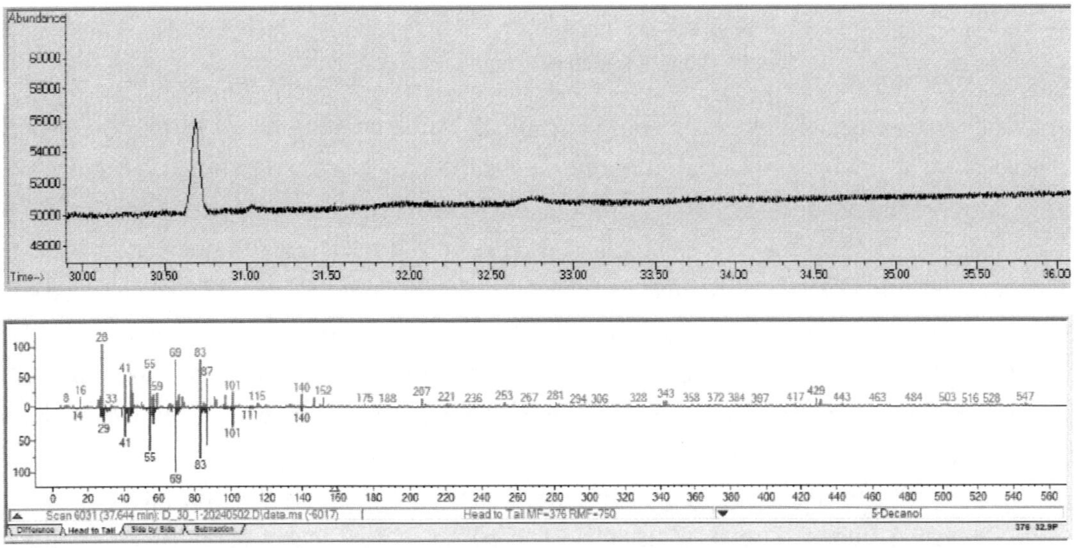

Figure 16: Gas chromatogram (upper panel) of the FID signal and (lower panel) the mass spectrum associated to the peak at the retention time 30.65 min (red) compared to the 5-decanol mass spectrum (blue).

The main alcohol identified in peak #1 with the GC-MS apparatus, besides ethanol from the eluent, is 5-decanol as shown in Figure 16. This indicates that peak #1 corresponds to the signal of 5-decane hydroperoxide. The retention time of peak #1 of 6.7 min in the HPLC-PcR chromatogram is consistent with our previous analyses on cumene hydroperoxide (7.2 min) and tert-butylhydroperoxide (7.5 min [1]). Thus, the most abundant hydroperoxides detected are indeed decane hydroperoxide isomers, as expected. Notably, our group [18] developed a detailed chemical kinetic model to simulate the autoxidation of *n*-decane. Our simulations at low conversion levels have confirmed that *n*-decane is consumed through the classical pathways postulated in the literature (Figure 17).

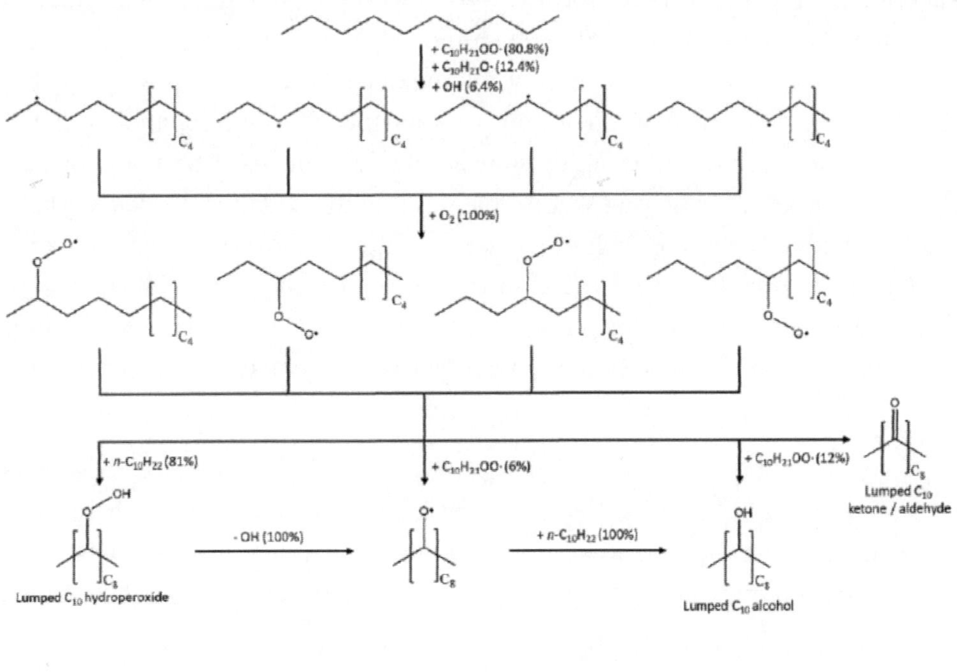

Figure 17: Reaction pathways of *n*-decane in the liquid phase at 413 K. ≈ 4 % conversion of *n*-decane (190 min residence time). (Adapted from Le et al. 2023)[18]

H-abstractions from *n*-decane by decane peroxy radicals is the only source of formation of decane hydroperoxides. The analysis of the simulations shows that, under the conditions of Figure 17, the branching ratios of the products is: 80.3% of *n*-decane hydroperoxides, 11.6% of C_{10} alcohols, 4% of C_{10} ketones and H_2O, 0.1% of decenes, and 0.05% of H_2O_2. Thus, the presence of H_2O_2i is expected in the chromatogram presented in Figure 14. In the simulation, H_2O_2 is formed by H-abstraction by HO_2• radical which comes from the oxidation reaction of decyl radicals: •$C_{10}H_{21}$ + O_2 → $C_{10}H_{22}$ + HO_2•. This reaction also explains the formation of decenes. Hydroperoxides from the oxidation of decenes could also be detected in our HPLC-PcR. The oxidation of C_{10} alcohols could also yield H_2O_2 in non-negligible amount, as we showed for 1-butanol. However, it is currently not possible to use simulations to verify these hypotheses as the current version of the kinetic model only includes lumped sub-mechanisms for the autoxidation of decenes, C_{10} alcohols and C_{10} ketones, that directly leads to smaller fragments that are decomposed in the reaction base. This demonstrates that further work on secondary mechanisms need to be done to take into account the new measurements of hydroperoxides made possible by our HPLC-PcR methodology.

CONCLUSION

This study presents a significant advancement in the understanding of hydroperoxide formation during the autoxidation of liquid fuels. This work identifies and quantifies the most abundant hydroperoxides found in aged liquid surrogate fuels. Different types of molecules were investigated i.e., an alkane: *n*-decane, an alcohol: *n*-butanol, and an aromatic: cumene. The liquid fuels were aged in the PetroOxy device that gives a quantitative data (IP) on their oxidation stabilities and the hydroperoxides thus formed were measured using different methods based on HPLC. The analysis of aged *n*-butanol with our newly developed HPLC-PcR methodology allowed the identification and quantification of hydrogen peroxide as the primary hydroperoxide formed in its oxidation, demonstrating an original reaction mechanism compared to alkanes. HPLC with a PDA detector has proven to be highly suited to study of the oxidation of cumene primarily due to the presence of its aromatic ring that is conserved in its oxidation products. The use of the classical HPLC, without the post-column reactor, allowed the direct quantification of the conversion of cumene at the IP time, which is nearly 4% at 18 min, along with its main product of oxidation (after calibration). When the post-column is coupled to the HPLC, two hydroperoxides are detected and quantified: the cumene hydroperoxide, which is expected, and methyl hydroperoxide, which is reported here for the first time and represents a novel contribution to the literature.

The innovative use of HPLC-PcR has demonstrated its capability to separate hydroperoxides in aged *n*-decane highlighting the formation of a multitude of different hydroperoxides that can be quantified. The collected fractions corresponding to each peak observed in the chromatogram were subjected to further analyses via a GC-MS apparatus. While this approach highlights the need for additional work on the accumulation technique of fractions, it successfully identified the most abundant hydroperoxide formed during n-decane autoxidation: 5-hydroperoxydecane. This successful recognition underscores the utility of this method. The analysis of the hydroperoxides chromatogram to identify each peak, required efforts both on the experimental (Orbitrap analyses, for example) and modeling sides.

ACKNOWLEDGEMENTS

This project has received funding from the European Research Council (ERC) under the European Union's Horizon 2020 research and innovation program (Project: 101003318). This work was partly supported by the French RENATECH network and its FEMTO-ST technological facility.

REFERENCES

(1) Benrabah, R.; El Sayah, Z.; Glaude, P.-A.; Arnoux, P.; Fournet, R.; Sirjean, B. Quantification of Hydroperoxides in Liquid Fuels: A Systematic Comparison of Titration and Absorption Methods and Their Coupling to HPLC. *Fuel* **2024**, *365*, 131218. https://doi.org/10.1016/j.fuel.2024.131218.

(2) Sicard, M.; Quignard, A. Carburéacteurs alternatifs aéronautiques et leur déploiement. *Techniques de l'Ingénieur* **2023**, *TRP4058 v1*. https://doi.org/10.51257/a-v1-trp4058.

(3) Sicard, M.; Quignard, A. Carburants aéronautiques. *Techniques de l'Ingénieur* **2021**, *TRP4054 V1*. https://doi.org/10.51257/a-v1-trp4054.

(4) Jia, T.; Zhang, X.; Liu, Y.; Gong, S.; Deng, C.; Pan, L.; Zou, J. J. A Comprehensive Review of the Thermal Oxidation Stability of Jet Fuels. *Chemical Engineering Science* **2021**, *229*, 116157. https://doi.org/10.1016/j.ces.2020.116157.

(5) *Standard - Standard Test Method for Hydroperoxide Number of Aviation Turbine Fuels, Gasoline and Diesel Fuels ASTM D3703 - Swedish Institute for Standards, SIS.* https://www.sis.se/en/produkter/external-categories/petroleum-products-lubricants-and-fossil-fuels-astm-vol-05/petroleum-products-and-lubricants-i-c1234--d3710-astm-vol-0501/astmd3703/ (accessed 2022-04-08).

(6) Alves-Fortunato, M.; Baroni, A.; Neocel, L.; Chardin, M.; Matrat, M.; Boucaud, C.; Mazarin, M. Gasoline Oxidation Stability: Deposit Formation Tendencies Evaluated by PetroOxy and Autoclave Methods and GDI/PFI Engine Tests. *Energy Fuels* **2021**, *35* (22), 18430–18440. https://doi.org/10.1021/acs.energyfuels.1c02466.

(7) Longobardi, F.; Contillo, F.; Catucci, L.; Tommasi, L.; Caponio, F.; Paradiso, V. M. Analysis of Peroxide Value in Olive Oils with an Easy and Green Method. *Food Control* **2021**, *130*, 108295. https://doi.org/10.1016/j.foodcont.2021.108295.

(8) Belhadj, N.; Benoit, R.; Dagaut, P.; Lailliau, M.; Serinyel, Z.; Dayma, G.; Khaled, F.; Moreau, B.; Foucher, F. Oxidation of Di-n-Butyl Ether: Experimental Characterization of Low-Temperature Products in JSR and RCM. *Combustion and Flame* **2020**, *222*, 133–144. https://doi.org/10.1016/j.combustflame.2020.08.037.

(9) Sahetchian, K. A.; Rigny, R.; Circan, S. Identification of the Hydroperoxide Formed by Isomerization Reactions during the Oxidation of *n*-Heptane in a Reactor and CFR Engine. *Combustion and Flame* **1991**, *85* (3), 511–514. https://doi.org/10.1016/0010-2180(91)90153-3.

(10) Battin-Leclerc, F.; Herbinet, O.; Glaude, P.-A.; Fournet, R.; Zhou, Z.; Deng, L.; Guo, H.; Xie, M.; Qi, F. Experimental Confirmation of the Low-Temperature Oxidation Scheme of Alkanes. *Angewandte Chemie International Edition* **2010**, *49* (18), 3169–3172. https://doi.org/10.1002/anie.200906850.

(11) Moshammer, K.; Jasper, A. W.; Popolan-Vaida, D. M.; Lucassen, A.; Diévart, P.; Selim, H.; Eskola, A. J.; Taatjes, C. A.; Leone, S. R.; Sarathy, S. M.; Ju, Y.; Dagaut, P.; Kohse-Höinghaus, K.; Hansen, N. Detection and Identification of the Keto-Hydroperoxide (HOOCH2OCHO) and Other Intermediates during Low-Temperature Oxidation of Dimethyl Ether. *J. Phys. Chem. A* **2015**, *119* (28), 7361–7374. https://doi.org/10.1021/acs.jpca.5b00101.

(12) Belhadj, N.; Benoit, R.; Dagaut, P.; Lailliau, M. Experimental Characterization of N-Heptane Low-Temperature Oxidation Products Including Keto-Hydroperoxides and Highly Oxygenated Organic Molecules (HOMs). *Combustion and Flame* **2021**, *224*, 83–93. https://doi.org/10.1016/j.combustflame.2020.10.021.

(13) West, Z. J.; Zabarnick, S.; Striebich, R. C. APPLIED CHEMISTRY Determination of Hydroperoxides in Jet Fuel via Reaction with Triphenylphosphine. *American Chemical Society, Division of Petroleum Chemistry, Preprints* **2005**, *49* (4), 481–483. https://doi.org/10.1021/ie0490379.

(14) Geleynse, S.; Brandt, K.; Garcia-Perez, M.; Wolcott, M.; Zhang, X. The Alcohol-to-Jet Conversion Pathway for Drop-In Biofuels: Techno-Economic Evaluation. *ChemSusChem* **2018**, *11* (21), 3728–3741. https://doi.org/10.1002/cssc.201801690.

(15) Hock, H.; Lang, S. Autoxydation von Kohlenwasserstoffen, IX. Mitteil.: Über Peroxyde von Benzol-Derivaten. *Berichte der deutschen chemischen Gesellschaft (A and B Series)* **1944**, *77* (3–4), 257–264. https://doi.org/10.1002/cber.19440770321.

(16) Benrabah, R.; Girot, E.; Commenge, J.-M.; Arnoux, P.; Fournet, R.; Glaude, P.-A.; Sirjean, B. An Innovative Microfluidic Reactor for Testing the Oxidation Stability of Fuels: Application to a SAF Surrogate and Comparison with the PetroOxy Test; 18th International Conference on Stability, Handling and Use of Liquid Fuels,Louisville, USA, 2024.

(17) de Saint Laumer, J.-Y.; Leocata, S.; Tissot, E.; Baroux, L.; Kampf, D. M.; Merle, P.; Boschung, A.; Seyfried, M.; Chaintreau, A. Prediction of Response Factors for Gas

Chromatography with Flame Ionization Detection: Algorithm Improvement, Extension to Silylated Compounds, and Application to the Quantification of Metabolites. *Journal of Separation Science* **2015**, *38* (18), 3209–3217. https://doi.org/10.1002/jssc.201500106.

(18) Le, M. D.; El Sayah, Z.; Benrabah, R.; Warth, V.; Glaude, P.-A.; Privat, R.; Fournet, R.; Sirjean, B. An Experimental and Detailed Kinetic Modeling of the Thermal Oxidation Stability of N-Decane as a Jet Fuel Surrogate Component. *Fuel* **2023**, *342*, 127754. https://doi.org/10.1016/j.fuel.2023.127754.

(19) Bhattacharya, A. Kinetic Modeling of Liquid Phase Autoxidation of Cumene. *Chemical Engineering Journal* **2008**, *137* (2), 308–319. https://doi.org/10.1016/j.cej.2007.04.043.

(20) Roohi, H.; Rajabi, M. Iodometric Determination of Hydroperoxides in Hydrocarbon Autoxidation Reactions Using Triphenylphosphine Solution as a Titrant: A New Protocol. *Industrial and Engineering Chemistry Research* **2018**, *57* (20), 6805–6814. https://doi.org/10.1021/ACS.IECR.7B05403.

(21) Baignee, A.; Howard, J. A.; Scaiano, J. C.; Stewart, L. C. Absolute Rate Constants for Reactions of Cumyloxy in Solution. *J. Am. Chem. Soc.* **1983**, *105* (19), 6120–6123. https://doi.org/10.1021/ja00357a024.

(22) Roux, B.; Simon, Y.; Poeuf, S.; Bouchez, M.; Lechevallier, M.; Glaude, P.-A. A.; Sirjean, B.; Fournet, R. Kinetic Study of the Pyrolysis of Cumene; 11th European Combustion Meeting, April 26-28, 2023, Rouen, France, 2023.

IASH 2024, the 18TH INTERNATIONAL CONFERENCE ON STABILITY, HANDLING AND USE OF LIQUID FUELS
Louisville, Kentucky, USA
September 8 – 12, 2024

(Technical Poster)

PREDICTION OF HYDROCARBONS IGNITION PERFORMANCES USING MACHINE LEARNING MODELING

Giacomo Flora*[1], Forood Karimzadeh[1], Moshan S.P. Kahandawala[1], Matthew J. DeWitt[1], Edwin Corporan[2]

[1]University of Dayton Research Institute, 300 College Park Ave, Dayton Ohio 45469 USA
[2]Air Force Research Lab, AFRL/RQTF, 1790 Loop Rd, Wright-Patterson AFB, OH 45433 USA

ABSTRACT

The excessive costs and long processing times associated with fuel certification could be mitigated by using computational models capable of predicting actual testing results. Quantitative Structure-Property Relationship (QSPR) models have been extensively utilized to correlate various physicochemical properties of organic compounds with molecular structure descriptors. The Derived Cetane Number (DCN) serves as a valuable pre-screening tool for fuels, and the abundant availability of DCN measurements for individual hydrocarbon species, surrogate blends, and practical fuels makes it an appealing candidate for the development of QSPR models. This study presents a novel computational methodology for determining the DCN of practical aviation fuels. The proposed approach integrates a novel Quantitative Structure-Property Relationship (QSPR) model designed to predict DCN for hydrocarbon species and mixtures with fuel composition analysis obtained through Two-Dimensional Gas Chromatography (GCxGC). The QSPR model used 20 independent variables computed based on selected hydrocarbon molecular descriptors, including functional groups and distance-based topological indexes. The multivariate regression analysis was used to train the QSPR model based on a dataset consisting of 48 individual hydrocarbon species and 157 surrogate mixtures. The model demonstrated robust predictive capabilities with high coefficients of determination (R^2) of 0.96 on the training dataset and 0.94 on the independent testing dataset. The latter consisted of 43 surrogate mixtures formulated both in-house and sourced from archived literature. The application of the QSPR model for practical jet fuels involves specifying the detailed jet fuel compositions using GCxGC analysis and a randomization algorithm based on a database featuring over 17,000 distinct hydrocarbon species. The overall model's performance on practical jet fuels aligns closely with its performance on the training and testing datasets, affirming its practical utility. To enhance prediction accuracy of the proposed computational approach for practical jet fuels, density was explored as a potential constraining property to narrow randomization results with those detailed composition having a density similar to the actual fuel. In this regard, a novel relationship was established to predict fuel densities based on compositional characteristics. Despite the promising results in density prediction, this study indicates that density alone is insufficient to effectively constrain randomized compositions for significantly improved DCN predictions, and thus, further development is required.

*Corresponding author: Giacomo Flora, Giacomo.Flora@udri.udayton.edu

IASH 2024, the 18TH INTERNATIONAL CONFERENCE ON STABILITY, HANDLING AND USE OF LIQUID FUELS
Louisville, Kentucky, USA
September 8 – 12, 2024

(Technical Poster)

INVESTIGATING THE SURVIVAL AND GROWTH OF MICROORGANISMS IN MICROCOSMS SIMULATING TEMPERATURE CYCLING IN AIRCRAFT FUEL TANKS

Lucy Owen, Gareth Williams*, Graham C. Hill

ECHA Microbiology, Cardiff, UK

ABSTRACT

Microbial growth in aircraft fuel tanks may impact operational safety through issues such as filter clogging, fuel quality indicator system malfunction or microbially influenced corrosion.

Fuel-associated microorganisms are typically mesophilic, with optimum growth at moderate temperatures (20-35°C), allowing proliferation to occur when aircraft are parked. However, aircraft fuel tanks experience temperatures as low as -40°C during flights. The impact of temperature fluctuations in aircraft fuel tanks between flights and parking on fuel-associated microbial communities is not well studied.

This investigation aimed to determine the survival and proliferation of microorganisms in jet fuel/water microcosms subjected to cycles of freezing and thawing.

Pure cultures of microorganisms (*Hormoconis resinae, Pseudomonas aeruginosa, Yarrowia lypolytica*) and an undefined jet fuel-derived consortium were exposed to temperature cycles representative of aircraft flight (-20°C) and parking conditions (25°C) up to 3 times in aqueous mineral salts medium (MSM) overlayed with jet fuel. The growth rate of surviving microorganisms was then measured during incubation at 25°C. Effects of aqueous phase solute concentration was determined by comparing survival and growth after temperature cycling in MSM and deionized water overlayed with jet fuel.

The results indicate that all test microorganisms can survive freeze-thaw temperature cycling in jet fuel. Whilst exposed to daily freeze cycles, microbial populations generally remained static or declined. However, as soon as freeze cycles ceased, recovery and proliferation to above initial levels was seen within 1 day for bacteria and ≤7 days for fungi. There were preliminary indications that some fungi could grow more rapidly after freeze-thaw cycling.

*Corresponding author: Gareth Williams, gwilliams@echamicrobiology.com

IASH 2024, the 18TH INTERNATIONAL CONFERENCE ON STABILITY, HANDLING AND USE OF LIQUID FUELS
Louisville, Kentucky, USA
September 8 – 12, 2024

(Technical Poster)

MOBILE DESULFURIZATION OF HIGH SULFUR JET FUELS

Lindsay Davis*, Giacomo Flora, Russell Tobias, Brian Hermano, Cody Tschantz, Juan Font-Rivera

University of Dayton Research Institute, 300 College Park Ave, Dayton Ohio 45469 USA

ABSTRACT

Sulfur-containing molecules are prevalent as native impurities in all petroleum distillates, leading to the formation of sulfur oxides (SOx) during fuel combustion. These oxides, combined with other combustion pollutants, contribute to environmental issues such as soot, global warming, and water pollution. Despite stringent heavy-duty highway engine standards, significant emissions of nitrogen oxides (NOx) and particulate matter (PM) persist, posing serious public health risks, including lung cancer, respiratory and cardiovascular diseases, and chronic bronchitis. Therefore, desulfurization of fuels is critically important in the petroleum industry. The US Environmental Protection Agency (EPA) limits sulfur levels in diesel fuel to 15 ppm by weight under the Ultra Low Sulfur Diesel (ULSD) specification. Compliance often necessitates commercial off-the-shelf (COTS) diesel engines to incorporate sulfur detection sensors, disabling engines if non-specification fuel is detected. Sulfur in fuels can also cause mechanical issues in engines, including coking, clogging, fouling, and corrosion. The Department of Defense (DoD) and Air Force (AF) extensively use these COTS engines, which amplifies the need for ULSD fuel, especially in regions outside the continental US where local sulfur standards may be less stringent, allowing for sulfur content to be over 15 ppm. To meet ULSD fuel demands, US and European refineries employ hydrodesulfurization (HDS), an energy-intensive and costly process suitable for large-scale operations. In regions where refinery upgrades are lacking, locally produced fuel often fails to meet ULSD standards, complicating DoD operations that require significant quantities of compliant fuel. In response to these challenges, an on-site deployable fuel desulfurization system is needed. Oxidative desulfurization (ODS) has emerged as a promising alternative, operating at low temperatures and pressures. ODS oxidizes sulfides into sulfoxides or sulfones, which are then extracted using liquid-liquid extraction techniques. The University of Dayton Research Institute (UDRI) has developed a deployable desulfurization technology using ODS by catalytic ozonation with solvent extraction, complemented by a solvent recovery unit to reduce operational costs. The effort undertaken by UDRI has been worked with the application of ULSD in tier four support equipment engines as used by the DoD. The current study demonstrates the applicability of the UDRI desulfurization technologies to jet fuels such as Jet-A. Optimization studies were conducted to identify optimal process parameters, including ozone-to-fuel ratio, weight hourly space velocity, number of extraction stages, and solvent recovery efficiencies. The results of the optimization demonstrated the suitability of this technology for mobile utilization. In this regard, UDRI has employed a model-based systems engineering (MBSE) methodology to design a deployable system to be fabricated in 2024 and undergo testing in 2025. The use of MBSE in the development of the final prototype design illustrates the applicability of the methodology to chemical processes and ultimately integration to DoD lifecycle operations.

*Corresponding author: Lindsay Davis, Lindsay.McDowall@udri.udayton.edu

IASH 2024, the 18TH INTERNATIONAL CONFERENCE ON STABILITY, HANDLING AND USE OF LIQUID FUELS
Louisville, Kentucky, USA
September 8 – 12, 2024

(Technical Poster)

DETERMINATION OF TRACE METALS IN AVIATION TURBINE FUEL VIA ICP-MS

Amanda M. Arts[*1], Zachary J. West[1], and Paul Wrzesinski[2]

[1]*University of Dayton Research Institute, 300 College Park Dr, Dayton, Ohio 45469-0043, United States*

[2]*Air Force Research Laboratory, 1790 Loop Rd, Wright-Patterson AFB, Ohio 45433-7131, United States*

ABSTRACT

Trace elements in jet fuel, such as Copper (Cu) and Zinc (Zn), can negatively impact jet fuel thermal oxidative stability at concentrations as low as tens of parts per billion (ppb). The impacts of these metals include increased oxidation and/or deposition rates which foul critical fuel system components. The determination of trace metals in organic solutions has always been difficult, even with advanced instrumentation. However, this work outlines the development of a method using inductively coupled plasma (ICP) with mass spectrometry (MS) detection to determine the amounts of trace elements in organic sample matrices. This technique can be used to analyze trace metals in organic samples that are middle distillate products of low volatility such as kerosene, aviation turbine fuel, synthetic aviation turbine fuel (SATF), or diesel. This trace metals research will aid in the understanding of safe and effective fuel handling, cleanliness, and stability.

*Corresponding author: Amanda M. Arts, Amanda.Arts@udri.udayton.edu

IASH 2024, the 18TH INTERNATIONAL CONFERENCE ON STABILITY, HANDLING AND USE OF LIQUID FUELS
Louisville, Kentucky, USA
September 8 – 12, 2024

REDUCING EMISSIONS WITH DIESEL DETERGENT ADDITIVES: INSIGHTS FROM ISO8178 TEST CYCLES

Oliver James Rumford-Warr [1]

[1] Fuelcare Limited, Mercury House, Shrewsbury Business Park, Shrewsbury, SY2 6LG, UK, sales@fuelcare.com

ABSTRACT

This study investigates the impact of a PIBSI-based detergent additive on emissions reduction in heavy-duty diesel engines. The research focuses on two heavy-duty diesel engines tested to the ISO 8178 cycle F[1] using EN590 diesel fuel. The primary objective is to identify potential emissions savings derived from reduced carbon buildup due to the cleaning properties of the detergent additive on injectors and fuel systems. The ISO8178 cycle F was chosen for its relevance in simulating real-world operating conditions of heavy-duty engines and specifically those found in U.K. rail applications, providing a robust framework for emissions testing. A PIBSI-based additive was evaluated to determine its role in mitigating emissions. Results from the testing indicated a reduction in regulated emissions and fuel usage. Improved injector cleanliness led to more efficient combustion, which in turn resulted in lower carbon deposits and reduced emissions. The study demonstrates that the inclusion of a detergent additive can have a meaningful impact on emission reductions, particularly in heavy- duty diesel applications such as U.K. rail or global marine. These findings underscore the potential environmental benefits of incorporating detergent additives in diesel fuels, especially in the context of stringent emissions regulations. By maintaining injector and fuel system cleanliness, such additives can enhance engine performance but also contributes to a reduction in harmful emissions. The research provides valuable insights for policymakers and industry stakeholders aiming to develop strategies for cleaner and more efficient diesel engine operations.

INTRODUCTION

The transportation sector, particularly rail transport, is under increasing pressure to reduce greenhouse gas (GHG) emissions and improve fuel efficiency. The customer (a U.K. Government entity), in partnership with Fuelcare, conducted a comprehensive evaluation of a PIBSI-based detergent fuel additive on MTU 183TD and Cummins NT855 heavy duty diesel engines. This study aimed to determine the effectiveness of detergents in enhancing engine performance, fuel economy, and crucially, reducing emissions through a series of tests aligned with ISO8178 test cycles.

EXPERIMENTAL

The test program was executed at Millbrook Leyland, an independent, accredited engine testing facility. The methodology involved several key steps to ensure accurate and reliable results:

- Fuel Preparation: A batch of EN590 Gas Oil was set aside and blended with a detergent additive at 1000ppm by volume.
- Engine Setup: The MTU 183TD and Cummins NT855 engines were installed in test beds and instrumented for monitoring operational parameters.
- Test Phases: The evaluation comprised baseline performance tests with clean and dirty injectors, followed by fuel trial conditioning with detergent additive for
- over 200 hours.
- ISO8178 Cycle F Analysis: This modified test cycle, which closely mimics real-world locomotive operations, was employed to measure emissions, fuel
- consumption, and performance metrics such as torque and power.

RESULTS AND DISCUSSION

The full results[2] from the testing were significant and demonstrated clear benefits of using Detergent Additives in diesel engines:

- Fuel Economy: MTU 183TD: A 3.41% improvement in Brake Specific Fuel Consumption (BSFC) was observed at 200 hours. Cummins NT855: A 2.66% improvement in BSFC was noted at 100 hours.
- Power and Torque: MTU 183TD: Corrected power increased by 21.40% and torque by 21.70% at 200 hours. Cummins NT855: Corrected power increased by 12.42% and torque by 14.02% at 100 hours.
- Emissions Reduction: The improved fuel economy translates into substantial CO2e savings for the associated fleet. For the MTU 183TD engine, the annual CO2e reduction was estimated at 1,311,607 kg, while for the Cummins NT855 engine, it was 893,080 kg.
- Injector and Engine Cleanliness: Pre and post-trial injector analyses showed significant reductions in carbonaceous deposits, contributing to improved fuel atomisation and combustion efficiency. Borescope inspections of engine cylinders before and after the trial revealed noticeable clean-up effects, supporting the statistical data on improved performance.

CONCLUSIONS

The study concluded that the use of Detergent Additives in diesel engines provides substantial benefits in terms of fuel economy, power output, and emissions reduction. The MTU 183TD and Cummins NT855 engines, after conditioning with Detergent Additives, demonstrated improved performance metrics across the board.

These findings are critical for rail operators looking to reduce their environmental footprint and operational costs. The integration of diesel detergent additives can play a pivotal role in achieving these goals. The continued use of such additives is recommended, particularly as fleets transition to include higher biodiesel blends, which can exacerbate engine deposition issues.

Overall, the Detergent Additives fuel additive has proven to be an effective solution for enhancing diesel engine performance, leading to lower fuel consumption and reduced GHG emissions, aligning with the operators commitment to sustainable operations.

REFERENCES

1. https://dieselnet.com/standards/cycles/iso8178.php - complete ISO 8178 Cycle F duty cycle.
2. https://www.fuelcare.com/fuel-detergent - full test data available at www.fuelcare.com.

IASH 2024, the 18TH INTERNATIONAL CONFERENCE ON STABILITY, HANDLING AND USE OF LIQUID FUELS
Louisville, Kentucky, USA
September 8 – 12, 2024

(Technical Poster)

DEVELOPMENT OF BIOCIDES FOR MITIGATION OF MICROBIAL FUEL CONTAMINATION USING BACTERIA PSEUDOMONAS PROTEGENS AND BACILLUS SUBTILIS

Amanda Schroeder*

University of Dayton Research Institute, 300 College Park Ave, Dayton Ohio 45469 USA

ABSTRACT

Aviation fuel infrastructure sustainment is expensive and requires frequent testing for hydrocarbon-degrading microbes to assure safeguarding of fuel quality and engineering systems. To prevent biofouling and biocorrosion, biological control agents appear to be promising alternatives for current chemical-industry derived biocidal treatments that are used for fuel tank and engineering system sustainment. In this study, 493 microbial fuel isolates from our in-house repository were screened for their potential antimicrobial compound production targeted towards other competing microbial species that are capable of growth in hydrocarbon fuels. Through the development of specialized antimicrobial screening assays including agar plug testing, and small-scale 96-well liquid culture assays, two fuel-isolate strains, e.g., #133 *Pseudomonas Protegens*, and # 232 *Bacillus Subtilis* were identified with promising antimicrobial activity against gram-positive and gram-negative bacteria as well as fungi and yeast. While the #232 culture filtrate exhibited potent antimicrobial activity in 96-well culture assays the mass spectrometry identification and purification of lipopeptide compounds were Macrolactin B, Gageostatin C, and Gageopeptin yielded variable results due to the unreliable metabolic growth behavior of *Bacillus*. In contrast, mass spectrometry analysis of culture filtrate from the #133 *(P. protegens)* microbial strain identified the siderophore Pyochelin as one of the potential compounds relevant for its potent antimicrobial activity with demonstrated antimicrobial effect on gram positive bacteria. Accordingly, when testing the #133 antimicrobial activity of crude culture filtrate in Jet A fuel microbial consortium culture with *Pseudomonas putida, Gordonia* sp., *Yarrowia lipolytica, Hormoconis resinae* a significant bacterial growth inhibition was observed. Flow cytometry live/dead staining confirmed antimicrobial compound secreted by *P. protegens* has bactericidal effect against *Gordonia* sp. and caused cell death and lysis. While the primary mode of action is unknown, the PI (Propidium Iodide) exclusion assay confirmed that the cell death is associated with membrane damage. Furthermore, antimicrobial compounds in *P. protegens* filtrate are heat-stable and active even at higher temperatures (e.g., 60° C).

*Corresponding author: Amanda Schroeder, Amanda.Barry@udri.udayton.edu

IASH 2024, the 18TH INTERNATIONAL CONFERENCE ON STABILITY, HANDLING AND USE OF LIQUID FUELS
Louisville, Kentucky, USA
September 8 – 12, 2024

(Technical Poster)

USE OF SHEEP MYELOID ANTIMICROBIAL PEPTIDE (SMAP-18) AND SIDEROPHORE PYOCHELIN FOR MITIGATION OF AEROSPACE FUEL-DEGRADING MICROBES

Amanda Schroeder*

University of Dayton Research Institute, 300 College Park Ave, Dayton Ohio 45469 USA

ABSTRACT

Sustainment of aviation fuel infrastructure is expensive and requires frequent testing for hydrocarbon-degrading microbes to assure safeguarding of fuel quality and engineering systems. Microbial contaminants in jet fuel, including Gram-positive and Gram-negative bacteria, yeast, and filamentous fungi, necessitate early detection to prevent biofouling and biocorrosion. This study tests the efficacy of two novel Jet fuel sustainment antimicrobial agents, the sheep myeloid antimicrobial peptide SMAP-18 and the iron chelating siderophore, Pyochelin. Both compounds have previously demonstrated antimicrobial properties. Through the development of specialized antimicrobial assays, that range from 96-well plate liquid culture testing to Jet A fuel cultures in small-scale 25 mL and 1 L total volume, this study evaluates the antimicrobial activity of SMAP-18 and Pyochelin against bacterial and fungal microbial consortia in fuel. The results show that repetitive dosing of combined SMAP-18 and Pyochelin is bactericidal and able to control bacterial growth in fuel. Moreover, the synergy observed between SMAP-18 and Pyochelin highlights their complementary mechanisms of action against microbial targets resulting in complete elimination of bacterial growth with a -7.9 log fold reduction through day 24.

*Corresponding author: Amanda Schroeder, Amanda.Barry@udri.udayton.edu

IASH 2024, the 18TH INTERNATIONAL CONFERENCE ON STABILITY, HANDLING AND USE OF LIQUID FUELS
Louisville, Kentucky, USA
September 8 – 12, 2024

(Technical Poster)

WORLD JET FUEL SURVEY (PRELIMINARY RESULTS)

Shane Kosir and Zachary West*

University of Dayton Research Institute, 300 College Park, Dayton, OH, USA 45469

Jet fuel is defined by a robust chemical/physical & performance specification; however, the exact chemical composition is dynamic and depends on numerous market factors such as: petroleum crude sources, production/refinery techniques, and impacts from transportation/handling in the fungible system to name a few. Additionally, the global aviation fuels industry is attempting to ramp up adoption of synthetic aviation turbine fuels (SATF), sometimes referred to as sustainable aviation fuels (SAF), to decarbonize. It has been almost two decades since the last major survey of commercial jet fuels, which pre-dates most synthetic jet fuel qualifications, and therefore, many industry stakeholders are interested in a contemporary look at jet fuel compositions and properties to use as a baseline for candidate new fuels. Our work aims to collect jet fuel samples from the point of origin and conduct both specification and non-specification testing on the subject samples. Select, preliminary results are reported.

*Corresponding author: Zachary West, zachary.west@udri.udayton.edu